T0121711

# LIQUID PIPELINE
# HYDRAULICS

## Second Edition

E. Shashi Menon, Ph.D., P.E

Pramila S. Menon, M.B.A.

Order this book online at www.trafford.com
or email orders@trafford.com

Most Trafford titles are also available at major online book retailers.

Printed in the United States of America.

ISBN: 978-1-4669-7739-6 (sc)
ISBN: 978-1-4669-7741-9 (hc)
ISBN: 978-1-4669-7740-2 (e)

Library of Congress Control Number: 2013901598

*Trafford rev. 04/11/2013*

**Trafford**
PUBLISHING®  www.trafford.com

**North America & international**
toll-free: 1 888 232 4444 (USA & Canada)
phone: 250 383 6864 ✦ fax: 812 355 4082

# Contents

# PREFACE

This book covers liquid pipeline hydraulics as it applies to transportation of liquids through pipelines in a single phase steady state environment. It will serve as a practical handbook for engineers, technicians and others involved in design and operation of pipelines transporting liquids. Currently, existing books on the subject are mathematically rigorous, theoretical and lack practical applications. Using this book, engineers can better understand and apply the principles of hydraulics to their daily work in the pipeline industry without resorting to complicated formulas and theorems. Numerous examples from the author's real life experience are included to illustrate application of pipeline hydraulics.

The application of hydraulics to liquid pipelines involve understanding of various properties of liquids, concept of pressure, friction and calculation of the energy required to transport liquid from point A to point B through a pipeline. You will not find rigorous mathematical derivation of formulas in this book. The formulas necessary for calculations are presented and described without using calculus or complex mathematical methods. If the reader is interested in how the formulas and equations are derived he should refer to any of the books and publications listed under the Reference section toward the end of this book.

This book covers liquid properties that affect flow through pipelines, calculation of pressure drop due to friction, horsepower required and the number of pump stations required for transporting a liquid through a pipeline. Topics covered include - basic equations necessary for pipeline design, commonly used formulas to calculate frictional pressure drop and necessary horsepower, the feasibility of improving an existing pipeline performance using drag reduction additives and power optimization studies. The use of pumps and valves in pipelines are addressed along with modifications necessary to improve pipeline throughput. Economic analysis and transportation tariff calculations are also included.

This book can be used for the analysis of both liquid pipeline gathering systems, plant or terminal piping as well as long distance trunk lines. The primary audience for the book is engineers and technicians working in the petroleum, water and process industry. This book could also be used as a text book for a college level course in liquid pipeline hydraulics.

Finally, I would like to dedicate this book to my father, who always believed I could write a technical book, but unfortunately did not live long enough to see it completed.

# 1

# Introduction

Pipeline Hydraulics deals with the flow of fluids in pipelines. Fluids are defined as liquids and gases. Specifically, this book deals with liquid flow in pipelines. Liquids are considered to be incompressible for most purposes. Today, several thousand miles of pipelines are used to transport crude oil and petroleum products such as gasoline and diesel from the refineries to storage tanks and delivery terminals. Similarly, thousands of miles of concrete and steel pipelines are used to transport water from reservoir to distribution locations throughout the country. Engineers are interested in the effect of pipe size, liquid properties, pipe length, etc. in determining the pressure required and horsepower necessary for transporting a liquid from point A to point B in a pipeline. It is clear that heavier liquids will require more pressure and hence more horsepower to transport a given quantity of liquid for a specified distance compared to a lighter liquid. In all these cases we are interested in determining the optimum pipe size required to transport given volumes of liquids economically and safely through these pipelines.

This book consists of 12 chapters that cover the practical aspects of liquid pipeline hydraulic and economics of pipelines used to transport liquids under steady state conditions, except Chapter 10, which introduces the reader to unsteady flow. For a more detailed analysis and study of Unsteady Flow and Transients, the reader should consult one of the books listed under the Reference section.

A Reference section, Appendices containing Tables and Charts and answers to selected problems are also included at the end of this book.

Chapter 2 covers units of measurements, properties of liquids such as density, gravity and viscosity that are important in liquid pipeline hydraulics. Chapter 3 discusses pressure, velocity, Reynold's number, friction factor and pressure drop calculations using various formulas. Several example problems are discussed and solved to illustrate the various methods currently used in pipeline engineering.

Chapter 4 is devoted to the strength analysis of pipes. It addresses allowable internal working pressures and hydrostatic test pressures and how they are calculated.

Chapter 5 extends the concepts developed in Chapter 3 by analyzing the total pressure and horsepower required to pump a liquid through long distance pipelines with multiple pump stations, including the transportation of high vapor pressure liquids, such as liquefied petroleum gas (LPG). Injection and delivery along a long pipeline and branch pipe analysis are also covered. The use of pipe loops to reduce friction and increase throughput is also analyzed.

Chapter 6 deals with optimizing pump station locations in a trunk line, minimizing pipe wall thickness using telescoping and pipe grade tapering. Open channel flow, slack line operation in hilly terrain and batching different products are also addressed in this Chapter.

Chapter 7 covers centrifugal pumps and positive displacement pumps applied to pipeline transportation. Centrifugal pump performance curves, Affinity laws

And the effect of viscosity are discussed as well as the importance of net positive suction head (NPSH). Operation of pumps in series and parallel and modification needed to operate pumps effectively are also discussed in this chapter

Chapter 8 discusses pump station design, minimizing energy loss due to pump throttling with constant speed motor driven pumps. The advantages of using variable speed drive (VSD) pumps are also explained and illustrated with examples.

Chapter 9 introduces the reader to thermal hydraulics, pressure drop calculations and temperature profiles in a buried heated liquid pipeline. The importance of thermal conductivity, overall heat transfer coefficient and how they effect heat loss to the surrounding are covered.

Chapter 10 introduces flow measurement devices used in measuring liquid flow rate in pipelines. Several of the more common instruments such as the venturi meter, flow nozzle and orifice meter are discussed and calculation methods explained.

Chapter 11 gives a basic introduction to unsteady flow and transient hydraulic analysis. This is an advanced concept that requires a separate book to cover fully the subject. Therefore, this chapter will serve as a starting point in understanding transient pipeline hydraulics. The reader should consult one of the publications listed in the Reference section for a more detailed study of unsteady flow and pipeline transients

Chapter 12 addresses economic aspects related to pipeline feasibility studies. In addition, the pipeline and pump station capital cost, annual operating cost and calculation of transportation tariff are discussed. Also covered in this chapter is the analysis of the optimum pipe size and pumping equipment required that produces the least cost. A discounted cash flow approach using the Present Value (PV) of investment is employed in determining the optimum pipe size for a particular application.

In each chapter, example problems are used to illustrate the concepts introduced. Problems for practice are also included at the end of each chapter. Answers to selected problems may be found in the Appendix B.

Appendix A consists of tables and charts containing Units and Conversions, Common Properties of Petroleum Fluids, etc.

In addition, for quick reference, formulas used in all chapters have been assembled and summarized in Appendix C.

# 2

# Properties of Liquids

In this chapter we will discuss the various units of measurement employed in Liquid Pipeline Hydraulics and proceed to cover the more important properties of liquids that affect hydraulic calculations. The importance of specific gravity, viscosity of pure liquids and mixtures will be analyzed and the concepts will be illustrated with sample problems. This chapter forms the foundation for all calculations involving pipeline pressure drops and horsepower requirements in subsequent chapters. In the Appendix, you will find tables listing properties of commonly used liquids such as water and petroleum products.

## 2.1  Units of Measurement

Before we discuss liquid properties it would be appropriate to identify the different units of measurement used in pipeline hydraulics calculations.

Over the years the English speaking world adopted so called "English Units" of measurement, while most other European and Asian countries adopted the "Metric System of Units".

The English system of units (referred to in USA as Customary US units) derives from the old Foot-Pound-Second (FPS) and Foot-Slug-Second (FSS) system that originated in England. The basic units are foot for length, slug for mass and second for measurement of time. In the past, the FPS system used pound for mass. Since Force, a derived unit, was also measured in pounds, there was evidently some confusion. To clarify the term pound-mass (lbm) and pound-force (lbf) were introduced. Numerically, the weight (which is a force due to gravity) of one pound mass was equal to one pound force. However, the introduction of slug for unit of mass resulted in the adoption of pound exclusively for unit of force. Thus, in the FSS system which is now used in the USA, the unit of mass is slug. The relationship between a slug, lbf and lbm will be explained later in this chapter.

In the Metric system, originally known as Centimeter-Gram-Second (CGS) system, the corresponding units for length, mass and time were centimeter, gram and second respectively. In later years, a modified metric units called Meter-Kilogram-Second (MKS) system emerged. In MKS units, the meter was used for the measurement of length and kilogram for the measurement of mass. The measurement for time remained the second for all systems of units.

The scientific and engineering communities, during the last four decades have attempted to standardize on a universal system of units worldwide. Through the International Standards Organization (ISO), a policy for an International System of Units (SI) was formulated. The SI units are also known as Systeme Internationale units.

The conversion from the older system of units to SI units has advanced at different rates in different countries. Most countries of Western Europe and all of Eastern Europe, Russia, India, China, Japan, Australia and South America have adopted the SI units completely. In North America, Canada and Mexico have adopted the SI units almost completely. However, engineers and scientists in these countries use both SI units and English units due to their business dealings with the United States. In the United States, SI units is used increasingly in colleges and the scientific community. However, the majority of work is still done using the English units referred to some times as Customary US units.

The Metric Conversion Act of 1975 accelerated the adoption of the SI system of units in the USA. The American Society of Mechanical Engineers (ASME), American Society of Civil Engineers (ASCE) and other Professional Societies and Organizations have assisted in the process of conversion from English to SI units using the respective Institutions publications. For example, ASME through the ASME Metric Study Committee published a series of articles in the Mechanical Engineering magazine to help engineers master the SI system of units.

In the USA, the complete changeover to SI has not materialized fast enough. Therefore in this transition phase, engineering students, practicing engineers, technicians and scientists must be familiar with the different systems of units such as English, Metric CGS, Metric MKS and the SI units. In this book we will use both English units (Customary US) as well as the SI system of units.

Units of measurement are generally divided into three classes as follows:

Base units
Supplementary units
Derived units

By definition, Base units are dimensionally independent. These are units of length, mass, time, electric current, temperature, amount of substance and luminous intensity.

Supplementary units are those used to measure plain angles and solid angles. Examples include radian and steradian.

Derived units are those that are formed by combination of base units, supplementary units and other derived units. Examples of derived units are those of force, pressure and energy.

## 2.1.1 Base Units

In the English (Customary US) system of units, the following base units are used

| | |
|---|---|
| Length | - foot (ft) |
| Mass | - slug (slug) |
| Time | - second (s) |
| Electric current | - ampere (A) |
| Temperature | - degree Fahrenheit (°F) |
| Amount of substance | - mole (mol) |
| Luminous intensity | - candela (cd) |

In SI units, the following base units are defined

| | |
|---|---|
| Length | - meter (m) |

| Mass | - kilogram (kg) |
| Time | - second (s) |
| Electric current | - ampere (A) |
| Temperature | - kelvin (K) |
| Amount of substance | - mole (mol) |
| Luminous intensity | - candela (cd) |

### 2.1.2  Supplementary Units

Supplementary Units in both English and SI system of units are

| Plain angle | - radian (rad) |
| Solid angle | - steradian (sr) |

The radian is defined as the plain angle between two radii of a circle with an arc length equal to the radius. Thus, it represents the angle of a sector of a circle with the arc length the same as its radius.

The steradian is the solid angle having its apex at the center of a sphere such that the area of the surface of the sphere that it cuts out is equal to that of a square with sides equal to the radius of this sphere.

### 2.1.3  Derived Units

Derived units are generated from a combination of base units, supplementary units and other derived units. Examples of derived units include those of area, volume, etc.

In English units the following derived units are used:

| Area | - square inches ($in^2$), square feet ($ft^2$) |
| Volume | - cubic inch ($in^3$), cubic feet ($ft^3$), gallons (gal) and barrels (bbl) |
| Speed/Velocity | - feet per second (ft/s), |
| Acceleration | - feet per second per second ($ft/s^2$) |
| Density | - slugs per cubic foot ($slugs/ft^3$) |
| Specific weight | - pound per cubic foot ($lb/ft^3$) |
| Specific volume | - cubic foot per pound ($ft^3/lb$) |
| Dynamic Viscosity | - pound second per square foot ($lb\text{-}s/ft^2$) |

| Kinematic Viscosity | - square foot per second (ft²/s) |
|---|---|
| Force | - pounds (lb) |
| Pressure | - pounds per square inch (lb/in²) |
| Energy/Work | - foot pound (ft lb) |
| Quantity of Heat | - British Thermal Units (Btu) |
| Power | - Horsepower (HP) |
| Specific Heat | - Btu per pound per °F (Btu/lb/ °F) |
| Thermal Conductivity | - Btu per hour per foot per °F (Btu/hr/ft/ °F) |

In SI units the following derived units are used:

| Area | - square meter (m²) |
|---|---|
| Volume | - cubic meter (m³) |
| Speed/Velocity | - meter/second (m/s) |
| Acceleration | - meter per second per second (m/s²) |
| Density | - kilogram per cubic meter (kg/m³) |
| Specific volume | - cubic meter per kilogram (m³/kg) |
| Dynamic Viscosity | - Pascal second (Pa.s) |
| Kinematic Viscosity | - square meters per second (m²/s) |
| Force | - Newton (N) |
| Pressure | - Newton per square meter or Pascal (Pa) |
| Energy/Work | - Newton meter or Joule (J) |
| Quantity of Heat | - Joule (J) |
| Power | - Joule per second or Watt (W) |
| Specific Heat | - Joule per kilogram per K (J/kg/K) |
| Thermal Conductivity | - Joule/second/meter/Kelvin (J/s/m/K) |

Many other derived units are used in both English and SI units. A list of the more commonly used units in Liquid Pipeline Hydraulics and their conversions are listed in Appendix A.1.

## 2.2   Mass, Volume, Density and Specific Weight

Several properties of liquids that affect liquid pipeline hydraulics will be discussed here. In steady state hydraulics of liquid pipelines, the following properties are important:

## 2.2.1    Mass

Mass is defined as the quantity of matter. It is independent of temperature and pressure. Mass is measured in slugs (slugs) in English units or kilograms (kg) in SI units. In the past mass was used synonymously with weight. Strictly speaking weight depends upon the acceleration due to gravity at a certain geographic location and therefore is considered to be a force. Numerically mass and weight are interchangeable in the older FPS system of units. For example, a mass of 10 lbm is equivalent to a weight of 10 pound force (lbf). To avoid this confusion, in English units, the slug has been adopted for unit of mass. One slug is equal to 32.17 lb. Therefore, if a drum contains 55 gal of crude oil and weighs 410 lb, the mass of oil will be the same at any temperature and pressure. Hence the statement "conservation of mass".

## 2.2.2    Volume

Volume is defined as the space occupied by a given mass. In the case of the 55 gallon drum above, 410 lb of crude oil occupies the volume of the drum. Therefore the crude oil volume is 55 gal. Consider a solid block of ice measuring 12 in on each side. The volume of this block of ice is 12 x 12 x 12 or 1 728 cubic inches or one cubic foot. The volume of a certain petroleum product contained in a circular storage tank 100 ft in diameter and 50 ft high, may be calculated as follows, assuming the liquid depth is 40 ft:

Liquid volumes = $(\pi / 4)$ x 100 x 100 x 40 = 314 160 ft$^3$

Liquids are practically incompressible, take the shape of their container and have a free surface. Volume of a liquid varies with temperature and pressure. However for liquids, being practically incompressible, pressure has negligible effect on volume. Thus, if the liquid volume measured at 50 psi is 1 000 gal, its volume at a 1 000 psi will not be appreciably different, provided the liquid temperature remained constant. Temperature, however, has a more significant effect on volume. For example, the 55 gal volume of liquid in a drum at a temperature of 60°F will increase to a slightly higher value (such as 56 gal) when the liquid temperature increases to 100°F. The amount of increase in volume per unit temperature rise depends on the coefficient of expansion of the liquid. When measuring petroleum liquids, for the purpose of custody transfer, it is customary to correct volumes to a fixed temperature such as 60 °F. Volume correction factors from American Petroleum Institute (API) publications are commonly used in the petroleum industry.

In the petroleum industry, it is customary to measure volume in gallons or barrels. One barrel is equal to 42 U.S. gallons. The Imperial gallon as used in UK is a larger unit, approximately 20% larger than the US gallon. In SI units volume is generally measured in cubic meters ($m^3$) or liters (L).

In a pipeline transporting crude oil or refined petroleum products, it is customary to talk about the "line fill volume" of the pipeline. The volume of liquid contained between two valves in a pipeline can be calculated simply by knowing the internal diameter of the pipe and the length of pipe between the two valves. By extension, the total volume or the line fill volume of the pipeline can be easily calculated.

As an example, if a 16 in pipeline, 0.250 in wall thickness is 5 000 ft long from one valve to another, the line fill for this section of pipeline is

$$\text{Line fill volume} = (\pi \, / \, 4) \times (16 - 2 \times 0.250)^2 \times 5\,000$$
$$= 943\,461.75 \text{ ft}^3 \text{ or } 168\,038 \text{ bbl}$$

Above calculation is based on conversion factors of:

      1 728 $in^3$ per $ft^3$
      231 $in^3$ per gallon
and   42 gallons per barrel.

In a later chapter we will discuss a simple formula for determining the line fill volume of a pipeline.

The volume flow rate in a pipeline is generally expressed in terms of cubic feet per second ($ft^3$/s), gallons per minute (gal/min), barrels per hour (bbl/hr) and barrels per day (bbl/day) in Customary English Units. In the SI Units, volume flow rate is referred to in cubic meters per hour ($m^3$/hr) and liters per second (L/s).

It must be noted that since the volume of a liquid varies with temperature, the inlet flow rate and the outlet volume flow rate may be different in a long distance pipeline, even with no intermediate injections or deliveries. This is due to the fact that the inlet flow rate may be measured at an inlet temperature of 70°F to be 5 000 bbl/hr and the corresponding flow rate at the pipeline terminus, 100 miles away may be measured at an outlet temperature different than the inlet temperature. The temperature difference is due to heat loss or gain between the pipeline liquid and the surrounding soil or ambient conditions. Generally, significant variation in temperature is observed when pumping crude oils or other products that are heated at the pipeline inlet. In refined petroleum products and other pipelines that are not heated, temperature variations along the pipeline are insignificant. In any case if the volume measured at the pipeline inlet is corrected to a standard temperature such as 60°F, the corresponding outlet volume can also be corrected to the same standard temperature. With temperature correction it can be assumed that the same flow rate exists throughout the pipeline from inlet to outlet provided of course there are no intermediate injections or deliveries along the pipeline.

By the principle of conservation of mass, the mass flow rate at inlet will equal that at the pipeline outlet since the mass of liquid does not change with temperature or pressure.

### 2.2.3    Density

Density of a liquid is defined as the mass per unit volume. Customary units for density are slugs/ft$^3$ in the English Units. The corresponding units of density in SI Units is kg/m$^3$. This is also referred to as mass density. The weight density is defined as the weight per unit volume. This term is more commonly called specific weight and will be discussed next.

Since mass does not change with temperature or pressure, but volume varies with temperature, we can conclude that density will vary with temperature. Density and volume are inversely related since density is defined as mass per unit volume. Therefore, with increase in temperature liquid volume increases while its density decreases. Similarly, with reduction in temperature, liquid volume decreases and its density increases.

### 2.2.4    Specific Weight

Specific weight of a liquid is defined as the weight per unit volume. It is measured in lb/ft$^3$ in English Units and N/m$^3$ in SI units.

If a 55 gal drum of crude oil weighs 410 lb (excluding weight of drum), the specific weight of crude oil is

(410/55) or 7.45 lb/gal.

Similarly, consider the 5 000 ft pipeline discussed in section 2.2.2 previously. The volume contained between the two valves was calculated to be 168 038 bbl. If we use specific weight calculated above, we can estimate the weight of liquid contained in the pipeline as

7.45 x 42 x 168 038 = 52 579 090 lb or 26 290 tons

Similar to density, specific weight varies with temperature. Therefore, with increase in temperature specific weight will decrease. With reduction in temperature, liquid volume decreases and its specific weight increases.

Customary units for specific weight are lb/ft$^3$ and lb/gal in the English Units. The corresponding units of specific weight in SI Units is N/m$^3$.

For example, water has a specific weight of 62.4 lb/ft$^3$ or 8.34 lb/gal at 60°F. A typical gasoline has a specific weight of 46.2 lb/ft$^3$ or 6.17 lb/gal at 60°F.

Although density and specific weight are dimensionally different, it is common to use the term density instead of specific weight and vice versa when calculating hydraulics of liquid pipelines. Thus you will find that the density of water and specific weight of water are both expressed as 62.4 lb/ft$^3$.

## 2.3    Specific gravity and API gravity

Specific gravity of a liquid is the ratio of its density to the density of water at the same temperature and therefore has no units (dimensionless). It is a measure of how heavy a liquid is compared to water.

The term relative density is also used to compare the density of a liquid with another liquid such as water. In comparing the densities it must be noted that both densities must be measured at the same temperature to be meaningful.

At 60°F, a typical crude oil has a density of 7.45 lb/gal compared to a water density of 8.34 lb/gal. Therefore, the specific gravity of crude oil at 60°F is

Specific gravity = 7.45/8.34 or 0.8933.

By definition, the specific gravity of water is 1.00 since the density of water compared to itself is the same. Specific gravity, like density, varies with temperature. As temperature increases, both density and specific gravity decrease. Similarly, decrease in temperature causes the density and specific gravity to increase in value. As with volume, pressure has very little effect on liquid specific gravity as long as pressures are within the range of most pipeline applications.

In the petroleum industry it is customary to use units of °API for gravity. The API gravity is a scale of measurement using API = 10 on the low end for water at 60°F. All liquids lighter than water will have API values higher than 10. Thus gasoline has an API gravity of 60 while a typical crude oil may be 35°API.

The API gravity of a liquid is a value determined in the laboratory comparing the density of the liquid versus the density of water at 60°F. If the liquid is lighter than water its API gravity will be greater than 10.

The API gravity versus the specific gravity relationship is as follows

$$\text{Specific gravity } Sg = 141.5/(131.5 + API) \qquad (2.1)$$
$$\text{or} \quad API = 141.5/Sg - 131.5 \qquad (2.2)$$

Substituting an API value of 10 for water in Equation (2.1), yields as expected the specific gravity of 1.00 for water. It is seen from the above equation that the specific gravity of the liquid cannot be greater than 1.076, in order to result in a positive value of API.

Another scale of gravity for liquids heavier than water is known as the Baume Scale. This scale is similar to the API scale with the exception of 140 and 130 being used in place of 141.5 and 131.5 respectively in Equations (2.1) and (2.2).

As an another example, assume the specific gravity of gasoline at 60°F is 0.736. Therefore, the API gravity of gasoline can be calculated from Equation (2.2) as follows:

API gravity = 141.5/0.736 - 131.5 = 60.76°API

If diesel fuel is reported to have an API gravity of 35, the specific gravity can be calculated from Equation (2.1) as follows:

Specific gravity = 141.5 / (131.5 + 35) = 0.8498

It must be noted that API gravity is always referred to at 60°F. Therefore in the Equations (2.1) and (2.2), specific gravity must also be measured at 60°F. Hence, it is meaningless to say that the API of a liquid is 35°API at 70°F.

API gravity is measured in the laboratory using the method described in ASTM D1298 using a properly calibrated glass hydrometer. Also refer to API Manual of Petroleum Measurements for further discussion on API gravity.

### 2.3.1    Specific gravity variation with temperature

It was mentioned earlier that the specific gravity of a liquid varies with temperature. It increases with decrease in temperature and vice versa.

For commonly encountered range of temperatures in liquid pipelines, the specific gravity of a liquid varies linearly with temperature. In other words, the specific gravity versus temperature can be expressed in the form of the following equation

$$S_T = S_{60} - a\,(T\text{-}60) \tag{2.3}$$

where

$S_T$     - Specific gravity at temperature T
$S_{60}$    - Specific gravity at 60°F

T     - Temperature, $°F$

a     - A constant that depends on the liquid

In Equation (2.3) the specific gravity $S_T$ at temperature T is related to the specific gravity at $60°F$ by a straight line relationship. Since, the term $S_{60}$ and a are unknown quantities, two sets of specific gravities at two different temperatures are needed to determine the specific gravity versus temperature relationship. If the specific gravity at $60°F$ and the specific gravity at $70°F$ are known we can substitute these values in Equation (2.3) to obtain the unknown constant a. Once the value of a is known, we can easily calculate the specific gravity of the liquid at any other temperature using Equation (2.3). An example will illustrate how this is done.

Some handbooks such as Hydraulic Institute Engineering Design book and the Crane Handbook (See Reference section) provide specific gravity versus temperature curves from which the specific gravity of most liquids can be calculated at any temperature.

**Example Problem 2.1**

The specific gravity of gasoline at 60 $°F$ is 0.736. The specific gravity at 70$°$ F is 0.729. What is the specific gravity at 50 $°F$?

**Solution**

Using Equation (2.3), we can write

$$0.729 = 0.736 - a (70-60)$$

Solving for a, we get

$$a = 0.0007$$

We can now calculate the specific gravity at $50°F$ using Equation (2.3) as

$$S_{50} = 0.736 - 0.0007(50-60) = 0.743$$

## 2.3.2    Specific gravity of blended liquids

Suppose a crude oil of specific gravity 0.895 at 70°F is blended with a lighter crude oil of specific gravity 0.815 at 70 °F, in equal volumes. What will be the specific gravity of the blended mixture? Common sense suggests that since equal volumes are used, the resultant mixture should have a specific gravity of the average of the two liquids or

$$(0.895 + 0.815)/2 = 0.855$$

This is indeed the case, since specific gravity of a liquid is simply related to the mass and the volume of each liquid.

When two or more liquids are mixed homogenously, the resultant liquid specific gravity can be calculated using weighted average method. Thus, 10% of Liquid A with specific gravity of 0.85 when blended with 90% of Liquid B that has a specific gravity of 0.89 results in a blended liquid with specific gravity of

$$(0.1 \times 0.85) + (0.9 \times 0.89) = 0.886$$

It must be noted that when performing the above calculations, both specific gravities must be measured at the same temperature.

Using the above approach, the specific gravity of a mixture of two or more liquids can be calculated from the following equation:

$$S_b = \frac{(Q_1 \times S_1) + (Q_2 \times S_2) + (Q_3 \times S_3) + \ldots}{Q_1 + Q_2 + Q_3 + \ldots} \qquad (2.4)$$

where

| | |
|---|---|
| $S_b$ | - Specific gravity of the blended liquid |
| $Q_1, Q_2, Q_3$   etc | - Volume of each component |
| $S_1, S_2, S_3$   etc | - Specific gravity of each component |

The above method of calculating the specific gravity of a mixture of two or more liquids cannot be directly applied when the gravities are expressed in °API values. If the component gravities of a mixture are given in °API we must first convert API values to specific gravities before applying Equation (2.4)

**Example Problem 2.2**

Three liquids A, B and C are blended together in the ratio of 15%, 20% and 65% respectively. Calculate the specific gravity of the blended liquid if the individual liquids have the following specific gravities at 70°F:

**Solution**

Specific gravity of Liquid A: 0.815
Specific gravity of Liquid B: 0.850
Specific gravity of Liquid C: 0.895

Using Equation (2.4) we get the blended liquid specific gravity as

$S_b = (15 \times 0.815 + 20 \times 0.850 + 65 \times 0.895)/100 = 0.874$

## 2.4  Viscosity

Viscosity is a measure of sliding friction between successive layers of a liquid that flows in a pipeline. Imagine several layers of liquid that constitute a flow between two fixed parallel horizontal plates. A thin layer adjacent to the bottom plate will be at rest or zero velocity. Each subsequent layer above this will have a different velocity compared to the layer below. This variation in the velocity of the liquid layers results in a velocity gradient. If the velocity is V at the layer that is located a distance of y from the bottom plate, the velocity gradient is approximately

Velocity gradient = V / y                                    (2.5)

If the variation of velocity with distance is not linear, using calculus we can write more accurately that

Velocity gradient $= \dfrac{dV}{dy}$ (2.6)

where $\underline{dV}$ represents the rate of change of velocity with distance or the
$\phantom{where}$ dy velocity gradient.

Newton's law states that the shear stress between adjacent layers of a
flowing liquid is proportional to the velocity gradient. The constant of
proportionality is known as the absolute (or dynamic) viscosity of the
liquid.

Shear stress = (Viscosity) (Velocity gradient)

Absolute Viscosity of a liquid is measured in lb-sec/ft$^2$ in English
Units and Pascal-sec in SI Units. Other commonly used units of absolute
viscosity are Poise and Centipoise (cP).

The kinematic viscosity is defined as the absolute viscosity of a
liquid divided by its density at the same temperature.

$\nu = \mu/\rho$ (2.7)

where

$\nu$ = Kinematic viscosity
$\mu$ = Absolute viscosity
$\rho$ = Density

The units of kinematic viscosity are ft$^2$/s in English Units and m$^2$/s
in SI Units. See Appendix A for conversion of units. Other commonly
used units for kinematic viscosity are Stokes and centistokes (cSt). In
the petroleum industry, two other units for kinematic viscosity are also
used. These are Saybolt Seconds Universal (SSU) and Saybolt Seconds
Furol (SSF). When expressed in these units, it represents the time taken
for a fixed volume of a liquid to flow through an orifice of defined
size. Both absolute and kinematic viscosities vary with temperature.
As temperature increases, liquid viscosity decreases and vice versa.

However, unlike specific gravity, viscosity versus temperature is not a linear relationship. We will discuss this in the next session.

Viscosity also varies somewhat with pressures. Significant variations in viscosity is found when pressures are several thousand psi. In most pipeline applications, viscosity of a liquid does not change appreciably with pressure.

For example, the viscosities of Alaskan North Slope (ANS) crude oil may be reported as 200 SSU at 60°F and 175 SSU at 70°F. Viscosity in SSU and SSF maybe converted to their equivalent in centistokes using the following equations:

Conversion from SSU to Centistokes

$$\text{Centistokes} = 0.226(\text{SSU}) - 195/(\text{SSU}) \qquad (2.8)$$
$$\text{for } 32 \le \text{SSU} \le 100$$

$$\text{Centistokes} = 0.220(\text{SSU}) - 135/(\text{SSU}) \qquad (2.9)$$
$$\text{for SSU} > 100$$

Conversion from SSF to Centistokes

$$\text{Centistokes} = 2.24(\text{SSF}) - 184/(\text{SSF}) \qquad (2.10)$$
$$\text{for } 25 < \text{SSF} \le 40$$

$$\text{Centistokes} = 2.16(\text{SSF}) - 60/(\text{SSF}) \qquad (2.11)$$
$$\text{for SSU} > 40$$

### Example Problem 2.3

Let us use the above Equations to convert viscosity of ANS crude oil from 200 SSU to its equivalent in Centistokes.

### Solution

Using Equation (2.9)

$$\text{Centistokes} = 0.220 \times 200 - 135 / 200 = 43.33 \text{ cSt}$$

The reverse process of converting from viscosity in cSt to its equivalent in SSU using Equations (2.8) and (2.9) is not quite so direct. Since, Equations (2.8) and (2.9) are valid for certain range of SSU values, we need to first determine which of the two equations to use. This is difficult since the equation to use depends on the SSU value, which itself is unknown. Therefore we will have to assume that the SSU value to be calculated falls in one of the two ranges shown and proceed to calculate by trial and error. We will have to solve a quadratic equation to determine the SSU value for a given viscosity in cSt. An example will illustrate this method

**Example Problem 2.4**

Suppose we are given a liquid viscosity of 15 cSt and we are required to calculate the corresponding viscosity in SSU.

**Solution**

Let us assume that the calculated value in SSU is approximately 5 x 15 = 75 SSU. This is a good approximation, since the SSU value generally is about 5 times the corresponding viscosity value in cSt. Since the assumed SSU value is 75, we need to use Equation (2.8) for converting between cSt and SSU.

Substituting 15 cSt in Equation (2.8) gives

$$15 = 0.226(SSU) - 195/(SSU)$$

replacing SSU with variable x, above Equation become, after transposition

$$15x = 0.226x^2 - 195$$

rearranging we get

$$0.226x^2 - 15x - 195 = 0$$

Solving for x, we get

x = [15 + (15 x 15 + 4 x 0.226 x 195)$^{1/2}$] / (2 x 0.226) = 77.5

Therefore, the viscosity is 77.5 SSU

## 2.4.1    Viscosity variation with temperature

The viscosity of a liquid decreases as the liquid temperature increases and vice versa. For gases, the viscosity increases with temperature. Thus, if the viscosity of a liquid at 60°F is 35 cSt, as the temperature increases to 100°F, the viscosity could drop to a value of 15 cSt. The variation of liquid viscosity with temperature is not linear, unlike specific gravity variation with temperature discussed in a previous section. The viscosity temperature variation is found to be logarithmic in nature.

Mathematically, we can state the following:

$$\text{Log}_e(v) = A - B(T) \tag{2.12}$$

Where

| | | | |
|---|---|---|---|
| v | - Viscosity of liquid, cSt | | |
| T | - Absolute temperature, °R or °K | | |
| T | = (t + 460) °R | if t is in °F | (2.13) |
| T | = (t + 273) °K | if t is in °C | (2.14) |

A and B are constants that depend on the specific liquid.

It can be seen from Equation (2.12) that a graphic plot of $\text{Log}_e(v)$ against the temperature T will result in a straight line with a slope of -B. Therefore, if we have two sets of viscosity versus temperature for a liquid we can determine the values of A and B by substituting the viscosity, temperature values in Equation (2.12). Once A and B are known we can calculate the viscosity at any other temperature using Equation (2.12). An example will illustrate this.

## Example Problem 2.5

Suppose we are given the viscosities of a liquid at 60°F and 100°F as 43 cSt and 10 cSt. We will use Equation (2.12) to calculate the values of A and B first.

$$Log_e(43) = A - B (60 + 460)$$

and

$$Log_e(10) = A - B (100 + 460)$$

## Solution

Solving the above two equations for A and B results in

$$A = 22.7232 \qquad B = 0.0365$$

Having found A and B, we can now calculate the viscosity of this liquid at any other temperature using Equation (2.12). Let us calculate the viscosity at 80°F

$$Log_e(v) = 22.7232 - 0.0365 (80 + 460) = 3.0132$$

Viscosity at 80°F = 20.35 cSt

In addition to Equation (2.12), several researchers have put forth various equations that attempt to correlate viscosity variation of petroleum liquids with temperature. The most popular and accurate of the formulas is the one known as the ASTM method. In this method, also known as the ASTM D341 chart method, a special graph paper with logarithmic scales is used to plot the viscosity of a liquid at two known temperatures. Once the two points are plotted on the chart and a line drawn connecting them, the viscosity at any intermediate temperature can be interpolated. To some extent, values of viscosity may also be extrapolated from this chart. This is shown in Figure 2.1

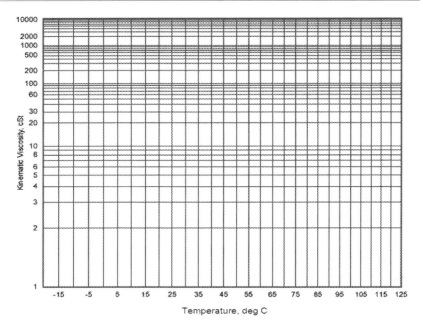

**Figure 2.1 ASTM D341 - Viscosity Temperature Chart**

In the following paragraphs we will discuss how to calculate the viscosity variations with temperature, using the ASTM method, without using the special logarithmic graph paper.

$$\text{Log Log } (Z) = A - B\text{Log}(T) \tag{2.15}$$

where

Log is the logarithm to base 10

Z      - depends on viscosity of the liquid $v$

$v$      - Viscosity of liquid, cSt

T      - Absolute temperature, °R or °K

A and B are constants that depend on the specific liquid.

The variable Z is defined as follows:

$$Z = (v + 0.7 + C - D) \tag{2.16}$$

where C and D are

$$C = \exp[-1.14883 - 2.65868(v)] \tag{2.17}$$

$$D = \exp[-0.0038138 - 12.5645(v)] \tag{2.18}$$

C, D and Z are all functions of the kinematic viscosity v.

Given two sets of temperature viscosity values $(T_1, v_1)$ and $(T_2, v_2)$ we can calculate the corresponding values of C, D and Z from Equations (2.17), (2.18) and (2.16).

We can then come up with two equations using the pairs of $(T_1, Z_1)$ and $(T_2, Z_2)$ values by substituting these values into Equation (2.15) as shown below:

$$\text{Log Log } (Z_1) = A - B\text{Log}(T_1) \tag{2.19}$$
$$\text{Log Log } (Z_2) = A - B\text{Log}(T_2) \tag{2.20}$$

From the above equations, the two unknown constants A and B can be easily calculated, since $T_1, Z_1$ and $T_2, Z_2$ values are known.

The following example will illustrate this approach for viscosity temperature variation.

**Example Problem 2.6**

A certain liquid has a temperature versus viscosity relationship as given below:

| Temperature, °F | 60 | 180 |
|---|---|---|
| Viscosity, cSt | 750 | 25 |

(a) Calculate the constants A and B that define the viscosity versus temperature correlation for this liquid using Equation (2.15).
(b) What is the estimated viscosity of this liquid at 85°F?

**Solution**

(a) At the first temperature 60°F

C, D and Z are calculated using Equations (2.17), (2.18) and (2.16)

$C_1 = \exp[-1.14883 - 2.65868 \times 750] = 0$

$D_1 = \exp[-0.0038138 - 12.5645 \times 750] = 0$

$Z_1 = (750 + 0.7) = 750.7$

Similarly at the second temperature of 180°F, the corresponding values of C, D and Z are calculated to be

$C_2 = \exp[-1.14883 - 2.65868 \times 25] = 0$

$D_2 = \exp[-0.0038138 - 12.5645 \times 25] = 0$

$Z_2 = (25 + 0.7) = 25.7$

Substituting in Equation (2.19) we get

Log Log (750.7) = A - BLog (60+460)

or

0.4587 = A - 2.716B

Log Log (25.7) = A - BLog (180+460)

or

0.1492 = A - 2.8062B

Solving for A and B, we get

A = 9.778
B = 3.4313

(b) At temperature of 85 °F using Equation (2.15) we get

Log Log (Z) = A - BLog (85+460)

Log Log (Z) = 9.778 - 3.4313 x 2.7364 = 0.3886

Z = 279.78

Therefore,

Viscosity at 85°F = 279.78 - 0.7 = 279.08 cSt

## 2.4.2    Viscosity of Blended Products

Suppose a crude oil of viscosity 10 cSt at 60°F is blended with a lighter crude oil of viscosity 30 cSt at 60°F, in equal volumes. What will be the viscosity of the blended mixture? We cannot average the viscosities as we did with specific gravities blending earlier. This is due to the non-linear nature of viscosity with mass and volumes of liquids.

When blending two or more liquids, the specific gravity of the blended product can be calculated directly, by using the weighted average approach as demonstrated in an earlier section. However, the viscosity of a blend of two or more liquids cannot be calculated by simply using the ratio of each component. Thus if 20% of Liquid A of viscosity 10 cSt is blended with 80% of Liquid B with a viscosity of 30 cSt, the blended viscosity is *not* the following

0.2 x 10 + 0.8 x 30 = 26 cSt

In fact, the actual blended viscosity would be 23.99 cSt as will be demonstrated in the following section

The viscosity of a blend of two or more products can be estimated using the following equation:

$$\sqrt{V_b} = \frac{Q_1 + Q_2 + Q_3 + \ldots}{(Q_1 / \sqrt{V_1}) + (Q_2 / \sqrt{V_2}) + (Q_3 / \sqrt{V_3})} \qquad (2.21)$$

where

| | | |
|---|---|---|
| $V_b$ | | - Viscosity of blend, SSU |
| $Q_1, Q_2, Q_3$ | etc | - Volumes of each component |
| $V_1, V_2, V_3$ | etc | - Viscosity of each component, SSU |

Since Equation (2.19) requires the component viscosities to be in SSU, we cannot use this equation to calculate the blended viscosity when viscosity is less than 32 SSU (1.0 cSt)

Another method of calculating the viscosity of blended products has been in use in the pipeline industry for over four decades. This method is referred to as the Blending Index method. In this method a Blending Index is calculated for each liquid based on its viscosity. Next the Blending Index of the mixture is calculated from the individual blending indices by using the weighted average of the composition of the mixture. Finally, the viscosity of the blended mixture is calculated using the Blending Index of the mixture. The equations used are described below:

$$H = 40.073 - 46.414 \, Log_{10} \, Log_{10} \, (V+B) \qquad (2.22)$$
$$B = 0.931 \, (1.72)^V \text{ for } 0.2 < V < 1.5 \qquad (2.23)$$
$$B = 0.6 \text{ for } V >= 1.5 \qquad (2.24)$$
$$Hm = [H1(pct1)+H2(pct2)+H3(pct3)+ \ldots]/100 \qquad (2.25)$$

where

| | |
|---|---|
| $H, H1, H2 \ldots$ | - Blending index of liquids. |
| $Hm$ | - Blending index of mixture. |
| $B$ | - Constant in Blending Index Equation. |
| $V$ | - Viscosity in centistokes. |
| $pct1, pct2, \ldots$ | - Percentage of liquids 1, 2, . . . . in blended mixture. |

**Example Problem 2.7**

Calculate the blended viscosity obtained by mixing 20% of Liquid A with a viscosity of 10 cSt and 80% of Liquid B with a viscosity of 30 cSt at 70°F.

## Solution

First, convert the given viscosities to SSU to use Equation (2.21). Viscosity of Liquid A is calculated using Equations (2.8) and (2.9)

$$10 = 0.226 \, (V_A) - \frac{195}{V_A}$$

Rearranging we get

$$0.226 \, V_A{}^2 - 10 \, V_A - 195 = 0$$

Solving the quadratic equation for $V_A$ we get

$$V_A = 58.90 \, \text{SSU}$$

Similarly Viscosity of liquid B is

$$V_B = 140.72 \, \text{SSU}$$

From equation (2.21), the blended viscosity is

$$\sqrt{V_{blnd}} = \frac{20 + 80}{(20 \, / \, \sqrt{58.9}) + (80 \, / \, \sqrt{140.72})} = 10.6953$$

Therefore the viscosity of the blend is

$$V_{blnd} = 114.39 \, \text{SSU}$$

or

Viscosity of blend = 23.99 cSt    after converting from SSU to cSt using Equation (2.9)

A graphical method is also available to calculate the blended viscosities of two petroleum products using ASTM D341-77. This method involves using a logarithmic chart with viscosity scales on the left and right side of the paper. The horizontal axis is for selecting the percentage of each product as shown in Figure 2.2 This chart is also available in handbooks such as Crane Handbook and the Hydraulic Institute Engineering Data Book. It must be noted that the viscosities of both products must be plotted at the same temperature.

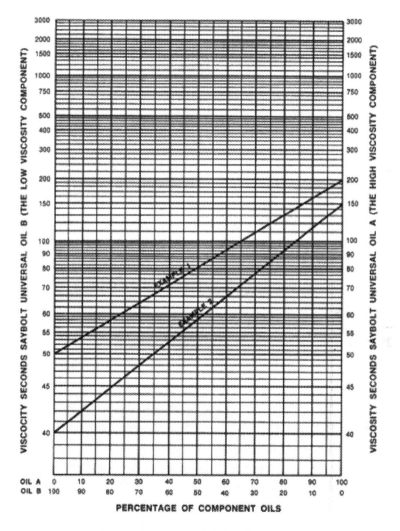

**Figure 2.2 Viscosity Blending chart**

Using this method, the blended viscosity of two products at a time is calculated and the process repeated for multiple products. Thus if three products are blended in the ratios of 10%, 20% and 70%, we would first calculate the blend using the first two liquids considering 10 parts of Liquid A mixed with 20 parts of Liquid B. This means that the blend would be calculated on the basis of one-third of Liquid A and two-third of Liquid B. Next this blended liquid will be mixed with Liquid C in the proportion of 30% and 70% respectively.

## 2.5   Vapor Pressure

Vapor Pressure of a liquid is defined as the pressure at a given temperature at which the liquid and vapor exist in equilibrium. The normal boiling point of a liquid can thus be defined as the temperature at which the vapor pressure equals

the atmospheric pressure. In the laboratory the vapor pressure is measured at a fixed temperature of 100°F and is then reported as the Reid Vapor Pressure. The vapor pressure of a liquid increases with temperature. Charts are available to determine the actual vapor pressure of a liquid at any temperature once its Reid vapor pressure is known. Refer to Crane Handbook for vapor pressure charts.

The importance of vapor pressure will be evident when we discuss the operation of centrifugal pumps on pipelines. To prevent cavitation of pumps, the liquid vapor pressure at the flowing temperature must be taken into account in the calculation of Net Positive Suction Head (NPSH) available at the pump suction. Centrifugal pumps are discussed in Chapter 7.

## 2.6   Bulk Modulus

The Bulk Modulus of a liquid is a measure of the compressibility of the liquid. It is defined as the pressure required to produce a unit change in its volume. Mathematically, bulk modulus is expressed as

$$\text{Bulk Modulus } K = VdP / dV \tag{2.26}$$

where dV is the change in volume corresponding to a change in pressure of dP.

The units of Bulk Modulus, K are psi or kPa. For most liquids the bulk modulus is approximately in the range of 250 000 to 300 000 psi. The fairly high number demonstrates the incompressibility of liquids.

Let us demonstrate the incompressibility of liquids by performing a calculation using Bulk Modulus. Assume the Bulk Modulus of a petroleum product is 250 000 psi. To calculate the pressure required to change the volume of a given quantity of liquid by one percent we would proceed as follows

From Equation (2.26), with some rearrangement,

Bulk Modulus = change in pressure / (change in volume / volume)

Therefore

250 000 = change in pressure / (0.01)

Therefore

change in pressure = 2 500 psi

It can be seen from the above a fairly large pressure is required to produce a very small (1%) change in the liquid volume. Hence we say that liquids are fairly incompressible.

Bulk modulus is used in line pack calculations and transient flow analysis. There are two bulk modulus values used in practice - isothermal and adiabatic. The bulk modulus of a liquid depends on temperature, pressure and specific gravity. The following empirical equations also known as ARCO formulas, may be used to calculate the bulk modulus.

## 2.6.1   Adiabatic Bulk Modulus

$$Ka = A + B(P) - C(T)^{1/2} - D(API) - E(API)^2 + F(T)(API) \qquad (2.27)$$

where

| | | |
|---|---|---|
| $A = 1.286 \times 10^6$ | $B = 13.55$ | $C = 4.122 \times 10^4$ |
| $D = 4.53 \times 10^3$ | $E = 10.59$ | $F = 3.228$ |

P — Pressure in psig
T — Temperature in °R
API — API gravity of liquid

## 2.6.2    Isothermal Bulk Modulus

$$Ki = A + B(P) - C(T)^{1/2} + D(T)^{3/2} - E(API)^{3/2} \qquad (2.28)$$

where

$A = 2.619 \times 10^6$     $B = 9.203$
$C = 1.417 \times 10^5$     $D = 73.05$
$E = 341.0$

P — Pressure in psig
T — Temperature in °R
API — API gravity of liquid

For a typical crude oil of 35°API gravity at 1 000 psig pressure and 80°F temperature, the Bulk Modulus calculated from above Equations (2.27) and (2.28) are:

Adiabatic Bulk Modulus = 231 426 psi
Isothermal Bulk Modulus = 181 616 psi

The bulk modulus of water at 70°F is 320 000 psi

## 2.7    Fundamental concepts of Fluid Flow

In this section we will discuss some fundamental concepts of fluid flow that will set the stage for the succeeding chapter. The basic principles of Continuity and Energy equations are introduced first.

### 2.7.1    Continuity

One of the fundamental concepts that must be satisfied in any type of pipe flow is the principle of Continuity of Flow. This principle states that the total amount of fluid passing through any section of a pipe is fixed. This may also be thought of as the principle of Conservation of Mass. Basically, it means that liquid is neither created nor destroyed as it flows through a pipeline. Since Mass is the product of the volume and density, we can write the following equation for Continuity

$$M = Vol \times \rho = Constant \qquad (2.29)$$

Where

M      - Mass flow rate at any point in the pipeline, slugs/s
Vol    - Volume flow rate at any point in the pipeline, ft³/s
ρ      - Density of liquid at any point in the pipeline, slugs/ft³

Since the volume flow rate at any point in a pipeline is the product of the area of cross section of the pipe and the average liquid velocity, we can rewrite Equation (2.29) as follows:

$$M = A \times V \times \rho = Constant \qquad (2.30)$$

Where

M      - Mass flow rate at any point in the pipeline, slugs/s
A      - Area of cross section of pipe, ft²
V      - Average liquid velocity, ft/s
ρ      - Density of liquid at any point in the pipeline, slugs/ft³

Since liquids are generally considered to be incompressible and therefore density does not change appreciably, the Continuity equation reduces to

$$AV = Constant \qquad (2.31)$$

### 2.7.2    Energy Equation

The basic principle of conservation of energy applied to liquid hydraulics is embodied in the Bernoulli's Equation. This equation simply states that the total energy of the fluid contained in the pipeline at any point is a constant. Obviously, this is an extension of the Principle of Conservation of Energy which states that energy is neither created nor destroyed, but transformed from one form to another.

Consider a pipeline shown in Figure 2.3 that depicts flow from point A to point B with elevation of point A being $Z_A$ and elevation at B being $Z_B$ above some chosen datum. The pressure in the liquid at point A is $P_A$ and that at B is $P_B$. Assuming a general case, where the pipe diameter at A may be different from that at B, we will designate the velocities at A and B to be $V_A$ and $V_B$ respectively. Consider a particle of the liquid of weight W at point A in the pipeline. This liquid particle at A may be considered to possess a total energy E that consists of three components:

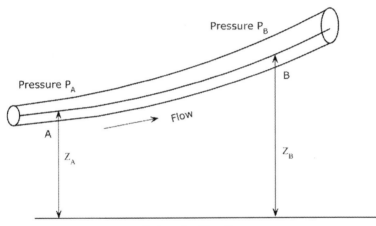

Pressure $P_B$

Pressure $P_A$

B

Flow

A

$Z_A$

$Z_B$

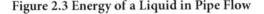

Datum for Elevations

**Figure 2.3 Energy of a Liquid in Pipe Flow**

Energy due to position, or potential energy = $W\,Z_A$
Energy due to pressure, or pressure energy = $WP_A/\gamma$
Energy due to velocity, or kinetic energy = $W\,(V_A/2g)^2$

Where $\gamma$ is the specific weight of liquid.
We can thus state that

$$E = W\,Z_A + WP_A/\gamma + W\,V_A^2/2g \qquad (2.32)$$

Dividing by W throughout, we get the total energy per unit weight of liquid as

$$H_A = Z_A + P_A/\gamma + V_A^2/2g \qquad (2.33)$$

Where $H_A$ is the total energy per unit weight at point A.

Considering the same liquid particle as it arrives at point B, the total energy per unit weight at B is

$$H_B = Z_B + P_B/\gamma + V_B^2/2g \qquad (2.34)$$

Due to conservation of energy

$$H_A = H_B$$

Therefore,

$$Z_A + P_A/\gamma + V_A^2/2g = Z_B + P_B/\gamma + V_B^2/2g \qquad (2.35)$$

Equation (2.35) is one form of the Bernoulli's Equation for fluid flow. In real world pipeline transportation, there is energy loss between point A and point B, due to friction in pipe. We include the energy loss due to friction by modifying Equation (2.35) as follows

$$Z_A + P_A/\gamma + V_A^2/2g = Z_B + P_B/\gamma + V_B^2/2g + \Sigma h_L \qquad (2.36)$$

Where $\Sigma h_L$ represents all the head losses between points A and B, due to friction

In the Bernoulli's equation (2.35), we must also include any energy added to the liquid, such as when there is a pump between points A and B. Thus the left hand side of the equation will have a positive term added to it that will represent the energy generated by a pump.

Equation (2.36) will be modified as follows to include a pump at point A that will add a certain amount of pump head to the liquid

$$Z_A + P_A/\gamma + V_A^2/2g + H_P = Z_B + P_B/\gamma + V_B^2/2g + \Sigma h_L \qquad (2.37)$$

where $H_P$ represents the pump head added to the liquid at point A

In the next chapter, we will further explore the concepts of pressure, velocity, flow rates and energy lost due to pipe friction.


## 2.8   Summary

In this chapter we discussed the more important properties of liquids that determine the nature of liquid flow in pipelines. The specific gravity and viscosity of liquids were explained along with how to calculate these properties in liquid mixtures and at various temperatures. We also introduced the basic concepts of liquid flow consisting of the continuity equation and the energy equation embodied in the Bernoulli's equation.


## 2.9   Problems

2.9.1   Calculate the specific weight and specific gravity of a liquid that weighs 312 lb, contained in volume of 5.9 ft3. Assume water weighs 62.4 lb/ft3.

2.9.2   The specific gravity of a liquid at 60oF and 100oF are reported to be 0.895 and 0.815 respectively. Determine the specific gravity of the liquid at 85oF. Assume linear relationship between gravity and temperature

2.9.3   The gravity of a petroleum product is 59oAPI. Calculate the corresponding specific gravity at 60oF.

2.9.4   The viscosity of a liquid at 70oF is 45 cSt. Express this viscosity in SSU. If the specific gravity at 70oF is 0.885, determine the absolute or dynamic viscosity

2.9.5   The viscosities of a crude oil at 60oF and 100oF are 40 cSt and 15 cSt respectively. Using the ASTM correlation method, calculate the viscosity of this product at 80oF.

2.9.6   Two liquids are blended to form a homogeneous mixture. The first liquid A has a specific gravity of 0.815 at 70 oF and a viscosity of 15 cSt at 70oF. At the same temperature, Liquid B has a specific gravity of 0.85 and viscosity of 25 cSt. If 20% of Liquid A is blended with 80% of Liquid B, calculated the specific gravity and viscosity of the blended product.

2.9.7   Using the viscosity blending chart, calculate the blended viscosity for two liquids as follows:

| Product | Percentage | Viscosity (SSU) |
| --- | --- | --- |
| Liquid A | 15 | 50 |
| Liquid B | 85 | 200 |

2.9.8   If Liquid A with a viscosity of 40 SSU is blended with Liquid B of viscosity 150 SSU. What percentage of each component would be required to obtain a blended viscosity of 46 SSU.

2.9.9   In Figure 2.3, consider the pipe to be 20 in diameter. The liquid is water with specific gravity = 1.00. The point A is at elevation 100 ft and B is at elevation 200 ft. The pressure at A is 500 psi and that at B is 400 psi. Specific weight of water is 62.34 lb/ft3 Write down the Bernoulli's equation for energy conservation between point A and B.

# 3

# Pressure Drop Due to Friction

In this chapter, we introduce the concept of pressure in a liquid and how it is measured. The liquid flow velocity in a pipe, types of flow and the importance of Reynold's number will be discussed. Depending upon the flow regime, such as laminar, critical or turbulent, methods will be discussed on how to calculate the pressure drop due to friction. Several popular formulas such as Colebrook-White and Hazen-Williams equations will be presented and compared. Also we will cover minor pressure losses in piping such as due to fittings and valves and those resulting from pipe enlargements and contractions. We will also explore drag reduction as a means of reducing energy loss in pipe flow.

## 3.1 Pressure

Hydrostatics is the study of hydraulics that deals with liquid pressures and forces resulting from the weight of the liquid at rest. Although this book is mainly concerned with flow of liquids in pipelines, we will address some issues related to liquids at rest in order to discuss some fundamental issues pertaining to liquids at rest and in motion.

The force per unit area at a certain point within a liquid is called the pressure p. This pressure at a certain depth h, below the free surface of the liquid consists of equal pressures in all directions. This is known as Pascal's Law. Consider an imaginary flat surface within the liquid located at a depth h, below the liquid surface as shown in Figure 3.1.

The pressure on this surface must act normal to the surface at all points along the surface because liquids at rest cannot transmit shear. The variation of pressure with the depth of the liquid is calculated by considering forces acting on a thin vertical cylinder of height Δh and a cross sectional area Δa as shown in Figure 3.1

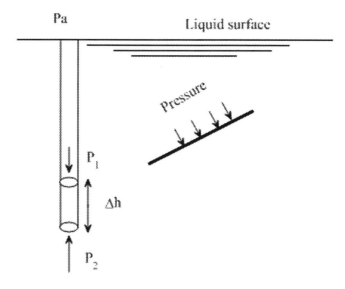

**Figure 3.1 Pressure in a Liquid**

Since the liquid is at rest, the cylindrical volume is in equilibrium due to the forces acting upon it. By the principles of Statics, the algebraic sum of all forces acting on this cylinder in the vertical and horizontal directions must equal zero. The vertical forces on the cylinder consists of the weight of the cylinder and the forces due to liquid pressure $P_1$ at the top and $P_2$ at the bottom as shown in Figure 3.1. Since the specific weight of the liquid γ does not change with pressure, we can write the following equation for the summation of forces in the vertical direction.

$$P_2 \, \Delta a = \gamma \, \Delta h \, \Delta a + P_1 \, \Delta a$$

Where the term γ Δh Δa represents the weight of the cylindrical element.

Simplifying above we get

$$P_2 = \gamma \Delta h + P_1 \tag{3.1}$$

If we now imagine that the cylinder is extended to the liquid surface, $P_1$ becomes the pressure at the liquid surface (atmospheric pressure $P_a$) and $\Delta h$ becomes h, the depth of the point in the liquid where the pressure is $P_2$. Replacing $P_2$ with P, the pressure in the liquid at depth h, Equation (3.1) becomes

$$P = \gamma h + P_a \tag{3.2}$$

From Equation (3.2) we conclude that the pressure in a liquid at a depth h increases with the depth. If the term $P_a$ (atmospheric pressure) is neglected we can state that the gauge pressure (based on zero atmospheric pressure) at a depth h is simply $\gamma h$.

Therefore, the gauge pressure is

$$P = \gamma h \tag{3.3}$$

Dividing both sides by $\gamma$ and transposing we can write

$$h = P / \gamma \tag{3.4}$$

In Equation (3.4) the term h represents the "pressure head" corresponding to the pressure P. It represents the depth in feet of liquid of specific weight $\gamma$ to produce the pressure P. Values of absolute pressure $(P + P_a)$ are always positive whereas the gauge pressure P may be positive or negative depending on whether the pressure is greater or less than the atmospheric pressure. Negative gauge pressure means that a partial vacuum exists in the liquid.

From the above discussion it is clear that the absolute pressure within a liquid consists of the head pressure due to the depth of liquid and the atmospheric pressure at the liquid surface. The atmospheric pressure at a geographic location varies with the elevation above sea level. Because the density of the atmospheric air varies with the altitude, a straight line relationship does not exists between the altitude and the atmospheric pressure (unlike the linear relationship between liquid pressure and depth). For most purposes, we can assume that the atmospheric pressure at sea level is approximately 14.7 psi. In SI units the atmospheric pressure is approximately 101 kPa.

The instrument used to measure the atmospheric pressure at a given location is called a "barometer". A typical barometer is shown in Figure 3.2.

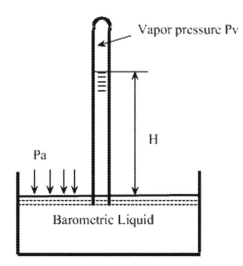

**Figure 3.2 Barometer for Measuring Pressure**

In such an instrument the tube is filled with a heavy liquid (usually mercury) and quickly inverting the tube and positioning it in the container full of the liquid as shown in Figure 3.2.

If the tube is sufficiently long, the level of liquid will fall slightly to cause a vapor space at the top of the tube just above the liquid surface. Equilibrium will be reached when the liquid vaporizes in the vapor space and creates a pressure $P_v$. Because of the density of mercury (approximately 13 times as heavy as water) and its vapor pressure being low, it is an ideal liquid for a barometer. If a liquid such as water were used, we would need a fairly long tube to measure the atmospheric pressure as we shall see shortly.

From Figure 3.2 the atmospheric pressure Pa exerted at the surface of the liquid is equal to the sum of the vapor pressure $P_v$ and the pressure generated by the column of the barometric liquid of height $H_b$.

$$P_a = P_v + \gamma \, H_b \qquad\qquad (3.5)$$

where

$P_a$     - Atmospheric pressure
$P_v$     - Vapor pressure of barometric liquid
$\gamma$     - Specific weight of barometric liquid
$H_b$     - Barometric reading

In the above equation, if pressures are in psi and liquid specific weight is in lb/ft³, the pressures must be multiplied by 144 to obtain the barometric reading in ft of liquid.

Equation (3.5) is valid for barometers with any liquid. Since, the vapor pressure of mercury is negligible we can rewrite Equation (3.5) for a mercury barometer as follows:

$$P_a = \gamma \, H_b \qquad\qquad (3.6)$$

Let us compare using water and mercury as barometric liquids to measure the atmospheric pressure.

**Example Problem 3.1**

Assume the vapor pressure of water at 70°F is 0.3632 psi and its specific weight is 62.3 lb/ft³. Mercury has a specific gravity of 13.54 and negligible vapor pressure. If the sea level atmospheric pressure is 14.7 psi. Determine the barometric heights for water and mercury.

**Solution**

From Equation (3.5) for water,

$$14.7 = 0.3632 + (62.3/144) \, H_b$$

The barometric height for water is

$$H_b = (14.7 - 0.3632) \times 144 / 62.3 = 33.14 \text{ ft}$$

Similarly, for mercury, neglecting the vapor pressure, using Equation (3.6) we get

14.7 x 144 = (13.54 x 62.3) $H_b$

The barometric height for mercury is

$H_b$ = (14.7 x 144) / (13.54 x 62.3) = 2.51 ft

It can be seen from the above that the mercury barometer requires a much shorter tube than a water barometer.

Manometers are instruments used to measure pressure in reservoirs, channels and pipes. Manometers are discussed further in the chapter titled Flow Measurement

The pressure in a liquid is measured in $lb/in^2$ (psi) in the English units or kiloPascal (kPa) in SI units. Since pressure is measured using a gauge and is relative to the atmospheric pressure at the specific location, it is also reported as psig (psi gauge). The absolute pressure in a liquid is the sum of the gauge pressure and the atmospheric pressure at the location. Thus,

Absolute pressure in psia = gauge pressure in psig + atmospheric pressure.

For example, if the pressure gauge reading is 800 psig, the absolute pressure in the liquid is:

Pabs = 800+14.7 = 814.7 psia

This is based on the assumption that atmospheric pressure at the location is 14.7 psia.

Pressure in a liquid may also be referred to in terms of feet (or meters in SI units) of liquid head. By dividing the pressure in $lb/ft^2$ by the liquid specific weight in $lb/ft^3$, we get the pressure head in ft of liquid. When expressed this way, the head represents the height of the liquid column required to match the given pressure in psig. For example, if the pressure in a liquid is 1 000 psig, the head of liquid corresponding to this pressure is calculated as follows:

Head = 2.31(psig)/Spgr ft (English units)            (3.7)

Head = 0.102(kPa)/Spgr m (SI units).            (3.8)

where

Spgr    - Liquid specific gravity

The factor 2.31 in the above equation comes from the ratio

$$\frac{144 \text{ in}^2 \,/\, \text{ft}^2}{62.34 \text{ lb/ft}^3}$$

where 62.34 lb/ft$^3$ is the specific weight of water.

Therefore, if the liquid specific gravity is 0.85, the equivalent liquid head is

Head = (1 000)(2.31) / 0.85 = 2 717.65 ft

This means that the liquid pressure of 1 000 psig is equivalent to the pressure exerted at the bottom of a liquid column, of specific gravity 0.85, 2 717.65 feet in height. If such a column of liquid had a cross-sectional area of one square inch, the weight of the column will be

2 717.65 (1/144) (62.34) (0.85) = 1 000 lb

where 62.34 lb/ft$^3$ is the density of water.

The above weight acts on area of one square inch. The calculated pressure is therefore 1 000 psig.

We can analyze head pressure due to a column of liquid in another way:

Consider a cylindrical column of liquid, of height H ft and area of cross section A in$^2$. If the top surface of the liquid column is open to the atmosphere, we can calculate the pressure exerted by this column of liquid at its base as

Pressure = $\dfrac{\text{Weight of liquid column}}{\text{Area of cross section}}$

or

Pressure = $\dfrac{\text{(Volume x Specific weight)}}{\text{Area of cross section}}$

= (AHγ) / (144 x A)

or

Pressure = Hγ / 144                                          (3.9)

Where

γ        - Specific weight of liquid, lb/ft$^3$

The factor 144 is used to convert from in$^2$ to ft$^2$.

If we use 62.34 lb/ft$^3$ for specific weight of water, the pressure for a water column from above Equation (3.9) is

Pressure = H x 62.34 / 144 = H/2.31

## 3.2   Velocity

Velocity of flow in a pipeline is the average velocity based on the pipe diameter and liquid flow rate. It may be calculated as follows:

Velocity = Flow rate/ area of flow

Depending on the type of flow (laminar, turbulent, etc), the liquid velocity in a pipeline at a particular pipe cross section will vary along the pipe radius. The liquid molecules at the pipe wall are at rest and therefore have zero velocity. As we approach the center line of the pipe, the liquid molecules are increasingly free and therefore have increasing velocity. The variation of velocity for laminar flow and turbulent flow are as shown in Figure 3.3. In laminar flow, the variation of velocity at a pipe cross section is parabolic. In turbulent flow there is an approximate trapezoidal shape to the velocity profile. Laminar flow is also known as viscous flow or streamline flow.

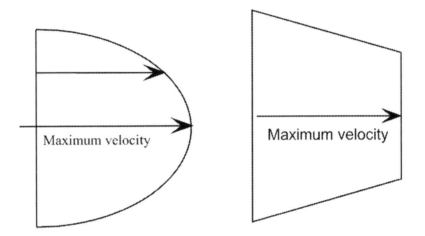

**Figure 3.3 Velocity variation in pipe for laminar and turbulent flow**

If the units of flow rate are bbl/day and pipe inside diameter is in inches the following equation for average velocity may be used.

$$V = 0.0119(bbl/day)/ D^2 \qquad (3.10)$$

where

V    - Velocity, ft/s
D    - Inside diameter, in

Other forms of the equation for velocity in different units are as follows:

$$V = 0.4085(gal/min)/ D^2 \qquad (3.11)$$
$$V = 0.2859(bbl/hr)/D^2 \qquad (3.12)$$

where

V     - Velocity, ft/s
D     - Inside diameter, in

In SI units, the velocity is calculated as follows:

$$V = 353.6777 \ (m^3/hr)/ \ D^2 \tag{3.13}$$

where

V     - Velocity, m/s
D     - Inside diameter, mm.

For example, liquid flowing through a 16-inch pipeline (wall thickness 0.250 inch) at the rate of 100 000 bbl/day, has an average velocity of:

$$0.0119(100 \ 000) \ / \ (15.5)^2 = 4.95 \ ft/s$$

This represents the average velocity at a particular cross-section of pipe. The velocity at the centerline will be higher than this, depending on whether the flow is turbulent or laminar.

## 3.3   Reynold's Number

Flow in a liquid pipeline may be smooth, laminar flow also known as viscous flow. In this type of flow the liquid flows in layers or laminations without causing eddies or turbulence. If the pipe is transparent and we inject a dye into the flowing stream it would flow smoothly in a straight line confirming smooth or laminar flow. As the liquid flow rate is increased, the velocity increases and the flow will change from laminar flow to turbulent flow with eddies and disturbances. This can be seen clearly when a dye is injected into the flowing stream.

An important dimensionless parameter, called Reynold's number is used in classifying the type of flow in pipelines.

Reynold's number of flow, R is calculated as follows:

$$R = VD\rho/\mu \qquad (3.14)$$

where

V — Average velocity, ft/s
D — Pipe internal diameter, ft
$\rho$ — Liquid density, slugs/ft$^3$
$\mu$ — Absolute viscosity, lb-s/ft$^2$
R — Reynold's number is a dimensionless value.

Since the kinematic viscosity $\nu = \mu / \rho$ the Reynold's number can also be expressed as

$$R = VD/\nu \qquad (3.15)$$

where

$\nu$ — Kinematic viscosity, ft$^2$/s

Care should be taken to ensure that proper units are used in Equation (3.14) and (3.15) such that R is dimensionless.

Flow through pipes are classified into three main flow regimes:

1. Laminar flow — R < 2 000
2. Critical flow — R > 2 000 and R < 4 000
3. Turbulent flow — R > 4 000

Depending upon the Reynold's number, flow through pipes will fall in one of the above three flow regimes. Let us first examine the concepts of Reynold's number. Sometimes an R value of 2 100 is used as the limit of laminar flow.

Using Customary units in the pipeline industry, the Reynold's number can be calculated using the following formula:

$$R = 92.24 \, Q/(vD) \tag{3.16}$$

where

Q    - Flow rate, bbl/day
D    - Internal diameter, in
v    - Kinematic viscosity, cSt.

The Equation (3.16) is simply a modified form of Equation (3.15) after performing conversions to commonly used pipeline units. R is still a dimensionless value.

Another version of Reynold's Number in English Units is as follows:

$$R = 3\,160 \, Q/(vD) \tag{3.17}$$

where

Q    - Flow rate, gal/min
D    - Internal diameter, in
v    - Kinematic viscosity, cSt.

A similar equation for Reynold's number in SI units is

$$R = 353\,678 \, Q/(vD) \tag{3.18}$$

where

Q    - Flow rate, m³/h
D    - Internal diameter, mm
v    - Kinematic viscosity, cSt.

As indicated earlier, if the Reynold's number is less than 2 000, the flow is considered to be laminar. This is also known as viscous flow. This means that the various layers of liquid flow without turbulence in the form of laminations. We will now illustrate the various flow regimes using an example.

Consider a 16 inch pipeline, 0.250 inch wall thickness transporting a liquid of viscosity 250 cSt. At a flow rate of 50 000 bbl/day the Reynold's number is using Equation (3.16)

R = 92.24(50 000) / (250 x 15.5) = 1 190

Since R is less than 2 000, this flow is laminar. If the flow rate is tripled to 150 000 bbl/day, the Reynold's number becomes 3 570 and the flow will be in the critical region. At flow rates above 168 040 bbl/day the Reynold's number exceeds 4 000 and the flow will be in the turbulent region. Thus, for this 16 inch pipeline and given liquid viscosity of 250 cSt, flow will be fully turbulent at flow rates above 168 040 bbl/day.

As the flow rate and velocity increase, the flow regime changes. With change in flow regime, the energy lost due to pipe friction increases. At laminar flow, there is less frictional energy lost compare to turbulent flow.

## 3.4   Flow Regimes

In summary, the three flow regimes may be distinguished as follows:

Laminar     - Reynold's number < 2 000
Critical - Reynold's number > 2 000 and Reynold's number < 4 000
Turbulent   - Reynold's number > 4 000

As liquid flows through a pipeline, energy is lost due to friction between the pipe surface and the liquid and due to the interaction between liquid molecules. This energy lost is at the expense of liquid pressure. See the Bernoulli's equation in Chapter 2. Hence we refer to the frictional energy lost as pressure drop due to friction.

The pressure drop due to friction in a pipeline depends on the flow rate, pipe diameter, pipe roughness, liquid specific gravity and viscosity. In addition, frictional pressure drop depends on the Reynold's number (and hence the flow regime). Our objective would be to calculate the pressure drop given these pipe and liquid properties and the flow regime.

The pressure drop in a given length of pipe, expressed in feet of liquid head (h), can be calculated using the Darcy-Weisbach equation as follows:

$$h = f(L/D)(V^2/2g) \tag{3.19}$$

where the pressure drop due to friction, h is expressed in ft of liquid head and the other symbols are defined below

f      - Darcy friction factor, dimensionless, usually a number 0.008 to 0.10

L      - Pipe length, ft

D      - Pipe internal diameter, ft

V      - Average liquid velocity, ft/s

g      - acceleration due to gravity, 32.2 ft/s² in English units.

In laminar flow, the friction factor f depends only on the Reynold's number. In turbulent flow f depends on pipe diameter, internal pipe roughness and Reynold's number, as we will see shortly.

**Example Problem 3.2**

Consider a pipeline transporting 4 000 bbl/hr of gasoline (Spgr = 0.736). Calculate the pressure drop in a 5 000 ft length of 16 in pipe (wall thickness 0.250 in) using the Darcy-Weisbach Equation.
Assume the friction factor is 0.02.

**Solution**

Using Equation (3.10)

Average liquid velocity = 0.0119(4 000 x 24) /(15.5) ² = 4.76 ft/s

using Darcy-Weisbach Equation (3.19)

Pressure drop = 0.02(5 000)(12/15.5)(4.76 ² /64.4) = 27.24 ft of head

Converting to pressure in psi, using Equation (3.7)

Pressure drop = 27.24(0.736)/2.31 = 8.68 psi.

In the above calculations, the friction factor f was assumed to be 0.02. However, the actual friction factor for a particular flow depends on various factors as explained previously. In the next section, we will see how the friction factor is calculated for the various flow regimes.

## 3.5 Friction Factor

For laminar flow, with Reynold's number R < 2 000, the Darcy friction factor f is calculated from the simple relationship:

$$f = 64/R \qquad\qquad (3.20)$$

It can be seen from Equation (3.20) that for laminar flow, the friction factor depends only on the Reynold's number and is independent of the internal condition of the pipe. Thus, regardless of whether the pipe is smooth or rough, the friction factor for laminar flow is a number that varies inversely as the Reynold's number.

Therefore, if the Reynold's Number R = 1 800, the friction factor becomes

$$f = 64 / 1\ 800 = 0.0356$$

It might appear that since f for laminar flow decreases with Reynold's number, using Darcy-Weisbach equation the pressure drop will decrease with increase in flow rate. This is not true. Since pressure drop is proportional to the velocity V squared (Equation 3.19), the influence of V is greater than that of f. Therefore, pressure drop will increase with flow rate in the laminar region.

To illustrate, consider the Reynold's number example in Section 3.3 discussed earlier. If the flow rate is increased from 50 000 bbl/day to 80 000 bbl/day, the Reynold's number R will increase from 1 190 to 1 904 (still laminar). The velocity will increase from $V_1$ to $V_2$ as follows

$V_1 = 0.0119 \ (50\ 000) \ / \ (15.5)^2 = 2.48$ ft/s
$V_2 = 0.0119 \ (80\ 000) \ / \ (15.5)^2 = 3.96$ ft/s

Friction factors at 50 000 bbl/day and 80 000 bbl/day flow rate are

$f_1 = 64/1\ 190 = 0.0538$
$f_2 = 64/1\ 904 = 0.0336$

Considering 5 000 ft length of pipe, the head loss due to friction using Darcy-Weisbach Equation (3.19)

$h_{L1} = 0.0538 \times (5\ 000 \times 12 \ / \ 15.5) \times (2.48^2 \ / \ 64.4) = 19.89$ ft
$h_{L2} = 0.0336 \times (5\ 000 \times 12 \ / \ 15.5) \times (3.96^2 \ / \ 64.4) = 31.67$ ft

Therefore from above it is clear in laminar flow even though the friction factor decreases with flow increase, the pressure drop still increases with increase in flow rate.

For turbulent flow, when the Reynold's number R > 4 000 the friction factor f depends not only on R, but also on the internal roughness of the pipe. As the pipe roughness increases, so does the friction factor. Therefore, smooth pipes have less friction factor compared to rough pipes. More correctly, friction factor depends on the relative roughness (e/D) rather than the absolute pipe roughness e.

Various correlations exist for calculating friction factor f. These are based on experiments conducted by scientists and engineers over the last 60 years or more. A good all purpose equation for the friction factor f in the turbulent region is called the Colebrook-White Equation as follows:

$1/ \sqrt{f} = -2 \ Log_{10}[(e/3.7D) + 2.51/(R \ \sqrt{f})]$         (3.21)
and applies only for Turbulent flow R > 4 000

where

| | |
|---|---|
| f | - Darcy friction factor, dimensionless |
| D | - Pipe internal diameter, inches |
| e | - Absolute pipe roughness, inches |
| R | - Reynold's number of flow, dimensionless |

In SI Units, the above equation for f remains the same, as long as the absolute roughness e and the pipe diameter D are both expressed in mm. All other terms in the equation are dimensionless.

It can be seen from Equation (3.21), that the calculation of f is not easy, since it appears on both sides of the equation. A trial and error approach needs to be used. We assume a starting value of f (say 0.02) and substitute it in the right hand side of Equation (3.21). This will yield a second approximation for f, which can then be used to re-calculate a better value of f, by successive iteration. Generally, three to four iterations will yield a satisfactory result for f correct to, within 0.001.

During the last two or three decades several formulas for friction factor for turbulent flow have been put forth by various researchers. All of these equations attempt to simplify calculation of the friction factor compared to the Colebrook-White Equation discussed above. Two such equations that are explicit equations in f, afford easy solution of friction factor compared to the implicit equation (3.21) that requires trial and error solution. These are called the Churchill Equation and Swamee-Jain Equation and will be discussed later on in this chapter.

In the critical zone, where Reynold's number is between 2 000 and 4 000 there is no generally accepted formula for determining the friction factor. This is because the flow is unstable in this region and therefore the friction factor is indeterminate. Most users calculate the value of f based upon turbulent flow.

To make matters more complicated, the turbulent flow region (R > 4 000) actually consists of three separate regions:

Turbulent flow in smooth pipes
Turbulent flow in fully rough pipes
Transition flow between smooth and rough pipes

For Turbulent flow in smooth pipes, pipe roughness has negligible effect on the friction factor. Therefore, friction factor in this region depends only on the Reynold's number as follows:

$$1/\sqrt{f} = -2 \, Log_{10} \, [2.51/(R \, \sqrt{f})] \tag{3.22}$$

For Turbulent flow in fully rough pipes, the friction factor f appears to be less dependent on Reynold's number as the latter increases in magnitude. It depends only on the pipe roughness and diameter. It can be calculated from the following equation:

$$1/\sqrt{f} = -2 \, Log_{10}[(e/3.7D)] \tag{3.23}$$

For the Transition region between turbulent flow in smooth pipes and turbulent flow in fully rough pipes, the friction factor f is calculated using the Colebrook-White equation given previously:

$$1/\sqrt{f} = -2 \, Log_{10}[(e/3.7D) + 2.51/(R \, \sqrt{f})] \tag{3.24}$$

As mentioned before, In SI Units, the above equation for f remains the same, if e and D are both in mm.

The friction factor equations discussed above are also plotted on the Moody diagram as shown in Figure 3.4. Relative roughness is defined as e/D. It is simply the result of dividing the absolute pipe roughness by the pipe inside diameter. The relative roughness term is a dimensionless parameter.

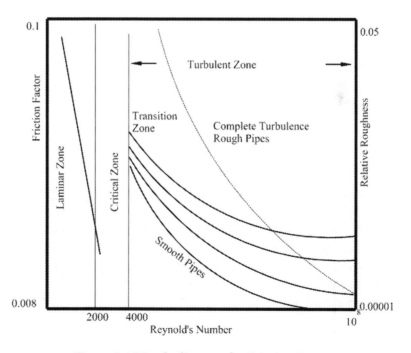

**Figure 3.4 Moody diagram for Friction Factor**

The Moody diagram represents the complete friction factor map for laminar and all turbulent regions of pipe flows. It is used commonly in estimating friction factor in pipe flow. If the Moody Diagram is not available, we must use trial and error solution of Equation (3.24) to calculate the friction factor.

To use the Moody diagram for determining the friction factor f we first calculate the Reynold's number R for the flow. Next, we find the location on the horizontal axis of Reynold's number for the value of R and draw a vertical line that intersects with the appropriate relative roughness (e/D) curve. From this point of intersection on the (e/D) curve, we go horizontally to the left and read the value of the friction factor f on the vertical axis on the left.

Before leaving the discussion of friction factor, we must mention an additional term called the Fanning friction factor. Some publication use this friction factor instead of the Darcy friction factor.

The Fanning friction factor is defined as follows:

$$f_f = f_d / 4 \qquad\qquad (3.25)$$

where

$f_f$     - Fanning friction factor
$f_d$     - Darcy friction factor

Unless otherwise specified, we will use the Darcy friction factor f throughout this book.

## Example Problem 3.3

Water flows through a 20 inch pipe at 5 700 gal/min. Calculate the friction factor using Colebrook-White Equation. Assume 0.375 inch pipe wall thickness and an absolute roughness of 0.002 inch. Use specific gravity of 1.00 and viscosity of 1.0 cSt. What is the head loss due to friction in 2 500 ft of pipe?

**Solution**

First we calculate the Reynold's number from Equation (3.17) as follows:

R = 3 160 x 5 700 / (19.25 x 1.0) = 935 688

The flow is fully turbulent and the friction factor f is calculated using Equation (3.21) as follows:

$1/\sqrt{f}$ = -2 $Log_{10}$[(0.002 / (3.7 x 19.25)) + 2.51/(935 688 $\sqrt{f}$)]

The above implicit equation for f must be solved by trial and error.

First assume a trial value of f = 0.02. Substituting in equation above, we get a successive approximations for f as follows:

f = 0.0133, 0.0136 and 0.0136

Therefore the solution is f = 0.0136

Using Equation 3.12,

velocity = 0.4085 (5 700) / $19.25^2$ = 6.28 ft/s

Using Equation (3.19), head loss due to friction is

h = 0.0136 x (2 500 x 12 / 19.25) x $6.28^2$ /64.4 = 12.98 ft

## 3.6   Pressure Drop Due to Friction

In the previous section, we introduced the Darcy-Weisbach equation as follows

$$h = f(L/D)(V^2/2g) \hspace{3cm} (3.26)$$

where the pressure drop h is expressed in ft of liquid head and the other symbols are defined below

f — Darcy friction factor, dimensionless
L — Pipe length, ft
D — Pipe internal diameter, ft
V — Average liquid velocity, ft/sec
g — Acceleration due to gravity, 32.2 ft/s$^2$ in English units.

A more practical equation, using customary pipeline units, is given below for calculating the pressure drop in pipelines.

Pressure drop due to friction per unit length of pipe, in English units is

$$Pm = 0.0605 \ f \ Q^2 \ (Sg/D^5) \qquad\qquad (3.27)$$

and in terms of transmission factor F

$$Pm = 0.2421(Q/F)^2 \ (Sg/D^5) \qquad\qquad (3.28)$$

where

Pm — Pressure drop due to friction, lb/in$^2$ per mile (psi/mi) of pipe length
Q — Liquid flow rate, bbl/day
f — Darcy friction factor, dimensionless
F — Transmission factor, dimensionless
Sg — Liquid specific gravity
D — Pipe internal diameter, in

The transmission factor F is directly proportional to the volume that can be transmitted through the pipeline and therefore has an inverse relationship with the friction factor f. The transmission factor F is calculated from the following equation.

$$F = 2/\sqrt{f} \qquad\qquad (3.29)$$

Since friction factor f ranges from 0.008 to 0.10 it can be seen from Equation (3.29) that the transmission factor F ranges from 6 to 22 approximately.

The Colebrook-White Equation (3.21) can be re-written in terms of the transmission factor F as follows:

$$F = -4 \, Log \, [(e/3.7D) + 1.255(F/R)] \qquad (3.30)$$
$$\text{for Turbulent flow R} > 4\,000$$

Similar to the calculation of the friction factor f using Equation (3.21), the calculation of transmission factor F from Equation (3.30) will also be a trial and error approach. We assume a starting value of F (say 10.0) and substitute it in the right hand side of Equation (3.30). This will yield a second approximation for F, which can then be used to re-calculate a better value, by successive iteration. Generally, three to four iterations will yield a satisfactory result for F.

In SI units, the Darcy equation(in pipeline units) for the pressure drop in terms of the friction factor is represented as follows:

$$P_{km} = 6.2475 \times 10^{10} \, f \, Q^2 \, (Sg/D^5) \qquad (3.31)$$

and the corresponding equation in terms of transmission factor F is written as follows:

$$P_{km} = 24.99 \times 10^{10} \, (Q/F)^2 \, (Sg/D^5) \qquad (3.32)$$

where

| | |
|---|---|
| $P_{km}$ | - Pressure drop due to friction in kiloPascal/km (kPa/km) |
| Q | - Liquid flow rate, $m^3/hr$ |
| f | - Darcy friction factor, dimensionless |
| F | - Transmission factor, dimensionless |
| Sg | - Liquid specific gravity |
| D | - Pipe internal diameter, mm |

In SI Units, the transmission factor F is calculated using Equation (3.30) as follows:

$$F = -4 \, Log[(e/3.7D) + 1.255(F/R)] \qquad (3.33)$$
for Turbulent flow R > 4 000

where

| | |
|---|---|
| D | - Pipe internal diameter, mm |
| e | - Absolute pipe roughness, mm |
| R | - Reynold's number of flow, dimensionless |

**Example Problem 3.4**

Consider a 100 mile pipeline, 16 inch diameter, 0.250 inch wall thickness, transporting a liquid (specific gravity of 0.815 and viscosity of 15 cSt at 70°F) at a flow rate of 90 000 bbl/day. Calculate the friction factor and pressure drop per unit length of pipeline using Colebrook-White Equation. Assume 0.002 in pipe roughness.

**Solution**

The Reynold's number is calculated first.

$$R = \frac{92.24 \times 90\,000}{15.5 \times 15} = 35\,706$$

Using Colebrook-White Equation (3.30), the Transmission Factor is

$$F = -4 \, Log[(0.002/ (3.7 \times 15.5)) + 1.255F/35\,706]$$

Solving above equation by trial and error, yields

$$F = 13.21$$

To calculate the friction factor f, we use Equation (3.29) after some transposition and simplification as follows:

Friction factor $f = 4/F^2 \quad = 4/(13.21)^2 \quad = 0.0229$

The pressure drop per mile is calculated using Equation (3.28)

$Pm = 0.2421(90\,000/13.21)^2\,(0.815/15.5^5) = 10.24\ psi/mi$

The total pressure drop in 100 mi length is then

Total pressure drop $= 100 \times 10.24 = 1\,024\ psi$

## 3.7  Colebrook-White Equation

In 1956 the US Bureau of Mines conducted experiments and recommended a modified version of the Colebrook-White equation. The modified Colebrook-White equation yields a more conservative transmission factor F. The pressure drop calculated using the modified Colebrook-White equation is slightly higher than that calculated using the original Colebrook-White equation. This modified Colebrook-White equation, in terms of Transmission factor F is defined as follows:

$$F = -4\,Log\,[(e/3.7D) + 1.4125(F/R)] \tag{3.34}$$

In SI units, the Transmission factor equation above remains the same with e and D expressed in mm, other terms being dimensionless.

Comparing Equation (3.34) with Equation (3.30) or (3.33), it can be seen that the only change is in the substitution of the constant 1.4125 in place of 1.255 in the original Colebrook-White Equation. Some companies use the modified Colebrook-White Equation stated in Equation (3.34).

An explicit form of an equation to calculate the friction factor was proposed by Swamee and Jain. This equation does not require trial and error solution like the Colebrook-White equation. It correlates very closely to the Moody Diagram values. Refer to the Appendix for a version of the Swamee-Jain Equation for friction factor.

## 3.8 Hazen-Williams Equation

Hazen-Williams equation is commonly used in the design of water distribution lines and in calculation of frictional pressure drop in refined petroleum products such as gasoline, diesel etc. This method involves the use of Hazen-Williams C-factor instead of pipe roughness or liquid viscosity.

The pressure drop calculation using Hazen-Williams equation takes into account flow rate, pipe diameter and specific gravity as follows:

$$h = 4.73 \, L \, (Q/C)^{1.852} / D^{4.87} \tag{3.35}$$

where

| | |
|---|---|
| h | - Head loss due to friction, ft |
| L | - Length of pipe, ft |
| D | - Internal diameter of pipe, ft |
| Q | - Flow rate, ft$^3$/s |
| C | - Hazen-Williams coefficient or C-Factor, dimensionless |

Typical values of Hazen-Williams C-factor are given in Appendix A, Table A.8

In customary pipeline units, the Hazen-Williams equation can be re-written as follows:

In English Units

$$Q = 0.1482(C) \, (D)^{2.63}(Pm/Sg)^{0.54} \tag{3.36}$$

where

| | |
|---|---|
| Q | - Flow rate, bbl/day |
| D | - Pipe internal diameter, in |
| Pm | - Frictional pressure drop, psi/mi |
| Sg | - Liquid specific gravity |
| C | - Hazen-Williams C-factor |

Another form of Hazen-Williams equation, when the flow rate is in gal/min and head loss is measured in feet of liquid per thousand feet of pipe is as follows:

$$GPM = 6.7547 \times 10^{-3} \ (C) \ (D)^{2.63} \ (H_L)^{0.54} \tag{3.37}$$

where

GPM  - Flow rate, gal/min
$H_L$    - Friction loss, ft of liquid per 1000 ft of pipe

Other symbols remain the same.

In SI Units, the Hazen-Williams formula is as follows

$$Q = 9.0379 \times 10^{-8} \ (C)(D)^{2.63}(P_{km}/Sg)^{0.54} \tag{3.38}$$

where

Q      - Flow rate, $m^3/hr$
D      - Pipe internal diameter, mm
$P_{km}$   - Frictional pressure drop, kPa/km
Sg     - Liquid specific gravity
C      - Hazen-Williams C-factor

Historically, many empirical formulas have been used to calculate frictional pressure drop in pipelines. Hazen-Williams Formula has been used widely in the analysis of pipeline networks and water distribution systems, because of its simple form and ease of use. A review of the Hazen-Williams formula shows that the pressure drop due to friction depends on the liquid specific gravity, pipe diameter and the Hazen-Williams coefficient or C factor.

Unlike, Colebrook-White equation where the friction factor is calculated based on pipe roughness, pipe diameter and the Reynold's number, which further depends on liquid specific gravity and viscosity, the Hazen-Williams C factor appears to not take into account the liquid viscosity or pipe roughness. It could be argued that the C factor is in fact a measure of the pipe internal roughness. However, there does not seem to be any indication of how the C factor varies from laminar flow to turbulent flow.

We could compare the Darcy-Weisbach equation with the Hazen-Williams equation and infer that the C factor is a function of Darcy friction factor and Reynold's number. Based on this comparison it can be concluded that the C factor is indeed an index of relative roughness of the pipe. It must be remembered that the Hazen-Williams equation, though convenient from the standpoint of its explicit nature, must be regarded as an empirical equation and that it is difficult to apply to all fluids under all conditions. Nevertheless, in real world pipelines, with sufficient field data we could determine specific C factors for specific pipelines and fluids pumped.

**Example Problem 3.5**

A 3 inch (internal diameter) smooth pipeline is used to pump 100 gal/min of water. Using Hazen-Williams formula, calculate the head loss in 3 000 ft of this pipe. Assume C factor = 140.

**Solution**

Using Equation (3.37) substituting given values we get

$$100 = 6.7547 \times 10^{-3} \times 140 \, (3.0)^{2.63} \, (H_L)^{0.54}$$

Solving for the head loss we get

$$H_L = 26.6 \text{ ft per } 1000 \text{ ft}$$

Therefore head loss for 3 000 ft = 26.6 x 3 = 79.8 ft of water.

## 3.9  Shell-MIT Equation

The Shell-MIT equation, sometimes called the MIT equation, is used in calculation of pressure drop in heavy crude oil and heated liquid pipelines. Using this method, a modified Reynold's number $R_m$ is calculated first from the Reynold's number as follows:

$$R = 92.24(Q)/(Dv) \tag{3.39}$$
$$Rm = R/(7\,742) \tag{3.40}$$

where

R     - Reynold's number, dimensionless
Rm    - Modified Reynold's number, dimensionless
Q     - Flow rate, bbl/day
D     - Internal diameter, inches
v     - Kinematic viscosity, cSt.

Next, depending on the flow (laminar or turbulent) the friction factor is calculated from one of the following equation.

$$f = 0.00207/Rm \qquad\text{- Laminar flow} \tag{3.41}$$
$$f = 0.0018 + 0.00662(1/Rm)^{0.355} \quad\text{- Turbulent flow} \tag{3.42}$$

Note that this friction factor f in the above equations is not the same as the Darcy friction factor f discussed earlier. In fact, the friction factor f in above equation is more like the Fanning friction factor discussed previously.

Finally, the pressure drop due to friction is calculated using the equation

$$Pm = 0.241\,(f\,SgQ^2)/D^5 \tag{3.43}$$

where

Pm    - Frictional pressure drop, psi/mi
f     - Friction factor, dimensionless
Sg    - Liquid specific gravity
Q     - Flow rate, bbl/day
D     - Pipe internal diameter, in

In SI Units the MIT Equation is expresses as follows

$$Pm = 6.2191 \times 10^{10}\,(f\,SgQ^2)/D^5 \tag{3.44}$$

where

Pm   - Frictional pressure drop, kPa/km
f    - Friction factor, dimensionless
Sg   - Liquid specific gravity
Q    - Flow rate, $m^3/hr$
D    - Pipe internal diameter, mm

Comparing Equation (3.43) with Equations (3.27) and (3.28) and recognizing the relationship between transmission factor F and Darcy friction factor f, using Equation (3.29) it is evident that the friction factor f in Equation (3.43) is not the same as the Darcy friction factor. It appears to be one-fourth the Darcy friction factor.

## Example Problem 3.6

A 500 mm outside diameter, 10 mm wall thickness steel pipeline is used to transport heavy crude oil at a flow rate of 800 $m^3/hr$ at 100°C. Using the MIT equation calculate the friction loss per kilometer of pipe assuming internal pipe roughness of 0.05 mm. The heavy crude oil has a specific gravity of 0.89 at 100°C and a viscosity of 120 cSt at 100 °C.

## Solution

From Equation (3.18)

Reynold's number = 353 678 x 800 / (120 x 480) = 4 912

The flow is therefore turbulent

Modified Reynold's number = 4 912 / 7 742 = 0.6345
Friction factor = $0.0018 + 0.00662 (1 / 0.6345)^{0.355} = 0.0074$

Pressure drop from Equation 3.44 is

Pm = $6.2191 \times 10^{10} (0.0074 \times 0.89 \times 800 \times 800)/480^5$
   = 10.29 kPa/km

## 3.10 Miller Equation

The Miller Equation also known as the Benjamin Miller formula is used in hydraulics studies involving crude oil pipelines. This equation does not consider pipe roughness and is an empirical formula for calculating the flow rate from a given pressure drop. The equation can also be re-arranged to calculate the pressure drop from a given flow rate. One of the popular versions of this equation is as follows

$$Q = 4.06 \ (M) \ (D^5 P_m / Sg)^{0.5} \tag{3.45}$$

where M is defined as follows:

$$M = Log_{10}(D^3 SgP_m / cp^2) + 4.35 \tag{3.46}$$

and

Q    - Flow rate, bbl/day
D    - Pipe internal diameter, in
$P_m$    - Pressure drop, psi/mi
Sg    - Liquid specific gravity
cp    - Liquid viscosity, centipoise

In SI Units, the Miller Equation is as follows:

$$Q = 3.996 \text{ x } 10^{-6} \ (M) \ (D^5 P_m / Sg)^{0.5} \tag{3.47}$$

and M is defined as follows:

$$M = Log_{10}(D^3 SgP_m / cp^2) - 0.4965 \tag{3.48}$$

where

Q    - Flow rate, m³/hr
D    - Pipe internal diameter, mm
Pm    - Frictional pressure drop, kPa/km
Sg    - Liquid specific gravity
cp    - Liquid viscosity, centipoise

It can be seen from the above version of Miller equation that calculating the pressure drop $P_m$ from the flow rate Q is not straight forward. This is because the parameter M depends on the pressure drop $P_m$. Therefore, if we solve for $P_m$ in terms of Q and other parameters from equation (3.45) we get

$$P_m = (Q / 4.06M)^2 \, (Sg / D^5) \, (3.49)$$

where M is calculated from Equation (3.46)

To calculate $P_m$ from a given value of flow rate Q, we use a trial and error approach. First, we assume a value of $P_m$ to get a starting value of M from Equation (3.46). This value of M is then substituted in Equation (3.49) above to determine a second approximation for $P_m$. This value of $P_m$ will be used to generate a better value of M form Equation (3.46) which is then used to recalculate $P_m$. Once the successive values of $P_m$ are within an allowable tolerance, such as 0.01 psi/mile, the iteration can be terminated and the value of pressure drop $P_m$ is calculated.

## Example Problem 3.7

Using Miller equation determine the pressure drop in a 14 inch, 0.250 inch wall thickness, crude oil pipeline at a flow rate of 3 000 gal/min. The crude oil properties are: Specific gravity = 0.825 at 60°F and viscosity = 15 cSt at 60°F.

## Solution

Liquid viscosity in centipoise = 0.825 x 15 = 12.375 cP

First the parameter M is calculated from Equation (3.46) using an initial value of Pm = 10.0

$$M = Log_{10}(13.5^3 \text{ x } 0.825 \text{ x } 10.0 / 12.375^2) + 4.35$$
$$= 6.4724$$

Using this value of M in Equation (3.49), we get

$$P_m = [3\,000 \times 34.2857 / (4.06 \times 6.4724)]^2 (0.825 / 13.5^5)$$
$$= 28.19 \text{ psi/mi}$$

We were quite far off in our initial estimate of $P_m$.
Using this value of Pm, a new value of M is calculated as

$$M = 6.9225$$

Substituting this value of M in Equation (3.49), we get

$$P_m = 24.64$$

By successive iteration, we get the final value for Pm = 25.02 psi/mi

## 3.11 T.R. Aude Equation

Another pressure drop equation used in the pipeline industry that is popular among companies that transport refined petroleum products is the T.R. Aude equation, sometimes referred to simply as the Aude equation. This equation is named after the engineer that conducted experiments on pipelines in the 1950s.

The Aude Equation is used in pressure drop calculations for 8 inch to 12 inch pipelines. This method requires the use of the Aude K factor, representing pipeline efficiency. One version of this formula is described below:

$$P_m = [Q(z^{0.104}) (Sg^{0.448}) / (0.871(K) (D^{2.656}))]^{1.812} \tag{3.50}$$

where

$P_m$    - Pressure drop due to friction, psi/mi
Q    - Flow rate, bbl/hr
D    - Pipe internal diameter, in
Sg    - Liquid specific gravity

z     - Liquid viscosity, centipoise

K     - T.R. Aude K-factor, usually 0.90 to 0.95

In SI Units the Aude Equation is as follows:

$$P_m = 8.888 \times 10^8 \, [Q(z^{0.104})(Sg^{0.448}) / (K(D^{2.656}))]^{1.812} \qquad (3.51)$$

where

Pm    - Frictional pressure drop, kPa/km
Sg    - Liquid specific gravity
Q     - Flow rate, $m^3/hr$
D     - Pipe internal diameter, mm
z     - Liquid viscosity, centipoise
K     - T.R. Aude K-factor, usually 0.90 to 0.95

Since the Aude formula for pressure drop given above does not contain pipe roughness, it can be deduced that the K factor somehow must take into account the internal condition of the pipe. As with the Hazen-Williams C factor discussed earlier, the Aude K factor is also an experience based factor and must be determined by field measurement and calibration of an existing pipeline. If field data is not available, engineers usually approximate using a value such as K = 0.90 to 0.95. A higher value of K will result in lower pressure drop for a given flow rate or higher flow rate for a given pressure drop.

It must be noted that the Aude Equation is based on field data collected from 6 inch and 8 inch refined products pipelines. Therefore, use caution when applying this formula to larger pipelines.

## 3.12 Minor Losses

In most long distance pipelines, such as trunk lines, the pressure drop due to friction in the straight lengths of pipe, form the significant portion of the total frictional pressure drop. Valves and fittings contribute very little to the total pressure drop in the entire pipeline. Hence, in such cases, pressure losses through valves, fittings and other restrictions are generally classified as *minor losses*. Minor losses include energy losses resulting from rapid changes in the direction or magnitude of liquid velocity in the pipeline. Thus pipe enlargements, contractions, bends and restrictions such as check valves and gate valves are included in minor losses.

In short pipelines, such as terminal and plant piping, the pressure loss due to valves, fittings, etc may be a substantial portion of the total pressure drop. In such cases, the term Minor Losses is a misnomer.

Therefore, in long pipelines the pressure loss through bends, elbows, valves, fittings etc. are classified as "minor" losses and in most instances may be neglected without significant error. However, in shorter pipelines these losses must be included for correct engineering calculations. Experiments with water, at high Reynold's numbers have shown that the minor losses varied approximately as the square of the velocity. This leads to the conclusion that minor losses can be represented by a function of the liquid velocity head or kinetic energy ($V^2/2g$).

Accordingly, the pressure drop through valves and fittings is generally expressed in terms of the liquid kinetic energy $V^2/2g$ multiplied by a head loss coefficient K. Comparing this with the Darcy-Weisbach Equation for head loss in a pipe, we can see the following analogy. For straight pipe, the head loss h is $V^2/2g$ multiplied by the factor (fL/D). Thus, the head loss coefficient for straight pipe is fL/D.

Therefore, the pressure drop in a valve or fitting is calculated as follows:

$$h = K \, V^2/2g \qquad\qquad (3.52)$$

where

| | |
|---|---|
| h | - Head loss due to valve or fitting, ft |
| K | - Head loss coefficient for the valve or fitting, dimensionless |
| V | - Velocity of liquid through valve or fitting, ft/s |
| g | - Acceleration due to gravity, 32.2 ft/s² in English units |

The head loss coefficient K is, for a given flow geometry, considered practically constant at high Reynold's number. K increases with pipe roughness and with lower Reynold's numbers. In general the value of K is determined mainly by the flow geometry or by the shape of the pressure loss device.

It can be seen from Equation (3.52) that K is analogous to the term (fL/D) for straight length of pipe. Values of K are available for various types of valves and fittings in standard handbooks, such as Crane Handbook and Cameron Hydraulic Data. A table of K values, commonly used for valves and fittings is included in Appendix A

### 3.12.1   Gradual Enlargement

Consider liquid flowing through a pipe of diameter $D_1$. If at a certain point, the diameter enlarges to $D_2$, the energy loss that occurs due to the enlargement can be calculated as follows:

$$h = K (V_1 - V_2)^2 / 2g \tag{3.53}$$

where $V_1$ and $V_2$ are the velocity of the liquid in the smaller diameter and the larger diameter respectively. The value of K depends upon the diameter ratio $D_1/D_2$ and the different cone angle due to the enlargement. A gradual enlargement is shown in Figure 3.5

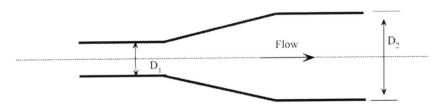

**Figure 3.5 Gradual Enlargement**

For a sudden enlargement K = 1.0 and the corresponding head loss is

$$h = (V_1 - V_2)^2 / 2g \tag{3.54}$$

### Example Problem 3.8

Calculate the head loss due to a gradual enlargement in a pipe that flows 100 gal/min of water from a 2 inch diameter to a 3 inch diameter with an included angle of 30 $^\circ$. Both pipe sizes are internal diameters.

The liquid velocities in the two pipe sizes are as follows:

$V_1 = 0.4085 \times 100 / 2^2 = 10.21$ ft/s
$V_2 = 0.4085 \times 100 / 3^2 = 4.54$ ft/s

Diameter ratio = 3/2 = 1.5

From charts, for diameter ratio = 1.5 and cone angle = $30^\circ$ the value of K is

K = 0.38

Therefore head loss due to gradual enlargement is

$h = 0.38 \times (10.21 - 4.54)^2 / 64.4 = 0.19$ ft

If the expansion was a sudden enlargement from 2 inch to 3 inch, the head loss would be

$h = (10.21 - 4.54)^2 / 64.4 = 0.50$ ft

### 3.12.2 Abrupt Contraction

For flow through an abrupt contraction, the flow from the larger pipe (diameter $D_1$ and velocity $V_1$) to a smaller pipe (diameter $D_2$ and velocity $V_2$) results in the formation of a vena contracta or throat, immediately after the diameter change. At the vena contracta, the flow area reduces to $A_c$ with increased velocity of $V_c$. Subsequently the flow velocity decreases to $V_2$ in the smaller pipe. From velocity $V_1$, the liquid first accelerates to velocity $V_c$ at the vena contracta and subsequently decelerates to $V_2$. This is shown in Figure 3.6

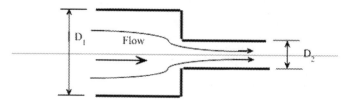

**Figure 3.6 Abrupt Contraction**

The energy loss due to the sudden contraction depends upon the ratio of the pipe diameters $D_2$ and $D_1$ and the ratio $A_c/A_2$. The value of the head loss coefficient K can be found using the Table 3.1, where $C_c = A_c/A_2$. The ratio $A_2/A_1$ can be calculated from the ratio of the diameters $D_2/D_1$.

A pipe connected to a large storage tank represents a type of abrupt contraction. If the storage tank is a large body of liquid, we can state that this is a limiting case of the abrupt contraction. For such a square edged pipe entrance from a large tank $A_2/A_1 = 0$. From Table 3.1 for this case K = 0.5 for turbulent flow.

Another type of pipe entrance from a large tank is called a re-entrant pipe entrance. If the pipe is thin walled and the opening within the tank is located more than one pipe diameter upstream from the tank wall, the K value will be close to 0.8.

If the edges of the pipe entrance in a tank are rounded or bell shaped, the head loss coefficient is considerably smaller. An approximate value for K for a bell-mouth entrance is 0.1.

Table 3.1 Head Loss Coefficient K for abrupt contraction

| $A_2/A_1$ | $C_c$ | K |
|-----------|-------|------|
| 0.0 | 0.617 | 0.50 |
| 0.1 | 0.624 | 0.46 |
| 0.2 | 0.632 | 0.41 |
| 0.3 | 0.643 | 0.36 |
| 0.4 | 0.659 | 0.30 |
| 0.5 | 0.681 | 0.24 |
| 0.6 | 0.712 | 0.18 |
| 0.7 | 0.755 | 0.12 |
| 0.8 | 0.813 | 0.06 |
| 0.9 | 0.892 | 0.02 |
| 1.0 | 1.000 | 0.00 |

### 3.12.3 Head Loss and L/D Ratio for Pipes and Fittings

We have discussed how minor losses can be accounted for using the head loss coefficient K in conjunction with the liquid velocity head. The Appendix A.9 lists K values for common valves and fittings.

Referring to the Table in Appendix A.9 for K values, we see that for a 16 inch Gate valve

$$K = 0.10$$

Therefore, compared to 16 inch straight pipe, we can write from Darcy-Weisbach Equation

$$\frac{fL}{D} = 0.10$$

or

$$\frac{L}{D} = 0.10f$$

If we assume f = 0.0125

We get     $\dfrac{L}{D} = 8$

This means that compared to straight pipe—16 inch diameter, a 16 inch gate valve has an (L/D) ratio of 8, that causes the same friction loss. The L/D ratio represents the equivalent length of straight pipe in terms of its diameter that will equal the pressure loss in the valve or fitting. In Appendix A.10, the (L/D) ratio for various valves and fittings are given.

Using the (L/D) ratio we can replace a 16 inch gate valve with 8x16 inch = 128 inch of straight 16 inch pipe. This length of pipe will have the same friction loss as the 16 inch gate valve. Thus we can use the K values or (L/D) ratios to calculate the friction loss in valves and fittings.

# 3.13 Internally Coated Pipes and Drag Reduction

In turbulent flow, pressure drop due to friction depends on the pipe roughness. Therefore, if the internal pipe surface can be smoothened, the frictional pressure drop can be reduced. Internally coating a pipeline with an epoxy will considerably reduce the pipe roughness, compared to uncoated pipe.

For example, if the uncoated pipe has an absolute roughness of 0.002 inch, coated pipe can reduce roughness to a value as low as 0.0002 inch. The friction factor f may therefore reduce from 0.02 to 0.01 depending on the flow rate, Reynold's number, etc. Since pressure drop is directly proportional to the friction factor in accordance with the Darcy-Weisbach equation, the total pressure drop in the internally coated pipeline in this example would be 50% of that in the uncoated pipeline.

Another method of reducing frictional pressure drop in a pipeline is by using Drag Reduction. Drag reduction is the process of reducing the pressure drop due to friction in a pipeline by continuously injecting a very small quantity (in parts per million or ppm) of a high molecular weight hydrocarbon, called the Drag Reduction Agent (DRA) into the flowing liquid stream. The DRA is effective, only in pipe segments between two pump stations. It degrades in performance as it flows through the pipeline for long distances. It also completely breaks up or suffers shear-degradation as it passes through pump stations, meters and other restrictions. DRA works only in turbulent flow and with low viscosity liquids. Thus, it works well with refined petroleum products(gasoline, diesel, etc) and light crude oils. It is ineffective in heavy crude oil pipelines, particularly in laminar flow. Currently, in the USA, two leading vendors of DRA products are Baker Petrolite and Conoco-Phillips.

To determine the amount of Drag Reduction using DRA we proceed as follows.

If the pressure drops due to friction with and without DRA are known, we can calculate the percentage drag reduction.

Percentage Drag Reduction = $100(DP_0 - DP_1)/ DP_0$ (3.55)

where

$DP_0$   - Friction drop in pipe segment without DRA, psi
$DP_1$   - Friction drop in pipe segment with DRA, psi

The above pressure drops are also referred to as untreated versus treated pressure drops. It is fairly easy to calculate the value of untreated pressure drop, using the pipe size, liquid properties, etc. The pressure drop with DRA is obtained using DRA vendor information. In most cases involving DRA, we are interested in calculating how much DRA we need to use to reduce the pipeline friction drop, and hence the pumping horsepower required. It must be noted that DRA may not be effective at the higher flow rate, if existing pump and driver limitations preclude operating at higher flow rates due to pump driver horsepower limitation.

Consider a situation where a pipeline is limited in throughput due to maximum allowable operating pressures (MAOP). Let us assume the friction drop in this MAOP limited pipeline is 800 psi at 100 000 bbl/day. We are interested in increasing pipeline flow rate to 120 000 bbl/day using DRA and we would proceed as follows:

Flow improvement desired = (120 000 - 100 000)/100 000 = 20%

If we calculate the actual pressure drop in the pipeline at the increased flow rate of 120 000 bbl/day (ignoring the MAOP violation) and assume we get the following pressure drop:

Frictional pressure drop at 120 000 bbl/day = 1 150 psi

and

Frictional pressure drop at 100 000 bbl/day = 800 psi

The percentage drag reduction is then calculated from Equation (3.55) as

Percentage Drag Reduction = 100(1150-800)/ 1150 = 30.43 %

In the above calculation, we have tried to maintain the same frictional drop (800 psi) using DRA at the higher flow rate as the initial pressure-limited case. Knowing the drag reduction percent required, we can get the DRA vendor to tell us how much DRA will be required to achieve the 30.43% drag reduction, at the flow rate of 120 000 bbl/ day. If the answer is 15 ppm of Brand x DRA, we can calculate the daily DRA requirement as follows:

Quantity of DRA required = $(15/10^6)(120\ 000)(42) = 75.6$ gal/day

If DRA costs $10 per gal, this equates to a daily DRA cost of $756. In this example, a 20% flow improvement requires a drag reduction of 30.43% and 15 ppm of DRA, costing $756 per day. Of course, these are simply rough numbers, used to illustrate the DRA calculations methods. The quantity of DRA required will depend on the pipe size, liquid viscosity, flow rate and Reynold's number, in addition to the percentage drag reduction required. Most DRA vendors will confirm that drag reduction is effective only in turbulent flow (Reynold's number > 4 000) and that it does not work with heavy (high viscosity) crude oil and other liquids.

Also, drag reduction cannot be increased indefinitely by injecting more DRA. There is a theoretical limit to the drag reduction attainable. For a certain range of flow rates, the percentage drag reduction will increase as DRA ppm is increased. At some point, depending on the pumped liquid, flow characteristics etc., the drag reduction levels off. No further increase in drag reduction is possible by increasing the DRA ppm. We would have reached the point of diminishing returns in this case.

In a later chapter on feasibility studies and pipeline economics, we will explore the subject of DRA further.

## 3.14 Summary

We defined pressure and how it is measured in both static and dynamic context. The velocity and Reynold's number calculations for pipe flow were introduced and the use of Reynold's number in classifying liquid flow as laminar, critical and turbulent were explained.

Existing methods of calculating the pressure drop due to friction in a pipeline using the Darcy-Weisbach equation were discussed and illustrated using examples. The importance of the Moody diagram was explained. Also, the trial and error solutions of friction factor from the Colebrook-White equation were covered. The use of Hazen-Williams, MIT and other pressure drop equations were discussed. Minor losses in pipelines due to valves, fittings, pipe enlargements and pipe contractions were analyzed. The concept of drag reduction as a means of reducing frictional head loss was also introduced.

## 3.15 Problems

3.15.1 Calculate the average velocity and Reynold's number in a 20 inch pipeline that transports diesel fuel at a flow rate of 250 000 bbl/day. Assume 0.375 inch pipe wall thickness and the diesel fuel properties as follows:

Specific gravity = 0.85 Kinematic viscosity = 5.9 cSt

What is the flow regime in this case?

3.15.2 In the above example, what is the value of the Darcy friction factor using Colebrook-White equation? If the modified Colebrook-White equation is used, what is the difference in friction factors? Calculate the pressure drop due to friction in a 5-mile segment of this pipeline. Use pipe roughness of 0.002 in.

3.15.3 Using Hazen-Williams equation with a C-factor of 125, calculate the frictional pressure drop per mile in the 20 inch pipeline described in Problem 3.15.1 above. Repeat the calculations using Shell-MIT and Miller equations.

3.15.4 A crude oil pipeline, 500 km long is 400 mm outside diameter and 8 mm wall thickness is used to transport 600 m³/hr of product from a crude oil terminal at San Jose to a refinery located at La Paz. Assuming the crude oil has a specific gravity of 0.895 and viscosity of 200 SSU at 20°C, calculate the total pressure drop due to friction in the pipeline. If the MAOP of the pipeline is limited to 10 MPa, how many pumping stations will be required to transport this volume, assuming flat terrain. Use the modified Colebrook-White equation and assume 0.05 mm for the absolute pipe roughness.

3.15.5  In Problem 3.15.4 the volume transported was 600 m³/hr. It is desired to increase flow rate using DRA in the bottleneck section of the pipeline.

(a) What is the maximum throughput possible with DRA?
(b) Summarize any changes needed to the pump stations to handle the increased throughput.
(c) What options are available to further increase pipeline throughput?

# 4

# Pipe Analysis

In this chapter we will focus our attention mainly on the strength capabilities of a pipeline. We will discuss different materials used to construct pipelines and how to calculate the amount of internal pressure that a given pipe can withstand.

We will determine the amount of internal pressure a particular size of pipe can withstand, based on the pipe material, diameter and wall thickness. Next, we establish the hydrostatic test pressure, the pipeline will be subject to, such that the previously calculated internal pressure can be safely tolerated.

We will also discuss the volume content or line fill volume of a pipeline and how it is used in batched pipelines with multiple products.

## 4.1 Allowable Operating Pressure & Hydrostatic Test Pressure

To transport a liquid through a pipeline, the liquid must be under sufficient pressure so that the pressure loss due to friction and the pressure required for any elevation changes can be accommodated. The longer the pipeline and higher the flow rate, friction drop will be higher requiring corresponding increase in liquid pressure at the beginning of the pipeline.

In gravity flow systems, flow occurs due to elevation difference without any additional pump pressure. Thus, a pipeline from a storage tank on a hill to a delivery terminus below may not need any pump pressure at the tank. However, the pipeline still needs to be designed to withstand pressure generated due to the static elevation difference.

The allowable operating pressure in a pipeline is defined as the maximum safe continuous pressure that the pipeline can be operated at. At this internal pressure the pipe material is stressed to some safe value below the yield strength of the pipe material. The stress in the pipe material consists of circumferential (or hoop) stress and longitudinal (or axial) stress. This is shown in Figure 4.1. It can be proven that the axial stress is one-half the value of the hoop stress. The hoop stress therefore controls the amount of internal pressure the pipeline can withstand. For pipelines transporting liquids, the hoop stress may be allowed to reach 72% of the pipe yield strength.

If pipe material has 60 000 psi yield strength, the safe internal operating pressure cannot exceed a value that results in a hoop stress of

0.72 x 60 000 = 43 200 psi.

In order to ensure that the pipeline can be safely operated at a particular maximum allowable operating pressure (MAOP) we must test the pipeline using water, at a higher pressure.

The hydrostatic test pressure is a pressure higher than the allowable operating pressure. It is the pressure at which the pipeline is tested for a specified period of time, such as four hours (for above ground piping) or eight hours (buried pipeline) as required by the pipeline design code or by county, city, state or federal government regulations. In the USA, Department of Transportation (DOT) Code Part 195 applies. Generally, for liquid pipelines the hydrostatic test pressure is 25% higher than the MAOP. Thus, if the MAOP is 1 000 psig, the pipeline will be hydrostatically tested at 1 250 psig.

Calculation of internal design pressure in a pipeline is based on Barlow's equation for internal pressure in thin-walled cylindrical pipes as discussed next.

## 4.2 Barlow's Equation for Internal Pressure

The hoop stress or circumferential stress in a thin-walled cylindrical pipe due to an internal pressure is calculated using the formula

$$S_h = PD/2t \qquad\qquad (4.1)$$

where

$S_h$     - Hoop stress, psi
P      - Internal pressure, psi
D      - Pipe outside diameter, in
t       - Pipe wall thickness, in

Similarly, the axial (or longitudinal) stress, $S_a$ is

$$S_a = PD/4t \qquad\qquad (4.2)$$

The above form the basis of Barlow's equation used to determine the allowable internal design pressure in a pipeline. As can be seen from Equation (4.1) and (4.2), the hoop stress is twice the longitudinal stress. The internal design pressure, therefore will be based on the hoop stress Equation (4.1).

Barlow's Equation (4.1) can be derived easily as follows:

Consider one half of a unit length of pipe as shown in Figure 4.1. Due to internal pressure P, the bursting force on one half the pipe is

P x D x 1

where the pressure P acts on a projected area D x 1. This bursting force is exactly balanced by the hoop stress $S_h$ acting along both edges of the pipe.

Therefore,

$S_h$ x t x 1 x 2 = P x D x 1

Solving for $S_h$ we get

$S_h = PD / 2t$

Equation (4.2) for axial stress $S_a$ is derived as follows

The axial stress $S_a$ acts on an area of cross section of pipe represented by $\pi Dt$. This is balanced by the internal pressure P acting on the internal cross sectional area of pipe $\pi D^2 / 4$. Equating the two we get

$S_a \times \pi Dt = P \times \pi D^2 / 4$

Solving for $S_a$, we get

$S_a = PD / 4t$

In calculating the internal design pressure in liquid pipelines, we modify Barlow's equation slightly. The internal design pressure in a pipe is calculated in English units as follows:

$$P = \frac{2\,T \times S \times E \times F}{D} \qquad\qquad (4.3)$$

where

P     - Internal pipe design pressure, psig
D     - Nominal pipe outside diameter, in
T     - Nominal Pipe wall thickness, in
S     - Specified Minimum Yield Strength (SMYS) of pipe material, psig
E     - Seam Joint Factor, 1.0 for seamless and Submerged Arc Welded (SAW) pipes. See Table in Appendix A.11
F     - Design Factor, usually 0.72 for liquid pipelines, except that a design factor of 0.60 is used for pipe, including risers, on a platform located off shore or on a platform in inland navigable waters, and 0.54 is used for pipe that has been subjected to cold expansion to meet the SMYS and subsequently heated, other than by welding or stress

relieving as a part of the welding, to a temperature higher than 900°F (482°C) for any period of time or over 600°F (316°C) for more than one hour.

The above form of Barlow's equation may be found in Part 195 of DOT code of Federal Regulations, Title 49 and ASME standard B31.4 for liquid pipelines.

In SI units, the Barlow's equation can be written as:

$$P = \frac{2\,T \times S \times E \times F}{D} \qquad (4.4)$$

where

P     - Pipe internal design pressure, kPa
D     - Nominal pipe outside diameter, mm
T     - Nominal pipe wall thickness, mm
S     - Specified Minimum Yield Strength (SMYS) of pipe material, kPa
E     - Seam Joint Factor, 1.0 for seamless and Submerged Arc Welded (SAW) pipes. See Table in Appendix A.11
F     - Design Factor, usually 0.72 for liquid pipelines, except that a design factor of 0.60 is used for pipe, including risers, on a platform located off shore or on a platform in inland navigable waters, and 0.54 is used for pipe that has been subjected to cold expansion to meet the SMYS and subsequently heated, other than by welding or stress relieving as a part of the welding, to a temperature higher than 900°F (482°C) for any period of time or over 600°F (316°C) for more than one hour.

In summary, Barlow's equation for internal pressure is based on calculation of the hoop stress (circumferential) in the pipe material. Within the stressed pipe material there are two stresses on a pipe element called hoop stress and axial or longitudinal stress. It can be shown that the controlling stress is the hoop stress being twice the axial stress as depicted in Figure 4.1

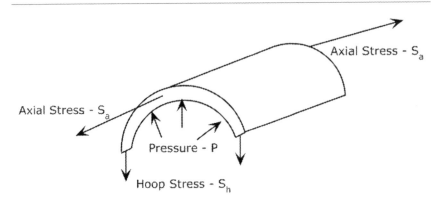

**Figure 4.1 Hoop stress and axial stress in pipe**

The strength of pipe material designated as Specified Minimum Yield Strength SMYS in Equations (4.3) and (4.4) depend on pipe material and grade. In the USA, steel pipeline material used in the Oil and Gas industry is manufactured in accordance with American Petroleum Institute (API) standards 5L and 5LX. For example, grades 5LX-42, 5LX-52, 5LX-60, 5LX-65, 5LX-70 and 5LX-80 are used commonly in pipeline applications. The numbers after 5LX above indicate the SMYS values in thousands of psi. Thus, 5LX-52 pipe has a minimum yield strength of 52 000 psi. The lowest grade of pipe material used is 5L Grade B which has an SMYS of 35 000 psi. In addition, seamless steel pipe designated as ASTM A106 and Grade B pipe are also used for liquid pipeline systems. These have an SMYS value of 35 000 psi.

It is obvious from Barlow's Equation (4.3), that for a given pipe diameter, pipe material and seam joint factor, the allowable internal pressure P is directly proportional to the pipe wall thickness. For example, 16 inch diameter pipe with a wall thickness of 0.250 inch made of 5LX-52 pipe has an allowable internal design pressure of 1 170 psi calculated as follows:

$$P = (2 \times 0.250 \times 52\ 000 \times 1.0 \times 0.72) / 16 = 1\ 170 \text{ psig}$$

Therefore if the wall thickness is increased to 0.375 inches the allowable internal design pressure increases to

$$(0.375/0.250) \times 1\ 170 = 1\ 755 \text{ psig}$$

On the other hand, if the pipe material is changed to 5LX-70, keeping the wall thickness at 0.250 inch, the new internal pressure is

(70 000 / 52 000) 1 170 = 1 575 psig

Note that we used the Barlow's equation to calculate the allowable internal pressure based upon the pipe material being stressed to 72% of SMYS. In some situations more stringent city or government regulations may require that the pipe be operated at a lower pressure. Thus, instead of using a 72% factor in Equation (4.3) we may be required to use a more conservative factor (lower number) in place of $F = 0.72$. As an example, in certain areas of the City of Los Angeles, liquid pipelines are only allowed to operate at a 66% factor instead of the 72% factor. Therefore, in the earlier example, the 16 inch/0.250 inch/X52 pipeline can only be operated at

1 170 (66 / 72) = 1 073 psig.

As mentioned before, in order to operate a pipeline at 1 170 psig, it must be hydrostatically tested at 25% higher pressure. Since 1 170 psig internal pressure is based on the pipe material being stressed to 72% of SMYS, the hydrostatic test pressure will cause the hoop stress to reach

1.25 (72) = 90% of SMYS.

Generally, the hydrostatic test pressure is specified as a range of pressures, such as 90% SMYS to 95% SMYS. This is called the hydrotest pressure envelope. Therefore, in the present example, the hydrotest pressure range is

1.25 (1 170) = 1 463 psig - lower limit (90% SMYS)
(95/90) 1 463 = 1 544 psig - higher limit (95% SMYS)

To summarize, a pipeline with an MAOP of 1 170 psig needs to be hydrotested at a pressure range of 1 463 psig to 1 544 psig. According to the design code, the test pressure will be held for a minimum four hour period for above ground pipelines and eight hours for buried pipelines.

In calculating the allowable internal pressure in older pipelines, consideration must be given to wall thickness reduction due to corrosion over the life of the pipeline. A pipeline that was installed 25 years ago with 0.250 inches wall thickness may have reduced in wall thickness to 0.200 inches or less due to corrosion. Therefore, the allowable internal pressure will have to be reduced in the ratio of the wall thickness, compared to the original design pressure.

## 4.3   Line Fill Volume and Batches

Frequently we need to know how much liquid is contained in a pipeline between two points along its length, such as between valves or pump stations.

For a circular pipe, we can calculate the volume of a given length of pipe by multiplying the internal cross sectional area by the pipe length. If the pipe inside diameter is D inches and the length is L feet, the volume of this length of pipe is

$$V = 0.7854 \ (D^2/144) \ L \tag{4.5}$$

where

$V$ - Volume, ft$^3$

Simplifying

$$V = 5.4542 \times 10^{-3} \ D^2 \ L \tag{4.6}$$

We will now restate this equation in terms of conventional pipeline units, such as the volume in bbl in a mile of pipe.

The quantity of liquid contained in a mile of pipe also called the line fill volume is calculated as follows:

$$V_L = 5.129(D)^2 \tag{4.7}$$

where

$V_L$    - Line fill volume of pipe, bbl/mile
$D$    - Pipe inside diameter, in

In SI Units we can express the line fill volume per km of pipe as follows:

$$V_L = 7.855 \times 10^{-4} D^2 \qquad\qquad (4.8)$$

where

$V_L$    - Line fill volume, m³/km
$D$    - Pipe inside diameter, mm

Using Equation (4.7), a pipeline 100 miles long, 16 inch diameter and 0.250 inch wall thickness, has a line fill volume of

$$5.129(15.5)^2 (100) = 123\ 224 \text{ bbl}$$

Many crude oil and refined product pipelines operate in a batched mode. Multiple products are simultaneously pumped through the pipeline as batches. For example, 50 000 bbl of product C will enter the pipeline followed by 30 000 bbl of product B and 40 000 bbl of product A. If the line fill volume of the pipeline is 120 000 bbl, an instantaneous snap shot condition of a batched pipeline is as shown in Figure 4.2

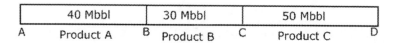

Figure 4.2 Batched Pipeline

**Example Problem 4.1**

A 50 mile pipeline consists of a 20 mile of 16 inch diameter, 0.375 inch wall thickness pipe followed by 30 miles of 14 inch diameter, 0.250 inch wall thickness pipe. Calculate the total volume contained in the 50 mile long pipeline.

## Solution

Using Equation (4.7) we get

For the 16 inch pipeline

Volume per mile = $5.129 (15.25)^2 = 1\ 192.81$ bbl/mi

For the 14 inch pipeline

Volume per mile = $5.129 (13.5)^2 = 934.76$ bbl/mi

Total line fill volume is

20 x 1 192.81 + 30 x 934.76 = 51 899 bbl.

## Example Problem 4.2

A pipeline 100 km long is 500 mm outside diameter and 12 mm wall thickness. If batches of three liquids A ($3\ 000$ m³), B ($5\ 000$ m³) and C occupy the pipe, at a particular instant, calculate the interface locations of the batches, considering the origin of the pipeline to be at 0.0 km.

## Solution

Using Equation (4.8) we get the line fill volume per km to be

$V_L = 7.855 \times 10^{-4} (500 - 24)^2 = 177.9754$ m³/km

The first batch A will start at 0.0 km and will end at a distance of

$3\ 000 / 177.9754 = 16.86$ km

The second batch B starts at 16.86 km and ends at

$16.86 + (5\ 000 / 177.9754) = 44.95$ km

The third batch C starts at 44.95 km and ends at 100 km

The total volume in pipe is

177.9754 x 100 = 17 798 m³

Thus the volume of the third batch C is

17 798 - 3 000 - 5 000 = 9 798 m³

It can thus be seen that line fill volume calculation is important when dealing with batched pipelines. We need to know the boundaries of each liquid batch, so that the correct liquid properties can be used to calculate pressure drops for each batch.

The total pressure drop in a batched pipeline would be calculated by adding up the individual pressure drop for each batch. Since intermixing of the batches is not desirable, batched pipelines must run in turbulent flow. In laminar flow, there will be extensive mixing of the batches, which defeats the purpose of keeping each product separate so that at the end of the pipeline each product maybe diverted into a separate tank. Some intermixing will occur at the product interfaces and this contaminated liquid is generally pumped into a slop tank at the end of the pipeline and may be blended with a less critical product. The amount of contamination that occurs at the batch interface depends on the physical properties of the batched products, batch length and Reynold's number. Several correlations have been developed for determining the contamination volumes and will be discussed in a subsequent chapter on batching.

## 4.4 Summary

In this chapter we discussed how allowable internal pressure in a pipeline is calculated depending on pipe size and material. We showed that for pipe under internal pressure the hoop stress in the pipe material will be a controlling factor. The importance of design factor in selecting pipe wall thickness was illustrated using an example. Based on Barlow's equation, the internal design pressure calculation as recommended by

ASME standard B31.4 and US Code of Federal Regulation, Part 195 of the DOT was illustrated. The need for hydrostatic testing pipelines for safe operation was discussed. The line fill volume calculation was introduced and its importance in batched pipelines was shown using an example.

## 4.5  Problems

4.5.1   Calculate the allowable internal design pressure at 72% design factor for an 18 inch pipeline, Wall thickness = 0.375 inch and Pipe Material is API 5L X-46. What is the hydrotest pressure range for this pipeline?

4.5.2   It has been determined that the design pressure for a storage tank piping system is 720 psi. If API 5L grade B pipe is used, what minimum wall thickness is required for 14 inch pipe?

4.5.3   In the previous problem, if the pressure rating were increased to ANSI 600 (1440 psi), calculate the pipe wall thickness required with 14 inch pipe, if high strength 5L X-52 pipe is used. What is the minimum hydrotest pressure for this system?

4.5.4   Determine the volume of liquid contained in a mile of 14 inch pipeline with wall thickness of 0.281 inch.

# 5

# Pressure and HP Required

In previous chapters we discussed how to calculate friction factors and pressure loss due to friction in a pipeline using various equations such as Colebrook-White, Hazen-Williams etc. We also analyzed the internal pressure allowable in a pipe and how to determine pipe wall thickness required for a specific internal pressure and pipe material, according to design code.

In this chapter we will analyze how much pressure will be required at the beginning of a pipeline to safely transport a given throughput to the pipeline terminus, taking into account the pipeline elevation profile and the pipeline terminus pressure required, in addition to friction losses. We will also calculate the pumping horsepower required and in many cases also determine how many pump stations are needed to transport the specified volume of liquid.

We will also examine pipes in series and parallel and how to calculate equivalent length of pipes in series and the equivalent diameter of parallel pipes. System head curves will be introduced along with flow injection and delivery along the pipeline. Incoming and outgoing branch connections will be studied as well as pipe loops.

## 5.1 Total Pressure Required

The total pressure $P_T$ required at the beginning of a pipeline, to transport a given flow rate from point A to point B will depend on

- Pipe diameter, wall thickness and roughness
- Pipe length
- Pipeline elevation changes from A to B
- Liquid specific gravity and viscosity
- Flow rate

If we increase the pipe diameter, keeping all other items above constant, we know that the frictional pressure drop will decrease and hence the total pressure $P_T$ will also decrease. Increasing pipe wall thickness or pipe roughness will cause increased frictional pressure drop and thus increase the value of $P_T$. On the other hand, if only the pipe length was increased, the pressure drop for the entire length of the pipeline will increase and so will the total pressure $P_T$.

How does the pipeline elevation profile affect $P_T$? If the pipeline were laid in a flat terrain, with no appreciable elevation difference between the beginning of the pipeline A and the terminus B, the total pressure $P_T$ will not be affected. But, if the elevation difference between A and B were substantial, and B was at a higher elevation than A, $P_T$ will be higher than that for the flat terrain pipeline.

Higher the liquid specific gravity and viscosity, higher will be the pressure drop due to friction and hence larger the value of $P_T$. Finally, increasing the flow rate will result in higher frictional pressure drop and therefore higher value for $P_T$.

In general, the total pressure required can be divided into three main components as follows:

- Friction head
- Elevation head
- Delivery pressure at terminus.

As an example, consider a pipeline from point A to point B operating at 4 000 bbl/hr flow rate. If the total pressure drop due to friction in the pipeline is 800 psi, the elevation difference from point A to point B is 500 ft (uphill flow) and the minimum delivery pressure required at the terminus B is 50 psi, we can state that the pressure required at A is the sum of the three components as follows:

Total pressure at A = 800 psi + 500 ft + 50 psi

If the liquid specific gravity is 0.85 and using consistent units of psi, the above equation reduces to

Total pressure = 800 + (500)(0.85/2.31) + 50 = 1 033.98 psi

Of course, this assumes that there are no controlling peaks or high elevation points between point A and point B. If an intermediate point C located half way between A and B had an elevation of 1 500 ft, compared to the elevation of point A (100 ft) and elevation of point B (600 ft), then the elevation of point C becomes a controlling factor. In this case, the calculation of total pressure required at A is a bit more complicated.

Assume that the example pipeline is 50 miles long, flowing at 4 000 bbl/hr and the pipe being of uniform diameter and thickness throughout its entire length. Therefore, the pressure drop per mile will be calculated as a constant value for the entire pipeline from the given values as

800/50 = 16 psi/mi.

The total frictional pressure drop between point A and the peak C located 25 miles away is

Pressure Drop from A to C = 16 x 25 = 400 psi.

Since, C is the mid point of the pipeline, an identical frictional pressure drop exists between C and B as follows

Pressure Drop from C to B = 16 x 25 = 400 psi.

Consider now, the portion of the pipeline from A to C, with a frictional pressure drop of 400 psi calculated above and an elevation difference between A and C of

1 500 ft - 100 ft = 1 400 ft

The total pressure required at A to get over the peak at C is the sum of the friction and elevation components as follows:

Total pressure = 400 + (1 400)(0.85/2.31) = 915.15 psi

It must be noted that this pressure of 915.15 psi at A will just about get the liquid over the peak at point C with zero gauge pressure. Sometimes it is desired that the liquid at the top of the hill be at some minimum pressure above the liquid vapor pressure at the flowing temperature. If the transported liquid were LPG, we would require a minimum pressure in the pipeline of 250 to 300 psi. On the other hand with crude oils and refined products with low vapor pressure, the minimum pressure required may be only 10 to 20 psi. In this example, we assume that we are dealing with low vapor pressure liquids and a minimum pressure of 10 psi is adequate at the high points in a pipeline.

Our revised total pressure at A will then be

Total pressure = 400 + (1 400)(0.85/2.31) + 10 = 925.15 psi

Therefore, starting with a pressure of 925.15 psi at A will result in a pressure of 10 psi at the highest point C after accounting for frictional pressure drop and elevation difference between A and the peak C.

Once the liquid reaches the high point C at 10 psi, it flows downhill from point C to the terminus B at the given flow rate, being assisted by gravity. Therefore, for the section of the pipeline from C to B the elevation difference helps the flow while the friction impedes the flow. The arrival pressure at the terminus B can be calculated by considering the elevation difference and frictional pressure drop between C and B as follows:

Delivery pressure at B = 10 + (1500 - 600) (0.85/2.31) - 400 = -58.83 psi

Since the calculated pressure at B is negative, it is clear that the specified minimum delivery pressure of 50 psi at B cannot be achieved with a starting pressure of 925.15 psi at A. Therefore, to provide the required minimum delivery pressure at B, the starting pressure at A must be increased to

Pressure at A = 925.15 + 50 + 58.83 = 1 033.98 psi

The above value is incidentally the same pressure we calculated at the beginning of this section, without considering the point C.

Hence the revised pressure at peak C = 10 + 50 + 58.83 = 118.83 psi.

Thus a pressure of 1033.98 psi at the beginning of the pipeline will result in a delivery pressure of 50 psi at B and clear the peak at C with a pressure of 118.83 psi. This is depicted in Figure 5.1 where

$$P_A = 1034 \text{ psi}$$
$$P_C = 119 \text{ psi}$$
$$\text{and} \quad P_B = 50 \text{ psi}$$

All pressure have been rounded off to the nearest one psi.

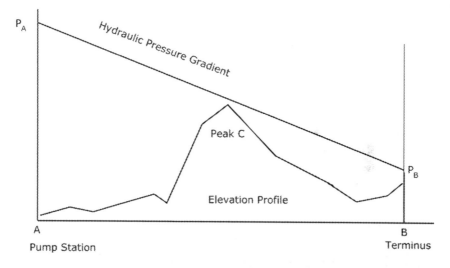

**Figure 5.1 Hydraulic Gradient**

Although the higher elevation point at C appeared to be controlling, calculation showed that the pressure required at A depended more on the required delivery pressure at the terminus B. In many cases this may not be true. The pressure required will be dictated by the controlling peak and therefore the arrival pressure of the liquid at the pipeline terminus may be higher than the minimum required.

Consider a second example in which the intermediate peak will be a controlling factor. Suppose the above pipeline now operates at 2 200 bbl/hr and the pressure drop due to friction is calculated to be 5 psi/mi.

We will first calculate the pressure required at A, ignoring the peak elevation at C. The pressure required at A is

5 x 50 + (600 - 100) x 0.85/2.31 + 50 = 484 psi

Based on this 484 psi pressure at A, the pressure at the highest point C will be calculated by deducting from 484 psi the pressure drop due to friction from A to C and adjusting for the elevation increase from A to C as follows:

$P_C$ = 484 - (5 x 25) - (1 500 - 100) x 0.85/2.31 = -156 psi

This negative pressure is not acceptable, since we require a minimum positive pressure of 10 psi at the peak to prevent liquid vaporization. It is therefore clear that the pressure at A calculated above is in adequate. The controlling peak at C therefore dictates the pressure required at A. We will now calculate the revised pressure at A to maintain a positive pressure of 10 psi at the peak at C.

$P_A$ = 484 + 156 + 10 = 650 psi

Therefore, starting with a pressure of 650 psi at A provides the required minimum of 10 psi at the peak C. The delivery pressure at B can now be calculated as

$P_B$ = 650 - (5 x 50) - (600 - 100) x 0.85/2.31 = 216 psi

which is more than the required minimum terminus pressure of 50 psi.

This example illustrates the approach used in considering all critical elevation points along the pipeline to determine the pressure required to transport the liquid.

Next we will repeat this analysis for a higher vapor pressure product.

If we were pumping a high vapor pressure liquid that requires a delivery pressure of 500 psi at the terminus, we would calculate the required pressure at A as follows:

$$\text{Pressure at A} = (6 \times 50) + \frac{(500 \times 0.65)}{2.31} + 500 = 940.69 \text{ psi}$$

where the high vapor pressure liquid (Sg = 0.65) is assumed to produce a pressure drop of 6 psi/mi for the same flow rate.

The pressure at the peak will be:

$$\text{Pressure at C} = 940.69 - (6 \times 25) - (1500-100) \times \frac{0.65}{2.31} = 396.75 \text{ psi}$$

If the higher vapor pressure product requires a minimum pressure of 400 psi, it can be seen from above that we do not have adequate pressure at the peak C. We therefore, need to increase the starting pressure at A to

940.69 + (400 - 396.75) = 944 psi, rounded off.

With this change, the delivery pressure at the terminus B will then be

500 + (944 - 940.69) = 503 psi

This is illustrated in Figure 5.2

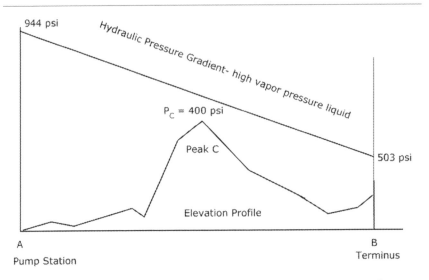

**Figure 5.2 Hydraulic Gradient - High Vapor Pressure Liquid**

## 5.2   Hydraulic Pressure Gradient

Generally, due to friction losses, the liquid pressure in a pipeline decreases continuously from the pipe inlet to the pipe delivery terminus. If there is no elevation difference between the two ends of the pipeline and the pipe elevation profile is essentially flat, the inlet pressure at the beginning of the pipeline will decrease continuously by the friction loss at a particular flow rate. When there is elevation differences along the pipeline, the decrease in pipeline pressure along the pipeline will be due to the combined effect of pressure drop due to friction and the algebraic sum of pipeline elevations. Thus starting at 1 000 psi pressure at the beginning of the pipeline, assuming 15 psi/mi pressure drop due to friction in a flat pipeline (no elevation difference) with constant diameter, the pressure would drop to

1 000 - 15 x 20 = 700 psi

at a distance of 20 mi from the beginning of the pipeline. If the pipeline is 60 mi long, the pressure drop due to friction in the entire line will be

15 x 60 = 900 psi

The pressure at the end of the pipeline will be

1 000 - 900 = 100 psi

Thus the liquid pressure in the pipeline has uniformly dropped from 1 000 psi in the beginning to 100 psi at the end of the 60 mi length. This pressure profile is referred to as the hydraulic pressure gradient in the pipeline.

The hydraulic pressure gradient is a graphical representation of the variation of pressures along the pipeline. It is shown along with the pipeline elevation profile. Since elevation is plotted in feet, it is convenient to represent the pipeline pressures also in feet of liquid head. This is shown in Figure 5.1, 5.2 and 5.3

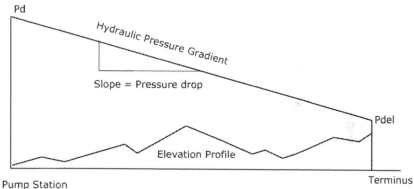

**Figure 5.3 Hydraulic Pressure Gradient**

In the example problem discussed in section 5.1, we calculated the pressure required at the beginning of the pipeline to be 1 034 psi for pumping crude oil at a flow rate of 4 000 bbl/hr. This pressure requires one pump station at the origin of the pipeline at point A.

Assume now, that the pipe length is 100 miles and the Maximum Allowable Operating Pressure (MAOP) is limited to 1 200 psi. Suppose the total pressure required at A is calculated to be 1 600 psi at a flow rate of 4 000 bbl/hr. We would then require an intermediate pump station between A and B to limit the maximum pressure in the pipeline to 1

200 psi. Due to the MAOP limit, the total pressure required at A will be provided in steps. The first pump station at A will provide approximately half the pressure followed by the second pump station located at some intermediate point providing the other half. This results in a saw-tooth like hydraulic gradient as shown in Figure 5.4

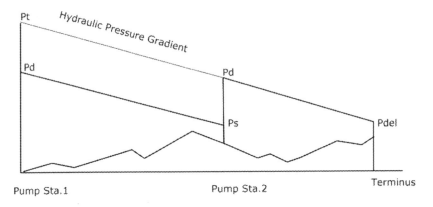

**Figure 5.4 Hydraulic Pressure Gradient - Two pump stations**

The actual discharge pressures at each pump station will be calculated considering pipeline elevations between A and B and the required minimum suction pressures at each of the two pump stations. An approximate calculation is described below, referring to Figure 5.4.

Let $P_s$ and $P_d$ represent the common suction and discharge pressure respectively, for each pump station and $P_{del}$ is the required delivery pressure at the pipe terminus B. The total pressure $P_t$ required at A can be written as follows.

$$P_t = P_{friction} + P_{elevation} + P_{del} \qquad (5.1)$$

where

Pt      - Total pressure required at A
$P_{friction}$  - Total frictional pressure drop between A and B
$P_{elevation}$ - Elevation head between A and B
$P_{del}$     - Required delivery pressure at B

Also, from Figure 5.4, based on geometry, we can state that

$$P_t = P_d + P_d - P_s \qquad (5.2)$$

Solving for $P_d$ we get

$$P_d = (P_t + P_s) / 2 \qquad (5.3)$$

where

$P_d$    - Pump station discharge pressure
$P_s$    - Pump station suction pressure

For example, if the total pressure calculated is 1 600 psi and the maximum pipeline pressure allowed is 1 200 psi we would need two pump stations. Considering minimum suction pressure of 50 psi, each pump station would have a discharge pressure of

$$P_d = (1\ 600 + 50) / 2 = 825 \text{ psi} \qquad \text{using Equation (5.3).}$$

Each pump station operates at 825 psi discharge pressure and the pipeline MAOP is 1 200 psi. It is clear that based on pipeline pressures alone we have the capability of increasing pipeline throughput further to fully utilize the 1 200 psi MAOP at each pump station. Of course, this would correspondingly require enhancing the pumping equipment at each pump station, since more horsepower (HP) will be required at the higher flow rate. We can estimate the increased throughput possible if we were to operate the pipeline at 1 200 psi MAOP level at each pump station as discussed next.

Let us assume that the pipeline elevation difference in this example contributes 300 psi to the total pressure required. This simply represents the station elevation head between A and B converted to psi. This component of the total pressure required ($P_t$= 1600 psi) depends only on the pipeline elevation and liquid specific gravity and therefore does not vary with flow rate. Similarly, the delivery pressure of 50 psi at B is also independent of flow rate. We can then calculate the frictional component (which depends on flow rate) of the total pressure $P_t$ using Equation 5.1 as follows:

Frictional pressure drop = 1 600 - 300 - 50 = 1 250 psi.

Assuming 100 mile length, the frictional pressure drop per mile of pipe is

$$P_m = 1250/100 = 12.5 \text{ psi/mi}$$

This pressure drop occurs at 4 000 bbl/hr flow rate. From Chapter 3, pressure drop calculations, we know that the pressure drop per mile, $P_m$ varies as the square of the flow rate as long as the liquid properties and pipe size do not change. Using Equation (3.28) we can write

$$P_m = K (Q)^2 \qquad (5.4)$$

where

$P_m$    - Frictional pressure drop per mile of pipe
K     - A constant for this pipeline that depends on liquid properties and pipe diameter.
Q     - Pipeline flow rate

Note that the K value above is not the same as the head loss coefficient discussed in Chapter 3. Strictly speaking, K also includes a transmission factor F (or a friction factor f) which varies with flow rate. However, for simplicity, we will assume that K is the constant that encompasses liquid specific gravity and pipe diameter. A more rigorous approach will require an additional parameter in Equation (5.4) that would include the transmission factor F, which in turn depends on Reynold's number, pipe roughness etc.

Therefore, using Equation (5.4) we can write at the initial flow rate of 4 000 bbl/hr

$$12.5 = K (4\,000)^2 \qquad (5.5)$$

In a similar manner, we can estimate the frictional pressure drop per mile when flow rate is increased to some value Q to fully utilize the 1 200 psi MAOP of the pipeline.

Using Equation (5.3) if we allow each pump station to operate at 1 200 psi discharge pressure, we can write

$$1\ 200 = (P_t + 50)\ /2$$

or

$$P_t = 2\ 400 - 50 = 2\ 350\ \text{psi}$$

This total pressure will now consist of friction, elevation and delivery pressure components at the higher flow rate Q.

From Equation (5.1) we can write

$$2\ 350 = P_{friction} + P_{elevation} + P_{del} \qquad \text{at the higher flow rate Q}$$

or

$$2\ 350 = P_{friction} + 300 + 50$$

Therefore,

$$P_{friction} = 2\ 350 - 300 - 50 = 2\ 000 \qquad \text{psi at the higher flow rate Q}$$

Thus the pressure drop per mile at the higher flow rate Q is

$$P_m = 2\ 000\ /\ 100 = 20\ \text{psi/mi}$$

From Equation (5.4) we can write

$$20 = K\ (Q)^2 \tag{5.6}$$

where Q is the unknown higher flow rate in bbl/hr.
By dividing Equation (5.6) by Equation (5.5) we get the following:

$$20/12.5 = (Q/4000)^2$$

Solving for Q we get

$$Q = 4000 (20/12.5)^{1/2} = 5\ 059.64\ \text{bbl/hr}$$

Therefore, by fully utilizing the 1 200 psi MAOP of the pipeline with the two pump stations, we are able to increase the flow rate to approximately 5 060 bbl/hr. As previously mentioned, this will definitely require additional pumps at both pump stations to provide the higher discharge pressure. Pumps will be discussed in Chapter 7.

In the preceding sections we have considered a pipeline to be of uniform diameter and wall thickness for its entire length. In reality, pipe diameter and wall thickness change, depending on the service requirement, design code and the local regulatory requirements. Pipe wall thickness may have to be increased due to different Specified Minimum Yield Strength(SMYS) of pipe because a higher or lower grade of pipe was used at some locations. As mentioned previously, some cities or counties through which the pipeline traverses may require different design factors (0.66 instead of 0.72) to be used, thus necessitating a different wall thickness. If there are drastic elevation changes along the pipeline, the low elevation points may require higher wall thickness to withstand the higher pipe operating pressures. If the pipeline has intermediate flow delivery or injections, the pipe diameter may be reduced or increased for certain portions to optimize pipe use. In all these cases, we can conclude that the pressure drop due to friction will not be the same uniform value throughout the entire pipeline length. Injections and delivery along the pipeline and their impact on pressure required is discussed later in this chapter.

When pipe diameter and wall thickness change along a pipeline, the slope of the hydraulic gradient, as shown in Figure 5.3 will no longer be uniform. Due to varying frictional pressure drop (because of pipe diameter and wall thickness change), the slope of the hydraulic gradient will vary along the pipe length.

# 5.3 Series Piping

Pipes are said to be in series, if different lengths of pipes are joined end to end with the entire flow passing through all pipes, without any branching.

Consider a pipeline consisting of two different lengths and pipe diameters joined together in series. A pipeline, 1 000 ft long, 16 inch diameter, connected in series with a pipeline, 500 ft long and 14 inch diameter would be an example of a series pipeline. At the connection point we will need to have a fitting, known as a reducer, that will join the 16 inch pipe with the smaller 14 inch pipe. This fitting will be a 16 inch x 14 inch reducer. The reducer causes transition in the pipe diameter smoothly from 16 inch to 14 inch. We can calculate the total pressure drop through this 16 inch /14 inch pipeline system, by adding the individual pressure drops in the 16 inch and the 14 inch pipe segment and accounting for the pressure loss in the 16 inch x 14 inch reducer.

If two pipes of different diameters are connected together in series, we can also use the equivalent length approach to calculate the pressure drop in the pipeline as discussed next.

A pipe is equivalent to another pipe or a pipeline system, when the same pressure loss due to friction occurs in the equivalent pipe compared to that of the other pipe or pipeline system. Since the pressure drop can be caused by an infinite combination of pipe diameter and pipe length, we must specify a particular diameter to calculate the equivalent length.

Suppose a pipe A of length $L_A$ and internal diameter $D_A$ is connected in series with a pipe B of length $L_B$ and internal diameter $D_B$. If we were to replace this two-pipe system with a single pipe of length $L_E$ and diameter $D_E$, we have what is known as the equivalent length of pipe. This equivalent length of pipe may be based on one of the two diameters $(D_A$ or $D_B)$ or a totally different diameter $D_E$.

The equivalent length $L_E$ in terms of pipe diameter $D_E$ can be written as

$$L_E / (D_E)^5 = L_A / (D_A)^5 + L_B / (D_B)^5 \qquad (5.7)$$

This formula for equivalent length is based on the premise that the total friction loss in the two-pipe system exactly equals that of the single equivalent pipe.

Equation (5.7) is based on Equation (3.28) for pressure drop, since pressure drop per unit length is inversely proportional to the fifth power of the diameter. If we refer to the diameter $D_A$ as the basis, this equation becomes, after setting $D_E = D_A$

$$L_E = L_A + L_B (D_A / D_B)^5 \tag{5.8}$$

Thus, we have an equivalent length $L_E$ that will be based on diameter $D_A$. This length $L_E$ of pipe diameter $D_A$ will produce the same amount of frictional pressure drop as the two lengths $L_A$ and $L_B$ in series. We have thus simplified the problem by reducing it to one single pipe length of uniform diameter $D_A$.

It must be noted that the equivalent length method discussed above is only approximate. Furthermore, if elevation changes are involved, it becomes more complicated, unless there are no controlling elevations along the pipeline system.

An example will illustrate this concept of equivalent pipe length.

Consider a pipeline 16 inch x 0.281 inch wall thickness pipeline, 20 miles long installed in series with a 14 inch x 0.250 inch wall thickness pipeline 10 miles long. The equivalent length of this pipeline is, using Equation (5.8),

$20 + 10 \times (16 - 0.562)^5 / (14 - 0.50)^5 = 39.56$ miles of 16 inch pipe.

The actual physical length of 30 miles of 16 inch and 14 inch pipes is replaced with a single 16 inch pipe 39.56 miles long, for pressure drop calculations. Note that we have left out the pipe fitting that would connect the 16 inch pipe with the 14 inch pipe. This would be a 16 x 14 reducer which would have its own equivalent length. To be precise, we should determine the equivalent length of the reducer from Table in Appendix A.10. and add it to the above length to obtain the total equivalent length, including the fitting.

Once an equivalent length pipe is determined, we can calculate the pressure drop based on this pipe size.

## 5.4 Parallel Piping

Pipes are said to be in parallel if they are connected in such a way that the liquid flow splits into two or more separate pipes and rejoins downstream into another pipe as illustrated in Figure 5.5

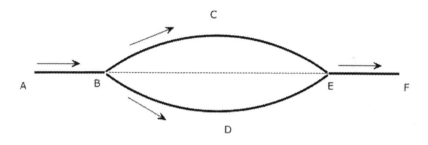

**Figure 5.5 Parallel Pipes**

In Figure 5.5 liquid flows through pipe AB and at point B, part of the flow branches off into pipe BCE, while the remainder flows through the pipe BDE. At point E, the flows recombine to the original value and the liquid flows through the pipe EF. Note that we are assuming that all pipes in Figure 5.5 are shown in plan view with no elevation changes.

In order to solve for the pressures and flow rates in a parallel piping system, such as the one depicted in Figure 5.5, we use the following two principles of pipes in parallel.

(a) Conservation of total flow
(b) Common pressure loss across each parallel pipe.

According to the principle (a), the total flow entering each junction of pipe must equal the total flow leaving the junction.

Or simply,

Total Inflow = Total Outflow

Thus, in Figure 5.5, all flows entering and leaving the junction B must satisfy the above principle. If the flow into the junction B is Q and the flow in branch BCE is $Q_{BC}$ and flow in the branch BDE is $Q_{BD}$, we have from above conservation of total flow,

$$Q = Q_{BC} + Q_{BD} \qquad (5.9)$$

Assuming $Q_{BC}$ and $Q_{BD}$ represent the flow out of the junction B.

The second principle of parallel pipes, defined as (b) above requires that the pressure drop across the branch BCE must equal the pressure drop across the branch BDE. This is simply due to the fact that point B represents the common upstream pressure for each of these branches, while the pressure at point E represents the common downstream pressure. Referring to these pressures as $P_B$ and $P_E$, we can state

Pressure drop in branch BCE = $P_B$ - $P_E$ $\qquad (5.10)$
Pressure drop in branch BDE = $P_B$ - $P_E$ $\qquad (5.11)$

Assuming that the flow $Q_{BC}$ and $Q_{BD}$ are in the direction of BCE and BDE respectively. If we had a third pipe branch between B and E, such as that shown by the dashed line BE in Figure 5.5, we can state that the common pressure drop $P_B$ - $P_E$ would be applicable to the third parallel pipe between B and E, as well.

We can re-write the Equations (5.9) and (5.11) above as follows for the three parallel pipe system.

$$Q = Q_{BC} + Q_{BD} + Q_{BE} \qquad (5.12)$$

and

$$\Delta P_{BCE} = \Delta P_{BDE} = \Delta P_{BE} \qquad (5.13)$$

where

$\Delta P$     - the pressure drop in respective parallel pipes.

Using the Equations (5.12) and (5.13) we can solve for flow rates and pressures in any parallel piping system. We will demonstrate the above using an example problem later in this chapter.

Similar to the equivalent length concept in series piping, we can calculate an equivalent pipe diameter for pipes connected in parallel.

Since each of the parallel pipes in Figure 5.5, has a common pressure drop indicated by Equation (5.13), we can replace all the parallel pipes between B and E with one single pipe of length $L_E$ and diameter $D_E$ such that the pressure drop through the single pipe at flow Q equals that of the individual pipes as follows:

Pressure drop in equivalent single pipe length $L_E$ and diameter $D_E$ at flow rate Q

$$= \Delta P_{BCE}$$

Assuming now that we have only the two parallel pipes BCE and BDE in Figure 5.5, ignoring the dashed line BE, we can state that

$$Q = Q_{BC} + Q_{BD}$$

and

$$\Delta P_{EQ} = \Delta P_{BCE} = \Delta P_{BDE}$$

The pressure $\Delta P_{EQ}$ for the equivalent pipe can be written as, using Equation (3.28)

$$\Delta P_{EQ} = K (L_E) (Q)^2 / D^5_E$$

where K is a constant, that depends on the liquid properties.

The above two equation will then become

$$K L_E Q^2 / D^5_E = K L_{BC} Q^2_{BC} / D^5_{BC} = K L_{BD} Q^2_{BD} / D^5_{BD}$$

Simplifying we get

$$L_E Q^2 / D^5_E = L_{BC} Q^2_{BC} / D^5_{BC} = L_{BD} Q^2_{BD} / D^5_{BD}$$

Further simplifying the problem by setting

$$L_{BC} = L_{BD} = L_E$$

We get

$$Q^2 / D^5_E = Q^2_{BC} / D^5_{BC} = Q^2_{BD} / D^5_{BD}$$

Substituting for $Q_{BD}$ in terms for $Q_{BC}$ from Equation (5.9) we get

$$Q^2 / D^5_E = Q^2_{BC} / D^5_{BC} \qquad (5.14)$$

and

$$Q^2_{BC} / D^5_{BC} = (Q - Q_{BC})^2 / D^5_{BD} \qquad (5.15)$$

From the above Equations (5.14) and (5.15) we can solve for the two flows $Q_{BC}$, $Q_{BD}$ and the equivalent diameter $D_E$ in terms of the known quantities $Q$, $D_{BC}$ and $D_{BC}$.

A numerical example will illustrate the above method.

**Example Problem 5.1**

A parallel pipe system, similar to the one shown in Figure 5.5 is located in a horizontal plane with the following data

Flow rate Q           = 2 000 gal/min of water
Pipe branch BCE    = 12 inch diameter, 8 000 ft
Pipe branch BCE    = 10 inch diameter, 6 500 ft

Calculate the flow rate through each parallel pipe and the equivalent pipe diameter for a single pipe 5 000 ft long between B and E to replace the two parallel pipes.

**Solution**

$$Q_1 + Q_2 = 2\ 000$$
$$Q_1^{\ 2} L_1 / D_1^{\ 5} = Q_2^{\ 2} L_2 / D_2^{\ 5}$$

where suffix 1 and 2 refer to the two branches BCE and BDE respectively.

$$(Q_2/Q_1)^2 = (D_2/D_1)^5 (L_1/L_2)$$
$$= (10/12)^5 \times (8\ 000/\ 6\ 500)$$
$$Q_2/Q_1 = 0.7033$$

Solving we get

$$Q_1 = 1\ 174\ \text{gal/min}$$
$$Q_2 = 826\ \text{gal/min}$$

The equivalent pipe diameter for a single pipe 5 000 ft long is calculated as follows

$$(2\ 000)^2 (5\ 000) / D_E^{\ 5} = (1\ 174)^2 \times 8\ 000 / (12)^5$$

or

$$D_E = 13.52\ \text{inch}$$

Therefore, a 13.52 inch diameter pipe, 5 000 ft long between B and E will replace the two parallel pipes.

## 5.5  Transporting High Vapor Pressure Liquids

As mentioned previously, transportation of high vapor liquids such as Liquified Petroleum Gas (LPG) requires that a certain minimum pressure be maintained throughout the pipeline. This minimum pressure must be greater than the liquid vapor pressure at the flowing temperature. Otherwise liquid may vaporize causing two phase flow in the pipeline which the pumps cannot handle. If the vapor pressure of LPG at the flowing temperature is 250 psi, the minimum pressure anywhere in the pipeline must be greater than 250 psi. Conservatively, at high elevation points or peaks along the pipeline, we must insure that more than the minimum pressure is maintained. This is illustrated in Figure 5.2. Additionally, the delivery pressure at the end of the pipeline must also satisfy the minimum pressure requirements. Thus, the delivery pressure at the pipeline terminus for LPG may be 300 psi or higher to account for any meter station and manifold piping losses at the delivery point. Also, sometimes, with high vapor pressure liquids, the delivery point may be a pressure vessel or a pressurized sphere maintained at 500 to 600 psi and therefore may require even higher minimum pressures compared to the vapor pressure of the liquid. Hence, both the delivery pressure and the minimum pressure must be considered when analyzing pipelines transporting high vapor pressure liquids.

## 5.6   Horsepower Required

So far we have examined the pressure required to transport a given amount of liquid through a pipeline system. Depending on the flow rate and MAOP of the pipeline, we may need one or more pump stations to safely transport the specified throughput. The pressure required at each pump station will generally be provided by centrifugal or positive displacement pumps. Pump operation and performance will be discussed in Chapter 7. In this section we will calculate the horsepower required to pump a given volume of liquid through the pipeline regardless of the type of pumping equipment used.

### 5.6.1   Hydraulic Horsepower

Power required is defined as energy or work performed per unit time. In English units, energy is measured in ft-lb and power is expressed in Horsepower (HP). One HP is defined as 33 000 ft.lb/min or 550 ft.lb/s.

In SI units, energy is measured in Joules and power is measured in Joules/second (Watts). The larger unit Kilowatts (kW) is more commonly used. One HP is equal to 0.746 kW.

To illustrate the concept of work, energy and power required, imagine a situation that requires 150 000 gallons of water to be raised 500 ft to supply the needs of a small community. If this requirement is on a 24-hour basis we can state that the work done in lifting 150 000 gal of water by 500 ft is

(150 000/7.48) x 62.34 x 500 = 625 066 845 ft lb

where the specific weight of water is assumed to be 62.34 lb/ft$^3$ and 1 ft$^3$ = 7.48 gals.

Thus we need to expend 6.25 x 10$^8$ ft lb of energy over a 24-hour period to accomplish this task. Since one HP equals 33 000 ft lb/min, the power required in this case is

$$HP = \frac{6.25 \times 10^8}{24 \times 60 \times 33\ 000} = 13.2$$

This is also known as the hydraulic horsepower (HHP), since we have not considered pumping efficiency.

As a liquid flows through a pipeline, pressure loss occurs due to friction. The pressure needed at the beginning of the pipeline to account for friction and any elevation changes is then used to calculate the amount of energy required to transport the liquid. Factoring in the time element, we get the power required to transport the liquid.

**Example Problem 5.2**

Consider 4 000 bbl/hr being transported through a pipeline with one pump station operating at 1 000 psi discharge pressure. If the pump station suction pressure is 50 psi, the pump has to produce (1 000 - 50) or 950 psi differential pressure to pump 4 000 bbl/hr of the liquid. If the liquid specific gravity is 0.85 at flowing temperature, calculate the HP required at this flow rate.

**Solution**

Flow rate of liquid in lb/min is calculated as follows:

M = 4 000 bbl/hr (5.6146 ft³/bbl)(1 hr/60 min)(0.85)(62.34 lb/ft³)

where 62.34 is the specific weight of water in lb/ft³.

or

M = 19 834.14 lb/min

Therefore the HP required is

$$HP = \frac{(lb/min)\,(ft.\,head)}{33\,000}$$

or

HP = ((19 834.14) (950) (2.31)/0.85) /33 000 = 1 552

It must be noted that in the above calculation, no efficiency value has been considered. In other words, we have assumed 100% pumping efficiency. Therefore, the above HP calculated is referred to as Hydraulic Horsepower (HHP), based on 100% efficiency.

HHP = 1 552

## 5.6.2    Brake Horsepower

The Brake Horsepower takes into account the pump efficiency. If a pump efficiency of 75% is used we can calculate the brake horsepower (BHP) in the above example as follows:

Brake Horsepower = Hydraulic Horsepower / Pump Efficiency
BHP = HHP/0.75 = 1552 / 0.75 = 2 070

If an electric motor is used to drive the above pump, the actual motor horsepower required would be calculated as

Motor HP = BHP/Motor Efficiency

Generally induction motors used for driving pumps, have fairly high efficiencies, ranging from 95% to 98%. Using 98% for motor efficiency, we can calculate the motor HP required as follows:

Motor HP = 2 070 / 0.98 = 2 112

Since the closest standard size electric motor is 2 500 HP, this application will require a pump that can provide a differential pressure of 950 psi at a flow rate of 4 000 bbl/hr and will be driven by a 2 500 HP electric motor.

Pump companies measure pump flow rates in gal/min and pump pressures are expressed in terms of feet of liquid head. We can therefore convert the flow rate from bbl/hr to gal/min and the pump differential pressure of 950 psi can be converted to liquid head in ft.

$$4\ 000\ \text{bbl/hr} = \frac{4\ 000 \times 42}{60} = 2\ 800\ \text{gal/min}$$

$$950\ \text{psi} = \frac{950 \times 2.31}{0.85} = 2\ 582\ \text{ft}$$

The above statement for the pump requirement can then be re-worded as follows:

This application will use a pump that can provide a differential pressure of 2 582 ft of head at 2 800 gal/min and will be driven by a 2 500 HP electric motor. We will discuss pump performance in more detail in Chapter 7.

The formula for BHP required in terms of customary pipeline units is as follows:

$$BHP = Q\ P\ /\ (2\ 449E) \tag{5.16}$$

where

Q     - Flow rate, bbl/hr
P     - Differential Pressure, psi
E     - Efficiency, expressed as a decimal value less than 1.0

Two additional formulas for BHP are expressed in terms of flow rate in gal/min and pressure in psi or ft of liquid are as follows:

$$BHP = (GPM)(H)(Spgr) / (3\ 960E) \qquad (5.17)$$

and

$$BHP = (GPM)P / (1\ 714E) \qquad (5.18)$$

where

GPM - Flow rate, gal/min
H     - Differential head, ft
P      - Differential Pressure, psi
E      - Efficiency, expressed as a decimal value less than 1.0
Spgr - Liquid specific gravity, dimensionless

In SI units, power in kW can be calculated as follows:

$$\text{Power (kW)} = \frac{Q\ H\ Spgr}{367.46\ (E)} \qquad (5.19)$$

where

Q     - Flow rate, m$^3$/hr
H     - Differential head, m
Spgr - Liquid specific gravity
E      - Efficiency, expressed as a decimal value less than 1.0

and

$$\text{Power (kW)} = \frac{Q\ P}{3\ 600\ (E)} \qquad (5.20)$$

where

P     - Pressure, kPa
Q     - Flow rate, $m^3$/hr
E     - Efficiency, expressed as a decimal value less than 1.0

## Example Problem 5.3

A water distribution system requires a pump that can produce 2 500 ft head pressure to transport a flow rate of 5 000 gal/min. Assuming a centrifugal pump driven by an electric motor, calculate the hydraulic HP, pump BHP and the motor HP required. Pump efficiency = 82%. Motor efficiency = 96%.

## Solution

$$\text{Hydraulic HP} = \frac{(5\ 000 \times 62.34)}{7.48} \times \frac{2\ 500}{33\ 000}$$

HHP = 3 157

$$\text{Pump BHP required} = \frac{3\ 157}{0.82} = 3\ 850 \text{ HP}$$

$$\text{Motor HP required} = \frac{3\ 850}{0.96} = 4\ 010 \text{ HP}$$

## Example Problem 5.4

A water pipeline is used to move 320 L/s and requires a pump pressure of 750 m. Calculate the power required at 80% pump efficiency and 98% motor efficiency.

**Solution**

Using Equation (5.20)

Pump power required $= \dfrac{(320 \times 60 \times 60)}{1\,000} \times \dfrac{750 \times 1.0}{367.46 \times 0.80}$

$= 2\,939$ KW

Motor power required $= \dfrac{2\,939}{0.98} = 3\,000$ KW

## 5.7 Effect of Gravity and Viscosity

It can be seen from the previous discussions that the pump BHP is directly proportional to specific gravity of liquid. Therefore, if the HP for pumping water is 1000, the HP required when pumping a crude oil of specific gravity 0.85 is

Crude Oil HP = 0.85 (Water HP) = 0.85 x 1 000 = 850 HP

Similarly, when pumping a liquid of specific gravity greater than 1.0, the HP required will be higher. This can be seen from examining Equation (5.17).

We can therefore conclude that, for same pressure and flow rate, HP required increases with specific gravity of liquid pumped. The HP required is also affected by viscosity of the liquid pumped. Consider water with a viscosity of 1.0 cSt. If a particular pump generates a head of 2 500 ft at a flow rate of 3 000 gal/min and has a efficiency of 85%, we can calculate the water HP using Equation (5.17) as follows:

BHP $= \dfrac{(3\,000)(2\,500)(1.0)}{(0.85)(3\,960)} = 2\,228.16$

If liquid with a viscosity of 1 000 SSU is used with this pump we must correct the pump head, flow rate and efficiency values using the

Hydraulic Institute viscosity correction charts for centrifugal pumps. These will be discussed in more detail in Chapter 7. The net result is that the BHP required when pumping the high viscosity liquid will be higher than the above calculated value. It has been found that, with high viscosity liquids the pump efficiency degrades a lot faster than flow rate or head. Also considering pipeline hydraulics, we can say that high viscosities increase pressure required to transport a liquid and therefore increase HP required.

The effect of viscosity and specific gravity on pump performance will be discussed in more detail in later chapters.

## 5.8   System Head Curves

A System Head Curve, also known as System Curve, for a pipeline shows the variation of pressure required with flow rate. See Figure 5.6 for a typical System Head Curve. As the flow rate increases, the head required increases.

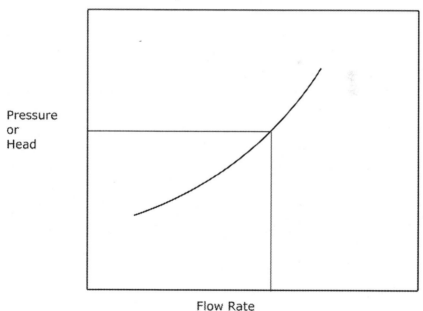

Figure 5.6 System Head Curve

Consider a pipeline of internal diameter D and length L used to transport a liquid of specific gravity Sg and viscosity v from a pump station at A to delivery point located at B. We can calculate the pressure required at A to transport the liquid at a particular flow rate Q. By varying flow rate Q we can determine the pressure required at A for each flow rate such that a given delivery pressure at B is maintained. For each flow rate Q, we would calculate the pressure drop due to friction for the length L of the pipeline, add the head required to account for elevation difference between A and B and finally, add the delivery pressure required at B as follows.

Pressure at A = Frictional pressure drop + Elevation head + Delivery pressure, using Equation (5.1).

Once the pressure at A is calculated for each flow rate we can plot a system head curve as shown in Figure 5.6

The vertical axis may be in feet of liquid or psi. The horizontal axis will be in units of flow such as gal/min or bbl/hr.

In Chapter 7 system head curves along with pump head curves will be reviewed in detail. We will see how the system head curve in conjunction with the pump head curve will determine the operating point for a particular pump-pipeline configuration.

Since system head curve represents the pressure required to pump various flow rates through a given pipeline, we can plot a family of such curves for different liquids as shown in Figure 5.7. The higher specific gravity and viscosity of diesel fuel requires greater pressures compared to gasoline. Hence the diesel system head curve is located above that of gasoline as shown in Figure 5.7.

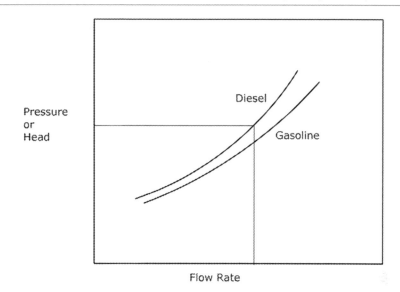

**Figure 5.7 System Head Curve - Different Products**

Note also that when there is no elevation difference involved the System Curve will start at the (0,0) point. This means at zero flow rate the pressure required is zero.

The shape of the system curve varies depending upon the amount of friction head compared to elevation head. Figure 5.8 and Figure 5.9 show two system curves that illustrate this. In Figure 5.8 there is comparatively less influence of the pipe elevations.

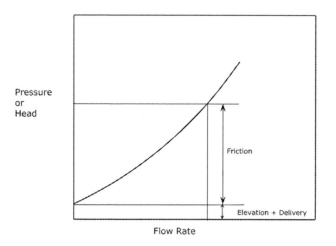

**Figure 5.8 - System Head Curve - High Friction**

Most of the system head required is due to the friction in the pipe. In comparison, Figure 5.9 shows a system head curve that consists mostly of the static head due to the pipe elevations.

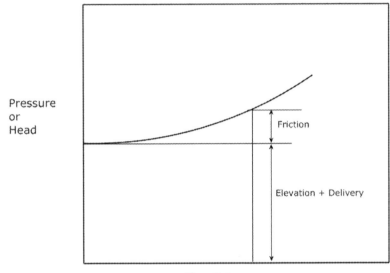

Figure 5.9 - System Head Curve - High Elevation

It can be seen from the above that the frictional head is a smaller component compared to the static elevation head.


## 5.9  Injections and Deliveries

In most pipelines liquid enters the pipeline at the beginning and would continue toward the end of the pipeline to be delivered at the terminus, with no deliveries or injection at any intermediate point along the pipeline. However, there are situation where liquid would be delivered off the pipeline (stripping) at some intermediate location and the remainder would continue towards the pipeline terminus. Similarly, liquid may enter the pipeline (injection) at some intermediate location thereby adding to the existing volume in the pipeline. These are called deliveries off the pipeline and injection into the pipeline. This is illustrated in Figure 5.10.

**Figure 5.10 Injection and Delivery**

Let us analyze the pressures required in a pipeline with injection and stripping. The pipeline AB in Figure 5.10 shows 6 000 bbl/hr entering the pipeline at A. At a point C, a new stream of liquid enters the pipeline at a rate of 1 000 bbl/hr. Further along the pipeline, at point D, a volume of 3 000 bbl/hr is stripped off the pipeline. Consequently, a resultant volume of (6 000 + 1 000 - 3 000) or 4 000 bbl/hr is delivered to the pipeline terminus at B. In order to calculate the pressures required at A for such a pipeline with injection and deliveries we proceed as follows.

First, the pipe segment between A and C that has a uniform flow of 6 000 bbl/hr is analyzed. The pressure drop in AC is calculated considering the 6 000 bbl/hr flow rate, pipe diameter and liquid properties. Next the pressure drop in the pipe segment CD with a flow rate of 7 000 bbl/hr is calculated taking into account the blended liquid properties by combining the incoming stream at C (6 000 bbl/hr) along the main line with the injection stream (1 000 bbl/hr) at C. Finally, the pressure drop in the pipe segment, DB is calculated considering a volume of 4 000 bbl/hr and the liquid properties in that segment, which would be the same as that of pipe segment CD. The total frictional pressure drop between A and B will be the sum of the three pressure drops calculated above. After adding any elevation head between A and B and accounting for the required delivery pressure at B we can calculate the total pressure required at point A for this pipeline system. This is illustrated in an example next.

**Example Problem 5.4**

In Figure 5.10 the pipeline from Point A to Point B is 48 miles long and is 18 inches nominal diameter, 0.281 inch wall thickness. It is constructed of 5LX-65 grade steel. At A, crude oil of specific gravity

0.85 and 10 cSt viscosity enters the pipeline at a flow rate of 6 000 bbl/hr. At C (milepost 22) a new stream of crude oil with a specific gravity of 0.82 and 3.5 cSt viscosity enters the pipeline at a flow rate of 1 000 bbl/hr. The mixed stream then continues to point D (milepost 32) where 3 000 bbl/hr is stripped off the pipeline. The remaining volume continues to the end of the pipeline at point B.

(a) Calculate the pressure required at A and the composition of the crude oil arriving at terminus B at a minimum delivery pressure of 50 psi. Assume elevations at A, C, D and B to be 100, 150, 250 and 300 ft respectively. Use Colebrook-White Equations for pressure drop calculations and assume a pipe roughness of 0.002 inch.
(b) How much pump HP will be required to maintain this flow rate at A, assuming 50 psi pump suction pressure at A and 80% pump efficiency?
(c) If a Positive Displacement (PD) pump is used to inject the stream at C, what pressure and HP are required at C?

**Solution**

The pressure drop due to friction for segment AC is calculated using Equation (3.27) as follows:

Reynold's number = 92.24 x 6 000 x 24 / (17.438 x 10) = 76 170
Friction Factor = 0.02
Pressure drop = 13.25 psi/mi
Frictional Pressure drop between A and C = 13.25 x 22 = 291.5 psi

Next, we calculate the blended properties of the liquid stream after mixing two streams at point C, by blending 6 000 bbl/hr of crude A (specific gravity of 0.85 and viscosity of 10 cSt) with 1 000 bbl/hr of crude B (specific gravity of 0.82 and viscosity of 3.5 cSt) using Equations (2.4) and (2.21) as follows:

Blended specific gravity at C = 0.8457
Blended Viscosity at C = 8.366 cSt

For pipe segment CD we calculate the pressure drop by using above properties at a flow rate of 7 000 bbl/hr.

Reynold's number = 92.24 x 7 000 x 24 / (17.438 x 8.366) = 106 222
Friction Factor = 0.0188
Pressure drop = 16.83 psi/mi
Frictional Pressure drop between C and D = 16.83 x 10 = 168.3 psi

Finally we calculate for pipe segment DB, the pressure drop by using above liquid properties at a flow rate of 4 000 bbl/hr.

Reynold's number = 92.24 x 4 000 x 24 / (17.438 x 8.366) = 60 698
Friction Factor = 0.021
Pressure drop = 6.13 psi/mi
Frictional Pressure drop between D and B = 6.13 x 16 = 98.08 psi

Therefore, the total frictional pressure drop between point A and point B is

291.5 + 168.3 + 98.08 = 557.9 psi

The elevation head between A and B consists of (150 - 100) ft between A and C and (300-150) between C and B. We need to separate the total elevation head in this fashion because of differences in liquid properties in pipe segments AC and CB. Therefore, the total elevation head is

[(150-100) x 0.85 / 2.31] + [(300 -150) x 0.8457 / 2.31] = 73.32 psi

Adding the delivery pressure of 50 psi, the total pressure required at A is therefore,

557.9 + 73.32 + 50 = 681.22 psi

Therefore, the pressure required at A is 681.22 psi and the composition of the crude oil arriving at terminus B is - specific gravity of 0.8457 and viscosity of 8.366 cSt.

The HP required at A is calculated using Equation (5.16) as follows:

BHP = 6 000 x (681.22 - 50) / (0.8 x 2 449) = 1 933 HP

To calculate the injection pump requirement at point C, we must first calculate the pressure in the pipeline at point C that the PD pump has to overcome.

The pressure at C = Pressure at A - pressure drop from A to C - elevation head A to C

$P_C$ = 681.22 - 291.5 - (150-100) x 0.85/2.31 = 371.3 psi

The HP required for the PD pump at C is calculated using Equation (5.16) as follows:

PD Pump HP required = (371.3 - 50) x 1 000 / (0.8 x 2 449) = 164

Assuming 50 psi suction pressure and 80% pump efficiency.

## 5.10 Pipe Branches

In the previous section we discussed a pipeline with an injection point and a delivery point between the pipeline inlet and pipeline terminus. These injections and deliveries may actually consist of branch pipes bringing liquid into the main line (incoming branch) and delivering liquid out of the pipeline (outgoing branch). This is illustrated in Figure 5.11.

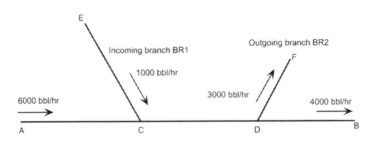

**Figure 5.11 Incoming and Outgoing Branches**

Sometimes we are interested in sizing the branch pipes for the required flow rates and pressures. For example in Figure 5.11 the incoming branch BR1 needs to be sized to handle 1 000 bbl/hr of liquid entering the pipeline at point C with the specified pressure $P_C$. We have to determine the pressure required at the beginning of the branch BR1 (point E) for a specified branch pipe diameter, liquid properties etc. so that the required pressure $P_C$ is achieved at the end of the branch at point C.

Similarly, for an outgoing branch BR2 as shown in Figure 5.11, we are interested in determining the branch pipe size such that the liquid stream flowing through the outgoing branch arrives at its destination F with a certain specified delivery pressure. We would use the starting pressure of BR2 as $P_D$ which represents the main line pressure at the junction with the outgoing branch. Using the value of $P_D$ and the required delivery pressure $P_F$ at F we can calculate the pressure drop per mile for BR2. Hence we can select a pipe size to handle the flow rate through BR2. An example next illustrates how branch piping are sized for injection and deliveries.

**Example Problem 5.5**

(a) Using the data in Example Problem 5.4 and Figure 5.11, determine the branch pipe sizes required for branch BR1 for the injection rate of 1 000 bbl/hr and an MAOP of 600 psi. Assume pipe branch BR1 to be 2.5 miles long and essentially flat elevation profile.
(b) Calculate the pipe size for the outgoing branch BR2 to handle the delivery of 3 000 bbl/hr at D and pressure of 75 psi at F. The branch BR2 is 4 miles long.
(c) What HP is required at beginning of branch BR1?

**Solution**

(a) Assume 6 inch pipe for branch BR1 and calculate pressure drop in 2.5 mile length of pipe as follows:

Using given liquid properties we calculate,

Pressure drop = 65.817 psi/mi for 1 000 bbl/hr in 6 inch pipe.

Pressure required at E to match junction pressure of 371.3 at C is

$P_E$ = 65.817 x 2.5 + 371.3 = 535.84 psi

Since this is less than the MAOP of 600 psi for BR1, 6 inch line is adequate.

(b) With a junction pressure of 166.39 psi available at D, a 10 inch pipe is inadequate to carry 3 000 bbl/hr of crude through the 4 miles long branch pipe BR2 and provide a 75 psi delivery at F.

Next, assuming a 12 inch pipe for BR2, 3,000 bbl/hr flow rate, specific gravity of 0.8457 and viscosity of 8.366 cSt, we get

Pressure drop in BR2 = 20.07 psi/mi

Therefore, we get a delivery pressure of

$P_F$ = 166.39 - 20.07 x 4 = 86.11 psi

which is higher than the minimum 50 psi delivery pressure required.

Therefore, branch pipe BR2 must be at least 12 inch nominal diameter.

(c) The HP required at beginning of branch BR1 is calculated to be as follows using Equation (5.16):

HP = (535.84 - 50) x 1 000 / (2 449 x 0.8) = 248

based on 80% efficiency and 50 psi suction pressure at E.

## 5.11 Pipe Loops

A pipe loop is a length of parallel pipe installed between two points on a main pipeline as shown in Figure 5.12 We discussed parallel pipes and equivalent diameter earlier in this chapter. In this section we will discuss how looping an existing pipeline will reduce the pressure drop due to friction and thus require less pumping HP.

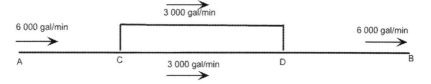

**Figure 5.12 Pipe Loop**

The purpose of the pipe loop is to split the flow through a parallel segment of the pipeline between the two locations resulting in a reduced pressure drop in that segment of the pipeline.

Consider pipeline from A to B with a loop installed from C to D as shown in Figure 5.12. The flow rate between A and C is 6 000 gal/min. At C where the loop is installed the flow rate of 6 000 gal/min is partially diverted to the loop section with the remainder going through the main line portion CD. If we assume that the diameter of the loop is the same as that of the main pipe, this will cause, 3 000 gal/min flow through the loop and an equal amount through the main line. Therefore, the section CD of the main line prior to looping had the full flow of 6 000 gal/min flowing through it resulting in a pressure drop of, say 25 psi/mi. With the loop installed, section CD has half the flow and therefore approximately one-fourth of the pressure drop (since the pressure drop varies, as the square of the flow rate) or 6.25 psi/mi, based on Equation (3.27). If the length CD were 10 miles, the total frictional pressure drop without the loop will be 250 psi. With the pipe loop the pressure drop will be 62.5 psi, a significant drop in pressure. Therefore if a pipeline section is bottle necked due to maximum allowable pressure, we can reduce the overall pressure profile by installing a loop in that pipe segment.

To illustrate this concept further, consider an example problem as discussed next.

**Example Problem 5.6**

A 16 inch crude oil pipeline (0.250 inch in wall thickness) is 30 miles long from point A to point B. The flow rate at the inlet A is 4 000 bbl/hr. The crude oil properties are specific gravity of 0.85 and viscosity of 10 cSt at a flowing temperature of 70 °F.

(a) Calculate the pressure required at A without any pipe loop. Assume 50 psi delivery pressure at the terminus B and a flat pipeline elevation profile.
(b) If a 10 mile portion CD, starting at milepost 10 is looped with an identical 16 inch pipeline. Calculate the reduced pressure at A.
(c) What is the difference in pump HP required at A between case (a) and (b) above. Use 80% pump efficiency and 25 psi pump suction pressure.

**Solution**

Using Colebrook-White Equation and assuming a pipe roughness of 0.002 inch we calculate the pressure drop per unit length using Equation (3.27) as follows:

Reynolds number = 92.24 x 4 000 x 24 / (15.5 x 10) = 57 129

Friction Factor f = 0.0213

and

Pressure drop = 11.28 psi/mi

(a) Therefore total pressure required at A = 30 x 11.28 + 50 = 388.40 psi without any pipe loop. With 10 miles of pipe loop the flow rate through the loop is 2 000 bbl/hr and the revised pressure drop at the reduced flow is calculated as

Pressure drop = 3.28 psi/mi

The pressure drop in the 10 mile loop = 10 x 3.28 = 32.8 psi
The pressure drop in the 10 mile section AC = 10 x 11.28 = 112.8 psi
The pressure drop in the last 10 mile section DB = 10 x 11.28 = 112.8 psi

Therefore

(b) Total pressure required at A = 112.8 + 32.8 + 112.8 + 50 = 308.40 psi with the pipe loop.

Therefore, using the 10 mile loop causes a reduction of 80 psi in the pressure required at point A.

(c) HP required without the loop is

HP = 4 000 x (388.40 - 25) / (2 449 x 0.8) = 742

HP required considering the loop is

HP = 4 000 x (308.40 - 25) / (2 449 x 0.8) = 579

Thus installing the pipe loop results in a reduction of

(742 - 579) / 742 = 22%

in pumping HP.

What would be the impact if the looped section of pipe were smaller than the main line? If we installed a smaller pipe in parallel, the flow will still be split through the loop section, but will not be equally divided. The smaller pipe will carry less flow rate than the larger main line pipe, such that the pressure drop through the main line from Point C to Point D will exactly equal the pressure drop through the pipe loop between C and D, since both the mainline pipe and the pipe loop have common pressures at the junction C and D.

To illustrate this, consider an 8 inch pipe looped with the 16 inch mainline. Assume $Q_8$ represents the flow rate through the 8 inch loop and $Q_{16}$ through the 16 inch mainline portion. We can write

$$Q_8 + Q_{16} = \text{Total flow} = 4\ 000 \text{ bbl/hr.} \qquad (5.21)$$

From Equation (3.27), we know that the frictional pressure drop in a pipe is directly proportional to the square of the flow rate and inversely proportional to the fifth power of the diameter. Therefore, we can write

Pressure Drop in 8 inch pipe = $K(Q_8)^2 / (8.625 - 0.5)^5$

considering 0.250 inch wall thickness and K is a constant.

or

$$\Delta P_8 = K\,(Q_8)^2 / 35\ 409 \qquad\qquad (5.22)$$

Similarly, for the 16 inch pipe

$$\Delta P_{16} = K\,(Q_{16})^2 / 894\ 661 \qquad\qquad (5.23)$$

Dividing Equation (5.22) by Equation (5.23), we get

$$\Delta P_8 / \Delta P_{16} = 25.27\,(Q_8 / Q_{16})^2$$

Since $\Delta P_8$ and $\Delta P_{16}$ are equal in a loop system, the above equation reduces to

$$Q_{16} = 5.03 \text{ x } Q_8 \qquad\qquad (5.24)$$

Solving Equation (5.21) and Equation (5.24) simultaneously, we get

$$Q_8 = 4\ 000 / 6.03 = 664 \text{ bbl/hr - flow through 8 inch loop}$$

and

$$Q_{16} = 4\ 000 - 664 = 3\ 336 \text{ bbl/hr - flow through 16 inch mainline}$$
portion.

We can now calculate the pressure drop and the total pressure required at A as we did in Example Problem 5.6 earlier.

Since the 10 mile looped portion of the 16 inch pipe has a flow rate of 3 336 bbl/hr we calculate the pressure drop in this pipe first

Reynolds number = 92.24 x 3 336 x 24 / (15.5 x 10) = 47 646
Friction Factor f = 0.0216

and

Pressure drop = 7.96 psi/mi

Therefore, the pressure drop in the loop section

= 7.96 x 10 = 79.6 psi

The pressure drop in section AC and section DB remain the same as before.

The total pressure required at A becomes

Pressure at A = 112.8 + 79.6 + 112.8 + 50 = 355.2 psi

Compare this with 388.4 psi (without loop) and 308.4 psi (with 16 inch loop).

We could also calculate the above using the Equivalent Diameter approach discussed earlier in this chapter. We will calculate an equivalent diameter pipe 10 miles long that will replace the 8 in/16 in loop.

Recalling our discussions of parallel pipes and equivalent diameter in section 5.4 and using Equations (5.14) and (5.15) we calculate the equivalent diameter $D_E$ as follows after some simplifications

$$D_E^{2.5} = (8.125)^{2.5} + (15.5)^{2.5}$$
$$D_E = 16.67 \text{ in}$$

Therefore, we can say that the two parallel pipes 8 inch and 16 inch together equal a single pipe that is 16.67 inch inside diameter. To show that the equivalent diameter concept is fairly accurate, we will calculate the pressure drop in this equivalent pipe and compare it to the two parallel pipes.

Reynolds number = 92.24 x 4 000 x 24 / (16.67 x 10) = 53 120
Friction Factor f = 0.0211

and

Pressure drop = 7.77 psi/mi

Compare this with 7.96 psi/mi we calculated earlier. It can be seen that the equivalent diameter method is within 2% of the more exact method and therefore is accurate for most practical purposes.

A practice problem with dissimilar pipe sizes as discussed above is included at the end of this Chapter as an exercise for the reader.

## 5.12 Summary

In this chapter we extended the pressure drop concept developed in Chapter 3, to calculate the total pressure required to transport liquid through a pipeline taking into account elevation profile of pipeline and required delivery pressure at the terminus. We analyzed pipes in series and parallel and introduced the concept of equivalent pipe length and equivalent pipe diameter. System head curve calculations were discussed to compare pressure required at various flow rates through a pipe segment. Flow injection and delivery in pipelines were studied and the impact on the hydraulic gradient discussed. We also sized branch piping connections and compared the advantage of looping a pipeline to reduce overall pressure drop and pumping horsepower. For a given pipeline system, the hydraulic horsepower, brake horsepower and motor horsepower calculations were illustrated using examples. A more comprehensive analysis of pumps and HP required will be covered in Chapter 7.

## 5.13 Problems

5.13.1 A pipeline 50 miles long, 16 inch outside diameter, 0.250 inch wall thickness is constructed of API 5LX-65 material. It is used to transport diesel and other refined products from the refinery at Carson to a storage tank at Compton. During Phase

I, a flow rate of 5 000 bbl/hr of diesel fuel is to be transported with one pump station located at Carson. The required delivery pressure at Compton is 50 psi. Assume a generally rolling pipeline elevation profile without any critical peaks along the pipeline. The elevation at Carson is 100 ft, and the storage tank at Compton is located on top of a hill at an elevation of 350 ft.

(a) Using diesel with specific gravity of 0.85 and a viscosity of 5.5 cSt at a flowing temperature of 70°F, calculate the total pressure required at Carson to transport 5 000 bbl/hr of diesel on a continuous basis. Use Hazen-Williams formula with C-factor = 125. The Carson pump suction pressure is 30 psi and the pump is driven by a constant speed electric motor.

(b) Determine the BHP and motor HP required at Carson assuming 82% pump efficiency and 96% motor efficiency.

(c) What size electric motor would be required at Carson?

(d) Assuming a maximum allowable operating pressure of 1 400 psi for the pipeline, how much additional throughput can be achieved in Phase II, if the pumps are modified at Carson?

5.13.2  A water pipeline is being built to transport 2 300 m³/hr from a storage tank at Lyon (elevation - 500 m) to a distribution facility 50 km away at Fenner (elevation - 850 m). What size pipe will be required to limit velocities to 3 m/s and an allowable pipeline pressure of 5.5 MPa.? No intermediate pump stations are to be used. Delivery pressure at Fenner is 0.3 MPa. Use Hazen-Williams formula with C-Factor = 110

5.13.3  In Problem 5.13.1, what throughputs can be achieved when pumping gasoline alone? Use specific gravity of 0.74 and viscosity of 0.65 for gasoline at the flowing temperature. Compare the pump head requirements when pumping diesel versus gasoline. Hazen-Williams formula with C-Factor = 145.

5.13.4  Calculate the BHP requirements for the above problem for both diesel and gasoline movements.

5.13.5   Consider a loop pipeline similar to the Example Problem 5.6. Instead of an the loop being 16 inch consider a smaller 10 inch pipe installed in parallel for the middle 10 mile section. How does the pressure at A change compared to the no loop case.

# 6

# Multi-Pump Station Pipelines

In the previous chapter we calculated total pressure required to pump a liquid through a pipeline from point A to point B at a specified flow rate. Three components of the total pressure required (friction head, elevation head and delivery pressure) were analyzed. Depending on the Maximum Allowable Operating Pressure (MAOP) of the pipeline we concluded that one or more pump stations may be required to handle the throughput.

In this chapter we will discuss pipeline hydraulics for multiple pump stations. Hydraulic balance and how to determine the intermediate booster pump stations will be explained. To efficiently utilize pipe material we will explore telescoping pipe wall thickness and grade tapering. Also covered will be slack line flow and more detail analysis of batched pipeline hydraulics.

## 6.1  Hydraulic Balance and Pump Stations Required

Suppose calculations indicate that at a flow rate of 5 000 gal/min, a 100 mile pipeline requires a pressure of 2 000 psi at the beginning of the pipeline. This 2 000 psi pressure may be provided in two steps of 1 000 psi each or three steps of approximately 670 psi each. In fact, due to the internal pressure limit of the pipe, we may not be able to provide one pump station at the beginning of the pipeline, operating at 2 000 psi, Most pipelines have an internal pressure limit of 1 000 to 1 440 psi based on pipe wall thickness, grade of steel, etc. as we found in Chapter 4. Therefore, in long pipelines the total pressure required to pump the liquid is provided in two or more stages by installing intermediate booster pumps along the pipeline.

In the example case with 2 000 psi requirement, at 1 400 psi pipeline MAOP, we would provide this pressure as follows. The pump station at the start of the pipeline will provide a discharge pressure of 1 000 psi, which would be consumed by friction loss in the pipeline and at some point (roughly halfway) along the pipeline the pressure will drop to zero. At this location we boost the liquid pressure to 1 000 psi using an intermediate booster pump station. We have assumed that the pipeline is essentially on a flat elevation profile.

This pressure of 1 000 psi will be sufficient to take care of the friction loss in the second half of the pipeline length. The liquid pressure will reduce to zero at the end of the pipeline. Since the liquid pressure at any point along the pipeline must be above the vapor pressure of the liquid at the flowing temperature, and the intermediate pumps require certain minimum suction pressure, we cannot allow the pressure at any point to drop to zero. Accordingly, we will locate the second pump station at a point where the pressure has dropped to a suitable suction pressure, such as 50 psi. The minimum suction pressure required is also dictated by the particular pump and may have to be higher than 50 psi, to account for any restrictions and suction piping losses at the pump station. For the present, we will assume 50 psi suction pressure is adequate for each pump station. Hence, starting with a discharge pressure of 1 050 psi (1 000 + 50) we will locate the second pump station (intermediate booster pump) along the pipeline where the pressure has dropped to 50 psi. This pump station will then boost the liquid pressure back up to 1 050 psi and will deliver the liquid to the pipeline terminus at 50 psi. Thus each pump station provides 1 000 psi differential pressure (Discharge pressure less suction pressure) to the liquid, together matching the total pressure requirement of 2 000 psi at 5 000 gal/min flow rate.

Note that in the above analysis, we ignored pipeline elevations and assumed that the pipeline profile is essentially flat. With elevations taken into account, the location of the intermediate booster pump will be different from that of a pipeline along a flat terrain.

Hydraulic balance is when each pump station supplies the same amount of energy to the liquid. Ideally pump station will be located at hydraulic centers. This will result in same HP added to the liquid at each pump station. For a single flow rate at the inlet of the pipeline (no intermediate injections or deliveries), the hydraulic centers will also result in same discharge pressures at each pump station. Due to topographic conditions it may not be possible to locate the intermediate pump station at locations desired for hydraulic balance. For example, calculations may show that three pump stations are required to handle the flow rate and that the two intermediate pump stations are to be located at milepost 50 and milepost 85. It is quite possible that the pump station location of milepost 50, when investigated in the field, may be in the middle of a swamp or a river. Hence we will have to relocate the subject pump station to a more suitable location after field investigation. If the revised location of the second pump station were at milepost 52, obviously, hydraulic balance would no longer be valid. Recalculations of hydraulics with the newly selected pump station locations will show hydraulic imbalance and all pump stations will not be operating at the same discharge pressure or providing the same amount of HP to the liquid at each pump station. Therefore, while it is desirable to have all pump stations balanced, it may not be practical. Balanced pump station locations afford the advantage of using identical pumps and motors and the convenience of maintaining a common set of spare parts (pump rotating elements, mechanical seal etc.) at a central operating district location.

In Chapter 5 we discussed how the location of an intermediate pump station can be calculated from given data on pump station suction pressure, discharge pressure etc. In Section 5.2 we presented a formula to calculate the discharge pressure for a two pump station pipeline system given the total pressure required for a particular flow rate. We will expand on that discussion by presenting a method to calculate the pump station pressures for hydraulic balance.

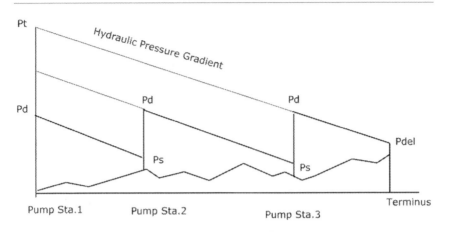

**Figure 6.1 - Hydraulic Gradient - Multiple Pump Stations**

Figure 6.1 shows a pipeline with varying elevation profile, but no significant controlling peaks along the pipeline. The total pressure $P_T$ was calculated for the given flow rate and liquid properties. The hydraulic gradient with one pump station at the total pressure $P_T$ is as shown. Since $P_T$ may be higher than the pipe MAOP, we will assume that three pump stations are required to provide the pressures needed within MAOP limits. Each pump station will discharge at pressure $P_D$. If $P_S$ represents the pump station suction pressure and $P_{del}$ the delivery pressure at the pipeline terminus, we can state the following, using geometry.

$$P_D + (P_D - P_S) + (P_D - P_S) = P_T \tag{6.1}$$

Since the above is based on one origin pump station and two intermediate pump stations, we can extend the above equation for N pump stations as follows

$$P_D + (N - 1)(P_D - P_S) = P_T \tag{6.2}$$

Solving for N we get

$$N = (P_T - P_S) / (P_D - P_S) \tag{6.3}$$

The above equation is used to estimate the number of pump stations required for hydraulic balance given the discharge pressure limit $P_D$ at each pump station.

Solving Equation (6.3) for the common pump station discharge pressure

$$P_D = (P_T - P_S) / N + P_S \qquad\qquad (6.4)$$

As an example, if the total pressure calculated were 2 000 psi and the suction pressure is 25 psi, the number of pump station required with 1 050 psi discharge pressure is

$$N = (2\ 000 - 50) / (1\ 050 - 50) = 1.95$$

Rounding up to the nearest whole number, we can conclude that two pump stations are required. Each pump station will operate at a discharge pressure $P_D$ from Equation (6.4)

$$P_D = (2\ 000 - 50) / 2 + 50 = 1\ 025\ \text{psi}$$

Once we calculate the discharge pressure required for hydraulic balance, as above, a graphical method can be used to locate the pump stations along the pipeline profile. First the pipeline profile (milepost versus elevation) is plotted and the hydraulic gradient superimposed upon it by drawing the sloped line starting at $P_T$ at A and ending at $P_{del}$ at D as shown in Figure 6.1. Note that the pressure must be converted to feet of head, since the pipe elevation profile is plotted in feet. Next starting at the first pump station A at discharge pressure $P_D$, a line is drawn parallel to the hydraulic gradient. The location B of the second pump station will be established at a point where the hydraulic gradient between A and B meet the vertical line at the suction pressure $P_S$. The process is continued to determine the location C of the third pump station.

In the above analysis, we have made several simplifying assumptions. We assumed that the pressure drop per mile was constant throughout the pipeline, meaning the pipe inside diameter was uniform throughout.

With variable pipe diameter or wall thickness, the hydraulic gradient slopes between pump station segments may not be the same.

## 6.2   Telescoping Pipe Wall Thickness

On examining a typical hydraulic gradient shown Figure 6.2, it is evident that under steady state operating conditions, the pipe pressure decreases from pump station to the terminus in the direction of flow. Thus, the pipeline segment immediately down stream of a pump station will be subject to higher pressures such as 1 000 to 1 200 psi while the tail end of that segment before the next pump station (or terminus) will be subject to lower pressures in the range of 50 to 300 psi. If we use the same wall thickness throughout the pipeline, we will be underutilizing the downstream portion of the piping. Therefore, a more efficient approach would be to reduce the pipe wall thickness as we move away from a pump station toward the suction side of the next pump station or the delivery terminus.

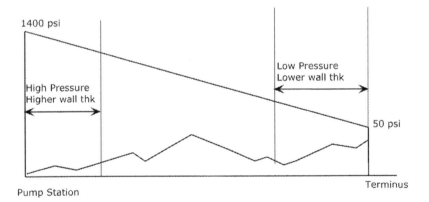

**Figure 6.2 - Telescoping Pipe Wall Thickness**

The higher pipe wall thickness immediately adjacent to the pump station will be able to withstand the higher discharge pressure and as the pressure reduces down the line, the lower wall thickness would automatically be designed to withstand the lower pressures as we approach the next pump station or delivery terminus. This process of varying the wall thickness to compensate for reduced pipeline pressures is referred to as telescoping pipe wall thickness.

A note of caution regarding the wall thickness tapering would be appropriate here. If a pipeline has two pump stations and the second pump station is shut down for some reason, the hydraulic gradient is as shown in Figure 6.3

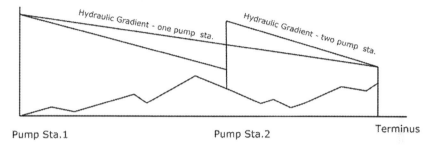

Figure 6.3 - Hydraulic Gradient - Pump Station Shutdown

It can be seen that portions of the pipeline on the upstream side of the second pump station will be subject to higher pressure than when the second pump station was online. Therefore, wall thickness reductions (telescoping) implemented upstream of a pump station, must be able to handle the higher pressures due to intermediate pump station shut down.

## 6.3   Change of Pipe Grade - Grade Tapering

Similar to changing pipe wall thickness to compensate for lower pressures as we approach the next pump station or delivery terminus, the pipe grade may also be varied. Thus the high pressure sections may be constructed of X-52 grade steel, whereas the lower pressure section may be constructed of X-42 grade pipe material thereby reducing the total cost. This process of varying the pipe grade is referred to as grade tapering. Sometimes a combination of telescoping and grade tapering is used to minimize pipe cost. It must be noted that such wall thickness variation and pipe grade reduction to match the requirements of steady state pressures may not always work. Consideration must be given to increased pipeline pressures under intermediate pump station shut down and upset conditions such as pump start up, valve closure, etc. These transient conditions cause surge pressures in a pipeline and therefore must be taken into account when selecting optimum wall thickness and pipe grade. This is illustrated in the modified hydraulic pressure gradient under transient conditions as depicted in Figure 6.4.

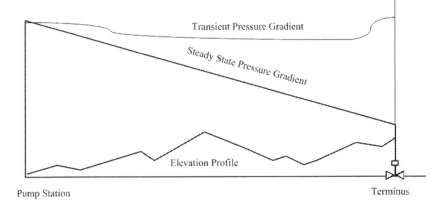

**Figure 6.4 - Hydraulic Gradient - Steady State versus Transient**

## 6.4   Slack Line and Open Channel Flow

Generally most pipelines flow full with no vapor space or a free liquid surface. However, under certain topographic conditions with drastic elevation changes, we may encounter pipeline sections that are partially full, called open channel flow or slack line conditions. Slack line operation may be unavoidable in some water lines, refined products and crude oil pipelines. Such a flow condition cannot be tolerated with high vapor pressure liquids and in batched pipelines. The latter would cause intermingling of batches with disastrous consequences.

Consider a long pipeline with a very high peak at some point between the origin A and the terminus B as shown in Figure 6.5

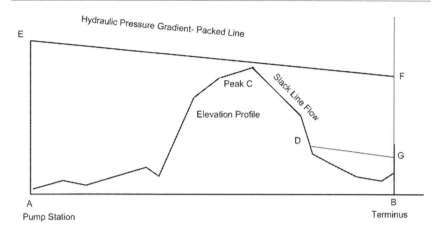

**Figure 6.5 - Hydraulic Gradient - Slack Line versus Packed Line**

Due to the high elevation point at C, the pressure required at A must be sufficient to take care of the friction loss between A and C, the pressure head due to elevation difference between A and C and the minimum pressure required at the top of the hill at C to prevent vaporization of liquid.

Once the liquid reaches the peak at C with the required minimum pressure, the elevation difference between C and B helps the liquid gain pressure as it flows down the hill from C to the terminus at B. The frictional pressure drop between C and B would have an opposite effect of the elevation and hence the resultant pressure at the terminus B will be the difference between the elevation head and the friction head. If the elevation head between C and B is sufficiently high compared to the frictional pressure drop between C and B, the final delivery pressure at B would be higher than the minimum required at the terminus. If the delivery at B is into an atmospheric storage tank, the hydraulic gradient will be modified as shown in the dashed line for slack line conditions.

The upper hydraulic gradient depicts a packed line condition where the delivery pressure at B represented by point F is substantially higher than that required for delivery into a tank represented by point G. The lower-hydraulic gradient shows that a portion of the pipeline between the peak at C and a point D will run in a partially full or slack line condition. Every point in the pipeline between C and D will be zero gauge pressure. From D to B the pipe will run full without any slack.

The slack line portion CD of the pipeline where the pipe is only partially full of the liquid is also referred to as open channel flow.

In this portion of the pipeline both liquid and vapor exists and therefore is an undesirable condition specially when pumping high vapor pressure liquids. Since a minimum pressure has to be maintained at the peak C, to prevent vaporization, the subsequent open channel flow in section CD of the pipeline defeats the purpose of maintaining a minimum pressure in the pipeline. In such instances, the pipeline must be operated in a packed condition (no slack line or open channel flow) by providing the necessary back pressure at B using a control valve, thus bringing the hydraulic gradient back to EF.

The control valve at B should have an upstream pressure equal to the pressure that will produce the packed line hydraulic gradient showed in Figure 6.5. In crude oil and refined products pipeline where a single product is transported, slack line can be tolerated. However, if the pipeline is operated in a batched mode with multiple products flowing simultaneously, slack line cannot be allowed, since intermingling and consequently degradation of the different batches would occur. A batched pipeline therefore must be operated as a tight line by using a control valve to create the necessary back pressure to pack the line.

In Figure 6.5, the back pressure valve would maintain the upstream pressure corresponding to point F on the hydraulic gradient. Downstream of the valve would see the lower pressure for delivering into a storage tank.

## 6.5   Batching Different Liquids

Batching is the process of transporting multiple products simultaneously through a pipeline with minimal mixing. Some commingling of the batches is unavoidable at the boundary or interface between contiguous batches.

For example, gasoline, diesel and kerosene may be shipped through a pipeline in a batched mode from a refinery to a storage terminal. Batched pipelines have to be run in turbulent mode or the velocities should be sufficiently high to ensure Reynold's number over 4 000. If flow were laminar (R < 2 100), the product batches would intermingle, thereby contaminating or degrading the products. Also, as discussed earlier, batched pipelines must be run in packed conditions (no slack line or open channel flow) to avoid contamination or intermingling of batches, in pipelines with drastic elevation changes such as the one shown in Figure 6.5.

In a batched pipeline, the total frictional pressure drop for a given flow rate will be calculated by adding the individual pressure drops for each product, considering its specific gravity, viscosity and the batch length. We will illustrate this using an example.

**Example Problem 6.1**

Consider a 16 inch pipeline, 0.250 inch wall thickness, 100 miles long from Douglas Refinery to Hampton Terminal used to ship three products at a flow rate of 4 000 bbl/hr as shown in Figure 6.6. The three batches shown represent an instantaneous snap shot condition.

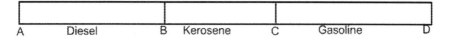

**Figure 6.6 - Batched Pipeline**

First we will calculate the total liquid volume in the pipeline. This is referred to as line fill volume.

The line fill volume can be calculated using the following equation:

$$\text{Line fill volume} = 5.129L(D)^2 \tag{6.5}$$

where

D    - Pipe inside diameter, in
L    - Pipe length, mi

In our example,

$$\text{Line fill} = 5.129(100)\,(15.5)^2 = 123{,}224 \text{ bbl}$$

Assuming the following physical properties and batch sizes for the three liquids, we can calculate the pressure drop for each liquid batch at the given flow rate.

| Product | Sp.Grav | Viscosity(cSt) | Batch size(bbl) |
|---------|---------|----------------|-----------------|
| Diesel | 0.85 | 5.9 | 50 000 |
| Kerosene | 0.82 | 2.7 | 30 000 |
| Gasoline | 0.74 | 0.7 | 43 224 |

Using Colebrook-White equation the frictional pressure drop in the different batch segments are calculated to be as follows:

| | | |
|---|---|---|
| Diesel: | 10.17 psi/mi | Batch length: 40.58 mi |
| Kerosene: | 8.58 psi/mi | Batch length: 24.35 mi |
| Gasoline: | 6.56 psi/mi | Batch length: 35.07 mi |

In the above, for each liquid, the pipeline length that represents the batch volume is shown. Thus the diesel batch will start at milepost 0.0 and end at milepost 40.58. Similarly the kerosene batch will start at milepost 40.58 and end at milepost (40.58+24.35) = 64.93. Finally, the gasoline batch will start at milepost 64.93 and end at milepost (64.93+35.07) = 100.0 for the snapshot configuration shown in Figure 6.6.

The batch lengths calculated above are based on 1 232.24 bbl per mile of 16 inch pipe calculated using Equation (6.5). The total frictional pressure drop for the entire 100 mile pipeline is obtained by adding up the individual frictional pressure drops for each product as follows:

$$\text{Total pressure drop} = 10.17(40.58) + 8.58(24.35) + 6.56(35.07) = 851.68 \text{ psi.}$$

In addition to the frictional pressure drop, the elevation head and the delivery pressure are combined to calculate the total pressure required at Douglas Refinery.

In a batched pipeline, there will be some intermixing of the two adjacent products, resulting in an interface of commingled product. The volume of this interface varies with the pipe size, Reynold's number and pipe length. Several correlations have been developed to estimate the amount of contamination in products at the interface.

As the batches arrive at the destination, density measuring instruments or densitometers (or gravitometers) monitor the density (or gravity) and flow is switched from one tank to another as appropriate. The contaminated volumes at the batch interface is diverted into a slop tank and later blended into a lower grade product.

When batching different products such as gasoline and diesel, flow rates vary as the batches move through the pipeline, due to the changing composition of liquid in the pipeline. We will discuss this further in Chapter 7, when centrifugal pump performance is combined with system head curves.

In order to economically operate the batched pipeline, by minimizing pumping cost, there exists an optimum batch size for the various products in a pipeline system. An analysis needs to be made over a finite period, such as a week or a month, to determine the flow rates and pumping costs considering various batch sizes. The combination of batch sizes that result in the least total pumping cost, consistent with shipper and market demands will then be the optimum batch sizes for the particular pipeline system.

## 6.6 Summary

We covered hydraulic balance in pipelines with multiple pump stations and learned that hydraulic balance may not always be possible due to topographic conditions. We demonstrated an approach to determine the number of pump stations from the total pressure required, minimum pump suction pressure and allowable pump discharge pressure based on pipeline MAOP. The advantages and cost implications of telescoping wall thickness and pipe grade tapering were discussed. Slack line and open channel flow in certain cases may occur but should be avoided on batched pipelines and high vapor pressure liquids. Also covered was the method of calculating hydraulics in a batched pipeline by analyzing a snapshot configuration of multiple products in a pipeline.

## 6.7 Problems

6.7.1 A pipeline 150 miles long from Beaumont pump station to a tank farm at Glendale is used to transport Alaskan North slope crude oil (ANS Crude). The pipe is 20 inches in outside diameter and constructed of X-52 steel. It is desired to operate the system at ANSI 600 pressure level (1 440 psi). The pipeline profile is such that there are two peaks located between Beaumont and Glendale. The first peak occurs at milepost 65.0 at an elevation of 1 500 ft. The second peak is located at milepost 110.0 at an elevation of 2 500 ft. Beaumont has an elevation of 350 ft and Glendale is situated at an elevation of 650 ft. During the initial phase of operation, 6 000 bbl/hr of ANS crude will be pumped at a temperature of 60°F and delivered to Glendale tankage at a pressure of 30 psi. The specific gravity and viscosity of ANS Crude at 60 °F may be assumed to be 0.895 and 43 cSt respectively.

(a) Determine the minimum wall thickness required to operate the pipeline system at a pressure of 1 400 psi.
(b) At a flow rate of 6 000 bbl/hr, how many pump stations would be required?
(c) During the second phase it is planned to expand the capacity of pipeline to 9 000 bbl/hr. How many additional pump stations would be required?
(d) Assuming 80% pump efficiency, calculate the total pumping HP required during the initial phase and under the expansion scenario. Use a minimum suction pressure of 50 psi at each pump station.

6.7.2 The pipeline described in Problem 6.7.1 is used to batch ANS crude along with a light crude (0.85 specific gravity and 15 cSt viscosity at 60 °F). Determine the optimum batch sizes to reduce pumping costs based on a 30 day operation. Consider a flow rate of 6 000 bbl/hr.

6.7.3 A batched pipeline, 60 km long is used to ship three grades of refined products. An instantaneous configuration shows an 8

000 m³ batch of Gasoline followed by 10 000 m³ of diesel and the remainder of the line consisting of batch of kerosene. The pipe size is 500 mm diameter and 10 mm wall thickness and essentially flat terrain. Assume 8 MPa for the operating pressure and pipeline delivery pressure of 300 kPa. Calculate the total pressure drop in the pipeline at a flow rate of 1800 m³/hr. The specific gravity and viscosities of the three products are at the flowing temperature of 20 °C are given below:

|  | Specific gravity | Viscosity(cSt) |
|---|---|---|
| Gasoline: | 0.74 | 0.65 |
| Kerosene: | 0.82 | 1.5 |
| Gasoline: | 0.85 | 5.9 |

# 7

# Pump Analysis

In this chapter we will discuss centrifugal pumps used in liquid pipelines. Though other types of pumps such as rotary, piston pumps etc. are sometimes used, the majority of pipelines today are operated with single and multi stage centrifugal pumps. We will cover the basic design of centrifugal pumps, their performance characteristics and how the performance may be modified by changing pump impeller speeds and trimming impellers. We will introduce Affinity Laws for centrifugal pumps, importance of Net Positive Suction Head (NPSH) and how to calculate horsepower requirements when pumping different liquids. Also covered will be the pump performance correction for high viscosity liquids using the Hydraulic Institute chart. We will also illustrate how the performance of two or more pumps in series or parallel configuration can be determined and how to estimate the operating point for pipeline by using the system head curve.

## 7.1   Centrifugal Pumps versus Reciprocating pumps

Pumps are needed to raise the pressure of a liquid in a pipeline so that the liquid may flow from the beginning of the pipeline to the delivery terminus at the required flow rate and pressure. As the flow rate increases more pump pressure will be required.

Over the years pipelines have been operated with centrifugal as well as reciprocating or Positive Displacement (PD) pumps. In this chapter we will concentrate mainly on centrifugal pumps, as these are used extensively in most liquid pipelines that transport water, petroleum products and chemicals. PD pumps are discussed as well, since they are used for liquid injection lines in oil pipeline gathering systems.

Centrifugal pumps develop and convert the high liquid velocity into pressure head in a diffusing flow passage. They generally have a lower efficiency than positive displacement (PD) pumps, such as reciprocating and gear pumps. However, centrifugal pumps can operate at higher speeds to generate higher flow rates and pressures. Centrifugal pumps also have lower maintenance requirements than PD pumps.

Positive Displacement (PD) pumps, such as reciprocating pumps operate by forcing a fixed volume of liquid from the inlet to outlet of the pump. These pumps operate at lower speeds than centrifugal pumps. Reciprocating pumps cause intermittent flow. Rotary screw pumps and gear pumps are also PD pumps, but operate continuously compared to Reciprocating pumps. PD pumps are generally larger in size and more efficient compared to centrifugal pumps, but require higher maintenance.

Modern pipelines are mostly designed with centrifugal pumps in preference to PD pumps. This is because, there is more flexibility in volumes and pressures with centrifugal pumps. In petroleum pipeline installations where liquid from a field gathering system is injected into a main pipeline, PD pumps may be used.

The performance of a centrifugal pump is depicted as a curve that shows variation in pressure head at different flow rates. The pump head characteristic curve shows the pump head developed on the vertical axis, while the flow rate is shown on the horizontal axis. This curve may be referred to as the H-Q curve or the Head-Capacity curve. Pump companies use the term capacity when referring to flow rate. Throughout this section we will use capacity and flow rate interchangeably when dealing with pumps.

Generally, pump performance curves are plotted for water as the liquid pumped. The head is measured in feet of water and flow rate is shown in gal/min. In addition to the head versus flow rate curve, two other curves are commonly shown on a typical pump performance chart. These are pump efficiency versus flow rate and pump brake horsepower BHP versus flow rate. Figure 7.1 shows typical centrifugal pump performance curves consisting of head, efficiency and BHP plotted against flow rate (or capacity). You will also notice another curve referred to as NPSH plotted against flow rate. NPSH will be discussed later in this chapter.

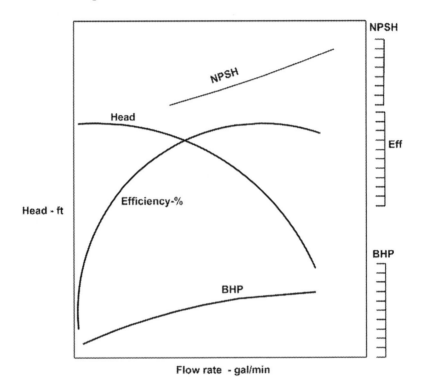

Figure 7.1 - Centrifugal Pump Performance

In summary, a typical centrifugal pump characteristic curves include the following four curves:

Head versus Flow rate
Efficiency versus Flow rate
BHP versus Flow rate
NPSH versus Flow rate

The above performance curves for a particular model pump are generally plotted for a particular pump impeller size and speed (example: 10 inch impeller, 3 560 RPM).

In addition to the above four characteristic curves, sometimes, you will encounter pump head curves drawn for different impeller diameters and iso-efficiency curves. When pumps are driven by variable speed electric motors or engines, the head curves may also be shown at various pump speeds. Pump performance at various impeller sizes and speeds will be discussed in detail later in this chapter.

A PD pump continuously pumps a fixed volume at various pressures. It is able to provide any pressure required at a fixed flow rate. This flow rate depends on the geometry of the pump such as bore, stroke, etc. A typical PD pump pressure volume curve is shown in Figure 7.2

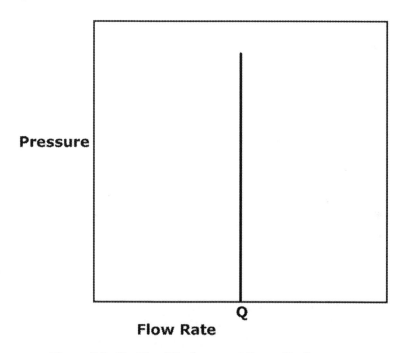

**Figure 7.2 - Positive Displacement Pump Performance**

## 7.2   Pump Head versus Flow Rate

For a Centrifugal pump, the head-capacity (flow rate) variation is shown in Figure 7.3.

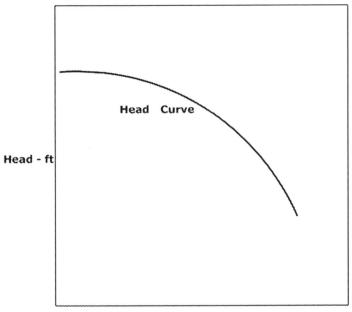

Figure 7.3 - Centrifugal Pump Performance - Head versus Flow Rate

We will refer to the Head versus Flow rate curve as the H-Q curve. It is generally referred to as a drooping head curve that starts of at the highest value (shut off head) at zero flow rate. The head decreases as the flow rate through the pump increases. The trailing point of the curve represents the maximum flow and the corresponding head that the pump can generate.

Pump head is always plotted in feet of head of water. Therefore the pump is said to develop the same head in feet of liquid, regardless of the liquid. If a particular pump develops 2 000 ft head at 3 000 gal/min flow rate, we can calculate the pressure developed in psi, when pumping water versus another liquid such as gasoline.

With water, pressure developed      = 2 000 x 1 /2.31      = 866 ft
With gasoline, pressure developed  = 2 000 x 0.74 /2.31  = 641 ft

This head versus flow rate relationship holds good for this particular pump that has a fixed impeller diameter D and operates at a fixed speed N. The diameter of the impeller generally can be increased within a certain range of sizes, depending on the internal cavity that houses the pump impeller. Thus, if the H-Q curve in Figure 7.3 is based on a 10 inch impeller diameter, similar curves can be generated for 9 inch impeller and 12 inch impellers representing the minimum and maximum sizes available for this particular pump. The minimum and maximum impeller sizes are defined by the pump vendor for the particular pump model. The 9 inch curve and 12 inch curve would be parallel to the 10 inch curve as shown in Figure 7.4. The variation of the pump H-Q curves with impeller diameter follow the Affinity Laws, discussed in a later section of this chapter.

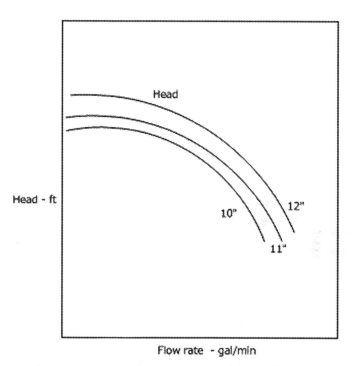

Figure 7.4 - Head versus Flow Rate - Different Impeller Sizes

Similar to H-Q variations with pump impeller diameter, curves can be generated for varying pump impeller speeds. If the impeller diameter is kept constant at 10 inch and the initial H-Q curve was based on a pump speed of 3 560 RPM (typical induction motor speed), by varying the pump speed, we can generate a family of parallel curve as shown in Figure 7.5

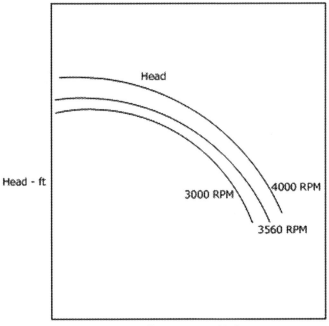

**Figure 7.5 - Head versus Flow Rate at Different Speeds**

It will be observed that the H-Q variation with pump impeller diameter and pump speed are similar. This is due to the Affinity Laws that the pump performance is based on. We will discuss Affinity Laws and prediction of pump performance at different diameters and speeds later in this chapter.

A pump may develop the head in stages. Thus single stage and multi stage pumps are used depending upon head required. A single stage pump may generate 200 ft head at 3 000 gal/min. A three stage pump of this design will generate 600 ft head at same flow rate and pump speed. An application that requires 2 400 ft head at 3 000 gal/min may be served by a multi stage pump with each stage providing 400 ft of head. De-staging is the process of reducing the active number of stages in a pump to reduce the total head developed. Thus the six stage pump discussed above may be de-staged to four stages if we need only 1 600 ft head at 3 000 gal/min.

## 7.3   Pump Efficiency versus Flow Rate

The variation of the efficiency E of a centrifugal pump with flow rate Q is as shown in Figure 7.6.

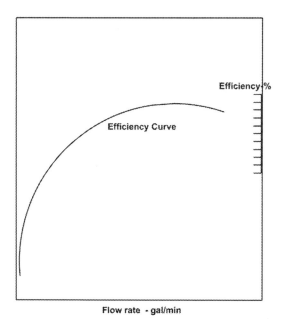

**Flow rate - gal/min**

**Figure 7.6 - Centrifugal Pump Performance - Efficiency versus Flow Rate**

It can be seen that the efficiency starts off at a zero value under shut off conditions (zero flow rate) and rises to a maximum value at some flow rate. After this point, with increase in flow rate, the efficiency drops off to some value lower than the peak efficiency. Generally centrifugal pump peak efficiencies are approximately 80-85%. It must be remembered in all these discussions we are referring to a centrifugal pump performance with water as the liquid being pumped. When heavy, viscous liquids are pumped, these water based efficiencies will be reduced by a correction factor. Viscosity corrected pump performance using the Hydraulic Institute Charts, is discussed later in this chapter. The flow rate at which the maximum efficiency occurs in a pump is referred to as the Best Efficiency Point (BEP). This flow rate and the corresponding head from the H-Q curve is thus the best operating point on the pump curve since the highest pumping efficiency is realized at this flow rate. In the combined H-Q and E-Q curves shown in Figure 7.7, the BEP is shown as a small triangle.

When choosing a centrifugal pump for a particular application we try to get the operating point as close as possible to the BEP. In order to allow for future increase in flow rates, the initial operating point is chosen slightly to the left of the BEP. This will ensure that with increase in pipeline throughput, the operating point on the pump curve will move to the right which would result in a slightly better efficiency.

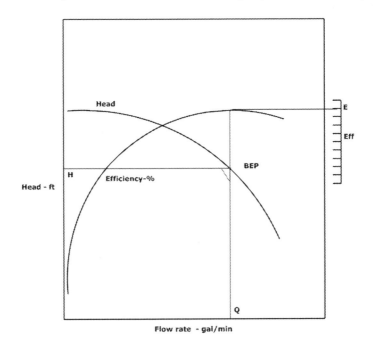

**Figure 7.7 - Best Efficiency Point**

## 7.4   BHP versus Flow rate

From the Head versus Flow rate (H-Q) curve and Efficiency versus Flow rate curves, we can calculate the Brake Horsepower (BHP) required at every point along the curve using Equation (5.17) as follows:

$$\text{Pump BHP} = \frac{Q \, H \, Sg}{3\,960 \, (E)} \tag{7.1}$$

where

Q    - Pump flow rate, gal/min
H    - Pump head, ft
E    - Pump efficiency as a decimal value, less than 1.0
Sg   - Liquid specific gravity (for water Sg = 1.0)

In SI units, power in kW can be calculated as follows:

$$\text{Power kW} = \frac{Q\,H\,Sg}{367.46\,(E)} \tag{7.2}$$

where

Q    - Pump flow rate, m³/hr
H    - Pump head, m
E    - Pump efficiency as a decimal value, less than 1.0
Sg   - Liquid specific gravity (for water Sg = 1.0)

For example, if the BEP on a particular pump curve occurs at a flow rate of 3 000 gal/min, 2 500 ft head at 85% efficiency, the BHP at this flow rate for water is calculated from Equation (7.1) as

Pump BHP = 3 000 x 2 500 x 1.0 / (3 960 x 0.85) = 2 228

This represents the BHP with water at 3 000 gal/min flow rate. Similarly, BHP can be calculated at various flow rates from zero to 4 000 gal/min (assumed maximum pump flow) by reading the corresponding Head and Efficiency values from the H-Q curve and E-Q curve and using Equation (7.1) as above. The BHP versus flow rate can than be plotted as shown in Figure 7.8

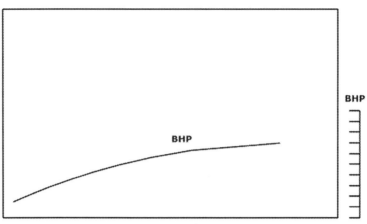

**Figure 7.8 - BHP versus Flow Rate**

BHP versus flow rate curve can also be shown on the same plot as H-Q and E-Q curves are shown in Figure 7.1.

As we discussed in an earlier chapter, the BHP calculated above is the brake horsepower demanded by the pump. An electric motor driving the pump will have an efficiency that ranges from 95% to 98%. Therefore the electric motor HP required is the pump BHP divided by the motor efficiency as follows

Motor HP = Pump BHP / Motor Efficiency
= 2228 / 0.96
= 2321 HP, based on 96% motor efficiency.

In determining the motor size required, we must first calculate the maximum HP demanded by the pump, which generally (not necessarily) will be at the highest flow rate. The BHP at this flow rate will then be divided by the electric motor efficiency to get the maximum motor HP required. The next available standard size motor will be selected for the application.

Suppose that for the above pump, 4 000 gal/min is the maximum flow rate through the pump and the combined head and efficiency are 1 800 ft and 76% respectively. We will calculate the maximum BHP required as

$$\text{Maximum pump BHP} = \frac{4\,000 \times 1\,800 \times 1.0}{3\,960 \times 0.76}$$

$$= 2\,392.34$$

$$\text{The Motor HP} = \frac{2\,392.34}{0.96}$$

$$= 2\,492$$

Therefore a 2 500 HP electric motor will be adequate for this case. Before we leave the subject of pump BHP and electric motor HP, we must mention that electric motors have built into their nameplate rating a service factor. The service factor for most induction motors is in the range of 1.10 to 1.15. This simply means that if the motor's rated HP is 2 000, the 1.15 service factor allows the motor to supply up to a maximum of 1.15 x 2 000 or 2 300 HP without burning up the motor windings. However, it is not advisable to use this extra 300 HP on a continuous basis as explained below.

At 1.15 service factor rating, the motor windings will be taking on an additional 15% electric current, compared to its rated current (Amps). This would translate to extra heat that would shorten the life of the windings. Even though the extra current is only 15%, the heat generated will be over 32%, since electrical heating is proportional to the square of the current ($1.15^2 = 1.3225$).

However, in an emergency, this additional 15% motor HP above the rated HP may be used. Continuous operation at the service factor rating is however discouraged.

## 7.5 NPSH versus Flow rate

In addition to the three pump curves, for head versus capacity (H-Q), efficiency versus capacity (E-Q) and BHP versus capacity (BHP-Q) discussed previously, the pump performance data will include a fourth curve for Net Positive Suction Head (NPSH) versus capacity. This curve is generally located above the head, efficiency and BHP curves as shown in Figure 7.1.

The NPSH curve shows the variation of the minimum net positive suction head at the impeller suction versus the flow rate. The NPSH increases at a faster rate as the flow rate increases.

NPSH is defined as the Net Positive Suction Head required at the pump impeller suction to prevent pump cavitation at any flow rate. It represents the resultant positive pressure at pump suction after accounting for frictional loss and liquid vapor pressure. NPSH is discussed in detail, later in this chapter. At present, we can say that as pump flow rate increases the NPSH required also increases. We will perform calculations to compare NPSH required (per pump curve) versus actual NPSH available based on pump suction piping and other parameters that affect the available suction pressure at the pump.

## 7.6   Specific Speed

The specific speed of a centrifugal pump is a parameter based on the impeller speed, flow rate and head at the best efficiency. It is used for comparing geometrically similar pumps and for classifying the different types of centrifugal pumps.

Specific speed may be defined as the speed at which a geometrically similar pump must be run such that it produces a head of 1 ft at a flow rate of 1 gal/min. Mathematically, specific speed is defined as

$$N_S = N Q^{1/2} / H^{3/4} \tag{7.3}$$

where

N$_S$     - Pump specific speed
N       - Pump impeller speed, RPM
Q       - Flow rate or capacity, gal/min
H       - Head, ft

Both Q and H are measured at the best efficiency point (BEP) for the maximum impeller diameter. The head H is measured per stage for a multi-stage pump.

Low specific speed is associated with high head pumps, while high specific speed is found with low head pumps.

Another related term, called the suction specific speed is defined as follows:

$$N_{SS} = N Q^{1/2} / (NPSH_R)^{3/4}$$  (7.4)

where

$N_{SS}$     - Suction specific speed
$N$       - Pump impeller speed, RPM
$Q$       - Flow rate or capacity, gal/min
$NPSH_R$ - NPSH required at BEP

When applying the above equations to calculate the pump specific speed and suction specific speed, use the full Q value for single or double suction pumps for $N_S$ calculation. For $N_{SS}$ calculation, use one-half the Q value for double suction pumps.

**Example Problem 7.1**

Calculate the specific speed of a 5-stage double suction centrifugal pump, 12 inch diameter impeller that when operated at 3 560 RPM generates a head of 2 200 ft at a capacity of 3 000 gal/min at the BEP on the head capacity curve. If the NPSH required is 25 ft, calculate the suction specific speed.

**Solution**

$N_S$    $= N Q^{1/2} / H^{3/4}$
       $= 3\ 560\ (3000)^{1/2} / (2200 / 5)^{3/4}$    $= 2030$

The suction specific speed is

$N_{SS}$   $= N Q^{1/2} / NPSH_R^{3/4}$
       $= 3560\ (3000 / 2)^{1/2} / (25)^{3/4}$      $= 12,332$

Centrifugal pumps are generally classified as

(a) Radial Flow
(b) Axial Flow
(c) Mixed Flow

Radial flow pumps develop head by centrifugal force. Axial flow pumps on the other hand develop the head due to propelling or lifting action of the impeller vanes on the liquid. Radial flow pumps are used when high heads are required, while the axial flow and mixed flow pumps are mainly used for low head - high capacity applications.

Table 7.1 lists the specific speed range for centrifugal pumps.

### Table 7.1 Specific Speeds of Centrifugal Pumps

| Description | Application | Specific Speed, Ns |
|---|---|---|
| Radial Vane | Low capacity/high head | 500 - 1 000 |
| Francis - Screw Type | Medium capacity/Medium head | 1000 - 4 000 |
| Mixed - Flow Type | Medium to high capacity, low to medium head | 4000 - 7000 |
| Axial - Flow Type | High capacity/low head | 7000 - 20,000 |

## 7.7 Affinity Laws - Variation with Impeller Speed and Diameter

Each family of pump performance curves for head, efficiency and BHP versus capacity is specific to a particular size of impeller and pump model. Thus Figure 7.1 depicts pump performance data for 10 inch diameter pump impeller. This particular pump may be able to accommodate a larger impeller, such as 12 inches in diameter. Alternatively, a smaller 9 inch impeller may also be fitted in this pump casing. The range of impeller sizes is dependent on the pump case design and will be defined by the pump vendor. Since centrifugal pumps generate pressure due to centrifugal force, it is clear that a smaller impeller diameter will generate less pressure at a given flow rate than a larger impeller at the same speed. If the performance data for a 10 inch impeller is available, we can predict the pump performance when using a larger or smaller size impeller by means of the Centrifugal Pump Affinity Laws.

The same pump with 10 inch impeller may be operated at a higher or lower speed to provide higher or lower pressure. Most centrifugal pumps are driven by constant speed electric motors. A typical induction motor in USA operates at 60 Hz and will have a synchronous speed of 3600 RPM. With some slip, the induction motor would probably run at 3560 RPM. Thus, most constant speed motor driven pumps have performance curves based on 3560 RPM. Some slower speed pumps operate at 1700 to 1800 RPM. If a variable speed motor is used, the pump speed can be varied from say, 3000 to 4000 RPM. This variation in speed produces variable head versus capacity values for the same 10 inch impeller. Given the performance data for a 10 inch impeller at 3560 RPM, we can predict the pump performance at different speeds ranging from 3 000 RPM to 4 000 RPM using the Centrifugal Pump Affinity Laws as described next.

The Affinity Laws for Centrifugal Pumps are used to predict pump performance for changes in impeller diameter and impeller speed. According to the Affinity Laws, for small impeller diameter changes, the flow rate (pump capacity) Q is directly proportional to the impeller diameter. The pump head H on the other hand is directly proportional to the square of the impeller diameter. Since the BHP is proportional to the product of flow rate and head (see Equation 7.1), BHP will vary as the third power of the impeller diameter.

The Affinity Laws are represented mathematically as follows:

For impeller diameter change

$$Q_2 / Q_1 = D_2 / D_1 \tag{7.5}$$
$$H_2 / H_1 = (D_2 / D_1)^2 \tag{7.6}$$

where

$Q_1, Q_2$ - Initial and final flow rates
$H_1, H_2$ - Initial and final heads
$D_1, D_2$ - Initial and final impeller diameters

From the H-Q curve corresponding to the impeller diameter $D_1$ we can select a set of H-Q values that cover the entire range of capabilities from zero to the maximum possible. Each value of Q and H on the corresponding H-Q curve for the impeller diameter $D_2$ can then be computed using the Affinity Law ratios per Equations (7.5) and (7.6).

Similarly, Affinity Laws state that for the same impeller diameter, if pump speed is changed, flow rate is directly proportional to the speed, while the head is directly proportional to the square of the speed. As with diameter change, the BHP is proportional to the third power of the impeller speed. This is represented mathematically as follows:

For impeller speed change

$$Q_2 / Q_1 = N_2 / N_1 \qquad\qquad (7.7)$$
$$H_2 / H_1 = (N_2 / N_1)^2 \qquad\qquad (7.8)$$

where

$Q_1, Q_2$ - Initial and final flow rates
$H_1, H_2$ - Initial and final heads
$N_1, N_2$ - Initial and final impeller speeds

Note that the Affinity Laws for speed change are exact. However, the Affinity Laws for impeller diameter change are only approximate and valid for small changes in impeller sizes. The pump vendor must be consulted to verify that the predicted values using Affinity Laws for impeller size changes are accurate or if any correction factors are needed. With speed and impeller size changes, the efficiency versus flow rate can be assumed to be the same.

An example using the Affinity Laws will illustrate how the pump performance can be predicted for impeller diameter change as well as for impeller speed change.

**Example Problem 7.2**

The head and efficiency versus capacity data for a centrifugal pump with a 10 inch impeller is as shown below.

| Q - gal/min | 0 | 800 | 1600 | 2400 | 3000 |
|---|---|---|---|---|---|
| H - ft | 3 185 | 3100 | 2900 | 2350 | 1800 |
| E - % | 0.0 | 55.7 | 78.0 | 79.3 | 72.0 |

The pump is driven by a constant speed electric motor at a speed of 3560 RPM.

(a) Determine the performance of this pump with an 11 inch impeller, using Affinity Laws.

(b) If the pump drive were changed to a variable frequency drive (VFD) motor with a speed range of 3000 RPM to 4000 RPM, calculate the new H-Q curve for the maximum speed of 4000 RPM with the original 10 inch impeller.

**Solution**

(a) Using Affinity Laws for impeller diameter changes, the multiplying factor for flow rate is

factor = 11/10 = 1.1

and the multiplier for head is $(1.1)^2 = 1.21$

Therefore, we will generate a new set of Q and H values for the 11 inch impeller by multiplying the given Q values by the factor 1.1 and the H values by the factor 1.21 as follows:

| Q - gal/min | 0 | 880 | 1760 | 2640 | 3300 |
|---|---|---|---|---|---|
| H - ft | 3854 | 3751 | 3509 | 2844 | 2178 |

The above flow rate and head values represent the predicted performance of the 11 inch impeller. The efficiency versus flow rate curve for the 11 inch impeller will approximately be the same as that of the 10 inch impeller.

(b) Using Affinity Laws for speeds, the multiplying factor for the flow rate is factor = 4000 / 3560 = 1.1236

and the multiplier for head is $(1.1236)^2 = 1.2625$

Therefore we will generate a new set of Q and H values for the pump at 4 000 RPM by multiplying the given Q values by the factor 1.1236 and the H values by the factor 1.2625 as follows:

| Q - gal/min | 0 | | 899 | 1798 | 2697 | 3371 |
|---|---|---|---|---|---|---|
| H - ft | | 4021 | 3914 | 3661 | 2967 | 2273 |

The above flow rates and head values represent the predicted performance of the 10 inch impeller at 4000 RPM. The new efficiency versus flow rate curve will approximately be the same as the given curve for 3560 RPM.

Thus, using Affinity Laws, we can determine whether we should increase or decrease the impeller diameter to match the requirements of flow rate and head for a specific pipeline application.

For example, suppose hydraulic calculations for a particular pipeline indicate that we need a pump that can provide a head of 2000 ft at a flow rate of 2 400 gal/min. The data for the pump in Example Problem 7.2 shows that a 10 inch impeller produces 2350 ft head at a flow rate of 2400 gal/min. If this is a constant speed pump, it is clear that in order to get 2000 ft head, the current 10 inch impeller needs to be trimmed in diameter to reduce the head from 2350 ft to the 2000 ft required. The Example Problem 7.4 will illustrate how the amount of impeller trim is calculated.

## 7.8   Effect of Specific Gravity and Viscosity on Pump Performance

The performance of a pump as reported by the pump vendor is always based on water as the pumped liquid. Hence the head versus capacity curve, the efficiency versus capacity curve, BHP versus capacity curve and NPSH required versus capacity curve are all applicable only when the liquid pumped is water (Specific gravity of 1.0) at standard conditions, usually 60°F. When pumping a liquid such as crude oil with specific gravity of 0.85 or a refined product such as gasoline with specific gravity of 0.74, both the H-Q and E-Q curves may still apply for these products if the viscosities are sufficiently low (less than 4.3 cSt for small pumps up to 100 gal/min capacity and less than 40 SSU for larger pumps up to 10 000 gal/min) according to Hydraulic Institute standards. Since BHP is a function of the specific gravity, from Equation (7.1), it is clear that the BHP versus flow rate will be different for other liquids compared to water. If the liquid pumped has a higher viscosity, in the range of (4.3 cSt to 3 300 cSt), the head, efficiency and BHP curves based on water must all be corrected for the high viscosity. The Hydraulic Institute has published viscosity correction charts that can be applied to correct the water performance curves to produce viscosity corrected curves. See Figure 7.9 for details of this chart. For any application involving high viscosity liquids, the pump vendor should be given the liquid properties. The viscosity corrected performance curves will be supplied by the vendor as part of the pump proposal. As an end user you may also use these charts to generate the viscosity-corrected pump curves.

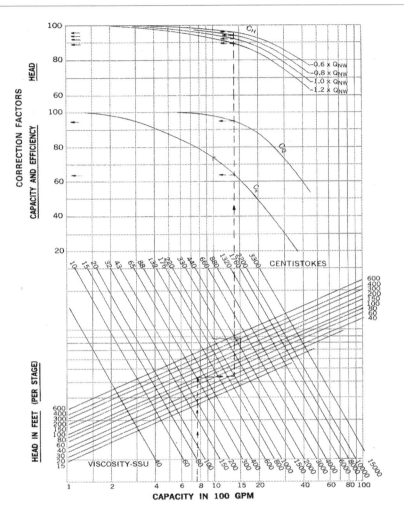

**Figure 7.9 - Viscosity correction chart**

The Hydraulic Institute method of viscosity correction require determining the Best Efficiency Point (BEP) values for Q, H and E from the water performance curve. This is called the 100% BEP point. Three additional sets of Q, H and E values are obtained at 60%, 80% and 120% of the BEP flow rate from the water performance curve. From these four sets of data, the Hydraulic Institute chart can be used to obtain the correction factors $C_q$, $C_h$ and $C_e$ for flow, head and efficiency for each set of data. These factors are used to multiply the Q, H and E values from the water curve, thus generating corrected values of Q, H and E for

60%, 80%, 100% and 120% BEP values. Example Problem 7.3 illustrates the Hydraulic Institute method of viscosity correction. Note that for multi-stage pumps, the values of H must be per stage.

**Example Problem 7.3**

The water performance of a single stage centrifugal pump for 60%, 80%, 100% and 120% of the BEP is as shown below:

| Q - gal/min | 450 | 600 | 750 | 900 |
|---|---|---|---|---|
| H - ft | 114 | 108 | 100 | 86 |
| E - % | 72.5 | 80.0 | 82.0 | 79.5 |

(a) Calculate the viscosity corrected pump performance when pumping oil with a specific gravity of 0.90 and a viscosity of 1000 SSU at pumping temperature.

**Solution**

(a) By inspection, the BEP for this pump curve is

$$Q = 750$$
$$H = 100$$

and

$$E = 82$$

we first establish the four sets of capacities to correspond to 60%, 80%, 100% and 120%. These have already been given as 450, 600, 750 and 900 gal/min. Since the head values are per stage, we can directly use the BEP value of head along with the corresponding capacity to enter the Hydraulic Institute Viscosity Correction chart at 750 gal/min on the lower horizontal scale. Going vertically from 750 gal/min to the intersection point on the line representing the 100 ft head curve and then horizontally to intersect the 1 000 SSU viscosity line and finally vertically up to intersect the three correction factor curves $C_e$, $C_q$ and $C_h$.

From the Hydraulic Institute chart of correction factors (Figure 7.9) we obtain the values of $C_q$, $C_h$ and $C_e$ for flow rate, head and efficiency as follows:

| | | | | |
|------|-------|-------|-------|-------|
| $C_q$ | 0.95 | 0.95 | 0.95 | 0.95 |
| $C_h$ | 0.96 | 0.94 | 0.92 | 0.89 |
| $C_e$ | 0.635 | 0.635 | 0.635 | 0.635 |

Corresponding to Q values of 450 (60% of $Q_{NW}$), 600 (80% of $Q_{NW}$), 750 (100% of $Q_{NW}$) and 900 (120% of $Q_{NW}$). The term $Q_{NW}$ is the BEP flow rate from the water performance curve.

Using these correction factors, we generate the Q, H and E values for the viscosity corrected curves by multiplying the water performance value of Q by $C_q$, H by $C_h$ and E by $C_e$ and obtain the following result.

| | | | | |
|---------|-------|-------|------|------|
| $Q_V$ | 427 | 570 | 712 | 855 |
| $H_V$ | 109.5 | 101.5 | 92.0 | 76.5 |
| $E_V$ | 46.0 | 50.8 | 52.1 | 50.5 |
| $BHP_V$ | 23.1 | 25.9 | 28.6 | 29.4 |

The last row of values for viscous BHP was calculated using Equation (7.1) as follows:

$$BHP_V = (Q_V) (H_V) \, 0.9 \, / \, (39.60 \times E_V) \tag{7.9}$$

where $Q_V$, $H_V$ and $E_V$ are the viscosity corrected values of capacity, head and efficiency tabulated above.

Note that when using the Hydraulic Institute chart, for obtaining the correction factors, two separate charts are available. One chart applies to small pumps up to 100 gal/min capacity and head per stage of 6 ft to 400 ft. The other chart applies to larger pumps with capacity between 100 gal/min and 10 000 gal/min and head range of 15ft to 600 ft per stage. Also, remember that when data is taken from a water performance curve, the head has to be corrected per stage, since the Hydraulic Institute charts are based on head in ft per stage rather than the total pump head. Therefore, if a 6 stage pump has a BEP at 2 500 gal/

min and 3 000 ft of head with an efficiency of 85%, the head per stage to be used with the chart will be 3 000 divided by 6 = 500 ft. The total head (not per stage) from the water curve can then be multiplied by the correction factors from the Hydraulic Institute Charts to obtain the viscosity corrected head for the 6 stage pump.

## 7.9  Pump Curve Analysis

So far we have discussed the performance of a single pump. When transporting liquid through a pipeline, we may need to use more than one pump at a pump station to provide the necessary flow rate or head requirement. Pumps may be operated in series or parallel configurations. Series pumps are generally used for higher heads and parallel pumps for increased flow rates.

For example, let us assume that pressure drop calculations indicate that the originating pump station on a pipeline requires 900 psi differential pressure to pump 3 000 gal/min of gasoline with specific gravity 0.736. Converting to customary pump units we can state that we require a pump that can provide the following:

$$\text{Head} = \frac{900 \times 2.31}{0.736} = 2825 \text{ ft}$$

at 3 000 gal/min flow rate. We have two options here. We could select from a pump vendor's catalog, one large pump that can provide 2 825 ft head at 3 000 gal/min or select two smaller pumps that can provide 1 413 ft at 3 000 gal/min. We would use these two smaller pumps in series to provide the required head, since pumps operated in series results in heads that are additive. Alternatively, we could also select two pumps that can provide 2825 ft at 1500 gal/min each. In this case these pumps would be operated in parallel. When pumps are operated in series, the same flow rate goes through each pump and the resultant head is the sum of the heads generated by each pump. In parallel operation, the flow rate is split between the pumps, while each pump produces the same head. Series and parallel pump configuration are illustrated in Figure 7.10

The choice of series or parallel pumps for a particular application depends on many factors, including pipeline elevation profile, as well as operational flexibility.

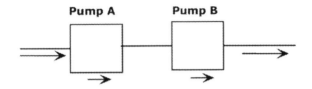

**Pump A**      **Pump B**

**Series - Same Flow through each Pump
Heads are additive**

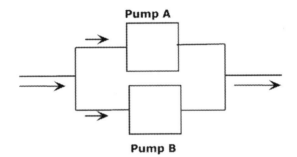

**Pump A**

**Pump B**

**Parallel - Same Head from each Pump
Flow rates are additive**

**Figure 7.10 Pumps in Series and Parallel**

Figure 7.11 shows the combined performance of two identical pumps in series, versus parallel configuration. It can be seen that parallel pumps are used when we need larger flows. Series pumps are used when we need higher heads than each individual pump. If the pipeline elevation profile is essentially flat, the pump pressure required is mainly to overcome the pipeline friction. On the other hand, if the pipeline has drastic elevation changes, the pump head generated is mainly for the static lift and to a lesser extent for pipe friction. In the latter case, if two pumps are used in series and one shuts down, the remaining pump alone will only be able to provide half the head and therefore will not be able to provide the necessary head for the static lift at any flow rate. If the pumps were configured in parallel, then shutting down one pump will still allow the other pump to provide the necessary head for the static lift at half the previous flow rate. Thus parallel pumps are generally used when elevation differences are considerable. Series pumps are used where pipeline elevations are not significantly high.

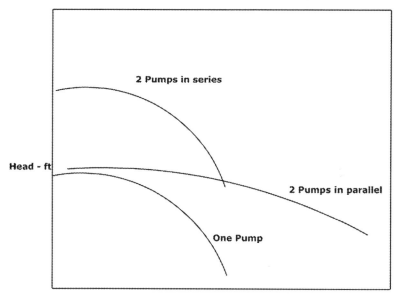

Figure 7.11 Pump Performance - Series and Parallel

## Example Problem 7.4

One large pump and one small pump are operated in series. The H-Q characteristic of the pumps are defined as follows:

Pump1

| Q - gal/min | 0 | 800 | 1600 | 2400 | 3000 |
|---|---|---|---|---|---|
| H - ft | 2389 | 2325 | 2175 | 1763 | 1350 |

Pump2

| Q - gal/min | 0 | 800 | 1600 | 2400 | 3000 |
|---|---|---|---|---|---|
| H - ft | 796 | 775 | 725 | 588 | 450 |

(a) Calculate the combined performance of Pump1 and Pump2 in series configuration.
(b) What changes (trimming impellers) must be made to either pump, to satisfy the requirement of 2000 ft of head at 2400 gal/min when operated in series?
(c) Can these pumps be configured to operate in parallel?

## Solution

(a) Pumps in series have the same flow through each pump and the heads are additive. We can therefore, generate the total head produced in series configuration by adding the head of each pump for each flow rate given as follows:

Combined performance of Pump1 and Pump2 in series:

| Q - gal/min | 0 | 800 | 1600 | 2400 | 3000 |
|---|---|---|---|---|---|
| H - ft | 3185 | 3100 | 2900 | 2351 | 1800 |

(b) It can be seen from the above combined performance that the head generated at 2400 gal/min is 2351 ft. Since the design requirement is 2000 ft at this flow rate, it is clear that the head needs to be reduced by trimming one of the pump impellers to produce the necessary total head. We will proceed as follows.

Let us assume that the smaller pump will not be modified and the impeller of the larger pump (Pump1) will be trimmed to produce the necessary head.

Modified head required of Pump1 = 2000 - 588 = 1412 ft.

Pump1 produces 1763 ft of head at 2400 gal/min. In order to reduce this head to 1412 ft, we must trim the impeller by approximately $(1412 / 1763)^{\frac{1}{2}} = 0.8949$ or 89.5% trim.

based on Affinity Laws. This is only approximate and we need to generate a new H-Q curve for the trimmed Pump1 so we can verify that the desired head will be generated at 2 000 gal/min. Using the method in Example Problem 7.2 we can generate a new H versus Q curve for a trim of 89.5%

Flow multiplier = 0.8949.
Head multiplier = $(0.8949)^2 = 0.8008$

Pump1 trimmed to 89.5% of present impeller diameter:

| Q - gal/min | 0 | | 716 | 1 432 | 2 148 | 2 685 |
|---|---|---|---|---|---|---|
| H - ft | | 1 913 | 1 862 | 1 742 | 1 412 | 1 081 |

It can be seen from above trimmed pump performance that the desired head of 1 412 ft will be achieved at a flow rate of 2 148 gal/min. Therefore, at the lower flow rate of 2 000 gal/min, we can estimate by interpolation that the head would be higher than 1 412 ft. Hence, slightly more trimming would be required to achieve the design point of 1 412 ft at 2 000 gal/min. By trial and error we arrive at a pump trim of 87.7% and the resulting pump performance for Pump1 at 87.7% trim is as follows.

Pump1 trimmed to 87.7% of present impeller diameter

| Q - gal/min | 0 | | 702 | 1 403 | 2 105 | 2 631 |
|---|---|---|---|---|---|---|
| H - ft | | 1 837 | 1 788 | 1 673 | 1 356 | 1 038 |

By plotting this curve, we can verify that the required head of 1 412 ft will be achieved at 2 000 gal/min.

(c) To operate satisfactorily, in a parallel configuration, the two pumps must have a common range of heads so that at each common head, the corresponding flow rates can be added to determine the combined performance. Pump1 and Pump2 are mismatched for parallel operation. Therefore, they cannot be operated in parallel.

## 7.10 Pump Head Curve versus System Head Curve

In a previous section in Chapter 5, we discussed development of the pipeline system head curves. In this section, we will see how, the system curve together with the pump head curve can predict the operating point (flow rate Q - head H) on the pump curve.

Since the system head curve for the pipeline is a graphic representation of the pressure required to pump a product through the pipeline at various flow rates (increasing pressure with increasing flow rate) and the pump H-Q curve shows the pump head available at various flow rates, when the head requirements of the pipeline match the available pump head, we have a point of intersection of the system head curve with the pump head curve as shown in Figure 7.12. This is the operating point for this pipe-pump combination.

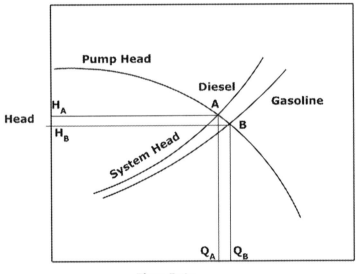

**Figure 7.12 Pump Curve - System Head Curve**

The point of intersection of the pump head curve and the system head curve for diesel (point A) indicates the operating point for this pipeline with diesel. Similarly, if gasoline were pumped through this pipeline, the corresponding operating point (point B) is as shown in the Figure 7.12.

Therefore, with 100% diesel in the pipeline, the flow rate would be $Q_A$ and the corresponding pump head would be $H_A$ as shown. Similarly, with 100% gasoline in the pipeline, the flow rate would be $Q_B$ and the corresponding pump head would be $H_B$ as shown in the Figure 7.12.

When batching the two products, a certain portion of the pipeline will be filled with diesel and the rest will be filled with gasoline. The flow rate will then be at some point between $Q_A$ and $Q_B$, since a new system curve located between the diesel curve and the gasoline curve will dictate the operating point.

If we had plotted the system head curve in psi instead of ft of liquid head, the Y-axis will be the pressure required (psi). The pump head curve also needs to be converted to psi versus flow rate by using Equation (3.7)

Psi = Head x Sg / 2.31

Therefore, using the same pump, we will have two separate H-Q curves for diesel and gasoline, due to different specific gravities. Such a situation is shown in Figure 7.13

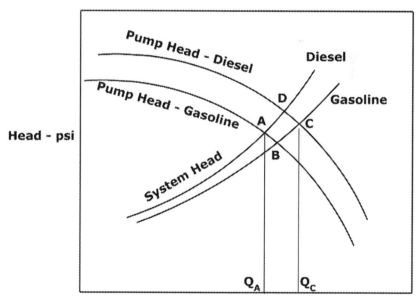

**Figure 7.13 Pump and System Curve - Batching**

Even though the pump develops the same head (in ft) with diesel or gasoline (or water), the pressure generated in psi will be different, and hence the two separate H-Q curves as indicated in Figure 7.13

In a batched pipeline, the operating point moves from D to A, A to B, B to C and C to D as shown in Figure 7.13. We start off with 100% diesel in the pipeline and the pump. This is point D in the Figure. When gasoline enters the pump, with diesel in the pipeline, the operating point moves from D to A. Then, as the gasoline batch enters the pipeline, the system head curve moves to the right until it reaches point B representing the operating point with 100% gasoline in the pipeline and gasoline in the pump. As diesel reaches the pump and the line is still full of gasoline, the operating point moves to C, where we have diesel in the pump and the pipeline filled with gasoline. Finally, as the diesel batch enters the pipeline, the operating point moves towards D. At D we have completed the cycle and both the pump and the pipeline are filled with diesel. Thus it is seen that in a batched operation the flow rates vary between $Q_A$ and $Q_C$ as in Figure 7.13

## 7.11 Multiple Pumps versus System Head Curve

When two or more pumps are operated in parallel on a pipeline system, we saw how the pump head curves added the flow rates at the same head to create the combined pump performance curve. Similarly, with series pumps, the heads are added up for the same flow rate resulting in the combined pump head curve.

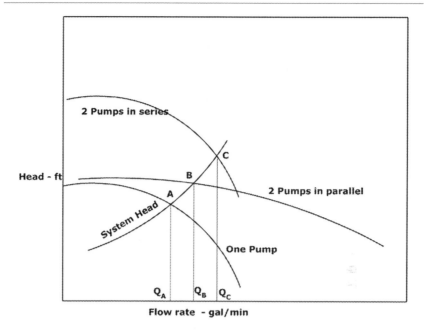

**Figure 7.14 - Multiple pumps and System Head Curve**

Figure 7.14 shows the pipeline system head curve superimposed on the pump head curves to show the operating point with one pump, two pumps in series and the same two pumps in parallel configurations. The operating points are shown as A, C and B with flow rates of $Q_A$, $Q_C$ and $Q_B$ respectively.

In certain pipeline systems, depending upon the flow requirements, we may be able to obtain higher throughput by switching from a series pump configuration to a parallel pump configuration. From Figure 7.14, it can be seen that a steep system head curve would favor pumps in series, while a relatively flat system head curve is associated with parallel pumps operation.

## 7.12 NPSH Required versus NPSH Available

As the pressure on the suction side of a pump is reduced to a value below the vapor pressure of the liquid being pumped, flashing can occur. The liquid vaporizes and the pump is starved of liquid. At this point the pump is said to cavitate due to insufficient liquid volume and pressure. The vapor can damage the pump impeller further reducing its ability to pump. To avoid vaporization of liquid, we must provide adequate positive pressure at the pump suction that is greater than the liquid vapor pressure.

NPSH for a centrifugal pump is defined as the Net Positive Suction Head required at the pump impeller suction to prevent pump cavitation at any flow rate. Cavitation will damage the pump impeller and render it useless. NPSH represents the resultant positive pressure at the pump suction. In this section, we will analyze a piping configuration from a storage tank to a pump suction, to calculate the available NPSH and compare it with NPSH required per pump vendor's performance curve. The NPSH available will be calculated by taking into account any positive tank head, including atmospheric pressure and subtracting the pressure drop due to friction in the suction piping and the liquid vapor pressure at the pumping temperature. The resulting value of NPSH for this piping configuration will represent the net pressure of the liquid at pump suction, above its vapor pressure. The value calculated must be more than the NPSH specified by the pump vendor at the particular flow rate.

Before we calculate the NPSH available in a typical pump-piping configuration, let us analyze the piping geometry associated with a pump taking suction from a tank and delivering liquid to another tank as shown in Figure 7.15

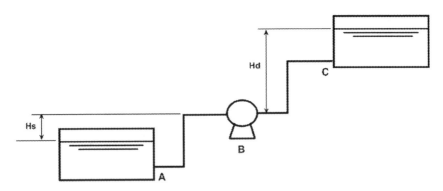

**Figure 7.15 Centrifugal Pump - Suction and Discharge Heads**

The vertical distance from the liquid level on the suction side of the pump center line is defined as the station suction head. More correctly it is the static suction lift ($H_S$) when the center line of the pump is above that of the liquid supply level as depicted in Figure 7.15. If the liquid supply level were higher than the pump center line, it will be called the static suction head on the pump.

Similarly the vertical distance from the pump center line to the liquid level on the delivery side is called the static discharge head ($H_D$) as shown in Figure 7.15.

The Total Static Head on a pump is defined as the sum of the static suction head and the static discharge head. It represents the vertical distance between the liquid supply level and the liquid discharge level. The static suction head, static discharge head and the Total Static Head on a pump are all measured in feet of liquid or meters of liquid in the SI system.

The friction head, measured in feet of liquid pumped, represents the pressure drop due to friction in both suction and discharge piping. It represents the pressure required to overcome the frictional resistance of all piping, fittings and valves on the suction side and discharge side of the pump as shown in Figure 7.15 Reviewing the piping system shown in Figure 7.15, it is seen that there are three sections of straight piping and two pipe elbows on the suction side between the liquid supply (A) and the center line of the pump. In addition, there would be at least two valves, one at the tank and the other at the inlet to the pump suction. An entrance loss will be added to account for the pipe entrance at the tank.

Similarly, on the discharge side of the pump between the center line of the pump (B) and the tank delivery point (C), there are three straight sections of pipe and two pipe elbows along with two valves. On the discharge of the pump there would also be a check valve to prevent reverse flow through the pump. An exit loss at the tank entry will also be added to account for the delivery pipe.

On the suction side of the pump, the available suction head $H_S$ will be reduced by the friction loss in the suction piping. This net suction head on the pump will be the available suction head at the pump center line.

On the discharge side the discharge head $H_D$ will be increased by the friction loss in the discharge piping. This is the net discharge head on the pump.

Mathematically, suction piping

Suction Head $\quad = H_S - H_{fs}$ $\qquad\qquad\qquad$ (7.10)

Discharge Head $\quad = H_D + H_{fd}$ $\qquad\qquad\qquad$ (7.11)

where

$H_{fs}$ $\quad$ - Friction loss in suction piping

$H_{fd}$ $\quad$ - Friction loss in discharge piping

If the suction piping is such that there is a suction lift (instead of suction head) the value of $H_S$ in Equation (7.10) will be negative. Thus a 20 ft static suction lift combined with a suction piping loss of 2 ft will actually result in an overall suction lift of 22 ft. The discharge head in the other hand will be the sum of the discharge head and friction loss in the discharge piping. Assuming 30 ft discharge head and 5 ft friction loss, the total discharge head will be 30 + 5 = 35 ft. In this example the total head of the pump is 22 + 35 = 57 ft.

**Example Problem 7.5**

A centrifugal pump is used to pump a liquid from a storage tank through 500 ft of suction piping as shown in Figure 7.16.

(a) Calculate the NPSH available at a flow rate of 3 000 gal/min.
(b) The pump vendor's data indicate NPSH required to be 35 ft at 3 000 gal/min and 60 ft at 4 000 gal/min. Can this piping system handle the higher flow rate without the pump cavitating?
(c) If cavitation is a problem in (b) above, what changes must be made to the piping system to prevent pump cavitation at 4 000 gal/min?

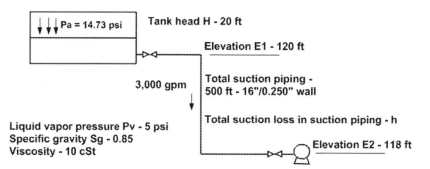

Figure 7.16 NPSH Calculation

## Solution

(a) NPSH available in ft of liquid head:

$$(P_a - P_v)(2.31/Sg) + H + E1 - E2 - h \qquad (7.12)$$

where

| | |
|---|---|
| $P_a$ | - Atmospheric pressure, psi |
| $P_v$ | - Liquid vapor pressure at flowing temperature, psi |
| Sg | - Liquid specific gravity |
| H | - Tank head, ft |
| E1 | - Elevation of tank bottom, ft |
| E2 | - Elevation of pump suction, ft |
| H | - Friction loss in suction piping, ft |

All items in Equation (7.12) are known except for the suction piping loss h.

The suction piping loss h is calculated at the given flow rate of 3 000 gal/min, considering 500 ft of 16 inch piping, pipe fitting, valves etc. in the given piping configuration. The total equivalent length of 16 inch pipe, including two gate valves and two elbows is:

Equivalent length of 16 inch pipe = 500 ft + 2 x 8 x (16/12) + 2 x 30 x (16/12) using an L/D ratio of 8 for the gate valves and 30 for each 90° elbow, from Appendix A.10

Therefore

$L_e = 500 + 21.33 + 80 = 601.33$ ft

Using Colebrook-White equation and assuming pipe roughness of 0.002 inch, we calculate the pressure drop at 3 000 gal/min as

$P_m = 12.77$ psi/mi

therefore

$h = 12.77 \times 2.31 \times 601.33/(0.85 \times 5280) = 3.95$ ft

Substituting h and other values in Equation (7.12) we get

$NPSH = (14.73 - 5) \times 2.31/0.85 + 20 + 120 - 118 - 3.95 = 44.49$ ft available

(b) At 4 000 gal/min the

Pressure loss $P_m = 21.43$ psi/mi

And        $h = 6.63$ ft

Available NPSH at 4 000 gal/min = 44.49 + 3.95 - 6.63 = 41.81 ft

Since available NPSH is less than NPSH required of 60 ft per pump vendor, the pump will cavitate.

(c) The extra head required to prevent cavitation = 60 - 41.81 = 18 ft

One solution is to locate the pump suction at an additional 18 ft or more below the tank. Another solution will be to provide a small vertical can pump that can serve as booster for the main pump. This pump will provide the additional head required to prevent cavitation.

## 7.13 Summary

We discussed centrifugal pumps and their performance characteristics as they apply to liquid pipeline hydraulics. Other types of pumps such as PD pumps used in injecting liquid into flowing pipelines were briefly covered. Pump performance at different impeller sizes and speeds, based on Affinity Laws were discussed and illustrated using examples. Trimming impellers or reducing speeds (VSD pumps) to match system pressure requirements were also explored.

The important parameter called NPSH was introduced and methods of calculations for typical pump configuration were shown. We also discussed how the water performance curves provided by a pump vendor must be corrected, using the Hydraulic Institute chart, when pumping high viscosity liquids. The performance of two or more pumps in series or parallel configuration was analyzed using an example. In addition, we demonstrated how the operating point can be determined by the point of intersection of a system head curve and the pump head curve. In batched pipelines the variation of operating point on a pump curve using the system head curves for different products was illustrated.

## 7.14 Problems

7.14.1 Two pumps are used in series configuration. Pump A develops 1000 ft head at 2 200 gal/min and Pump B generates 850 ft head at the same flow rate. At the design flow rate of 2 200 gal/min, the application requires a head of 1700 ft of head. The current impeller sizes in both are both 10 inch diameter.

(a) What needs to be done to prevent pump throttling at the specified flow rate of 2 200 gal/min?

(b) Determine the impeller trim size needed for Pump A

(c) If Pump A could be driven by a variable speed drive, what speed should it be run to match pipeline system requirement?

7.14.2 A pipeline 40 miles long, 10 inch nominal diameter and 0.250 inch wall thickness is used for gasoline movements. The static head lift is 250 ft from origin pump station to the delivery point. The delivery pressure is 50 psi and the pump suction pressure is 30 psi. Develop a system head curve for the pipeline for gasoline flow rates up to 2 000 gal/min. Use specific gravity of 0.736 and viscosity of 0.65 cSt at flowing temperature. Pipe roughness = 0.002 inch. Use Colebrook-White Equation.

7.14.3 In the previous problem for gasoline movements, the pump used has an Head versus Capacity and Efficiency versus Capacity as indicated below:

| Q - gal/min | 0 | 400 | 800 | 1 200 | 1 500 |
|---|---|---|---|---|---|
| H - ft | 3 185 | 3 100 | 2 900 | 2 350 | 1 800 |
| E - % | 0.0 | 55.7 | 78.0 | 79.3 | 72.0 |

(a) What gasoline flow rate will the system operate at with the above pump?

(b) If instead of gasoline, diesel fuel is shipped through the above pipeline on a continuous basis, what throughputs can be expected?

(c) Calculate the motor HP required in case (a) and (b) above

7.14.4 A tank farm consists of a supply tank A located at an elevation of 80 ft above mean sea level (MSL) and a delivery tank B at 90 ft above MSL. Two identical pumps are used to transfer liquid from tank A to tank B using interconnect pipe and valves. The pumps are located at elevation of 50 ft above MSL.

The suction piping from tank A is composed of 120 ft, 14 in diameter, 0.250 in wall thickness pipe and six 14 in, 90° elbows and two 14 in gate valves. On the discharge side of the pumps, the piping consists of 6 000 ft of 12.75 in diameter, 0.250 in wall thickness pipe, 12 in gate valves and eight 12 in 90° elbows. There is 10 in check valve on the discharge of eachpump. Liquid transferred has specific gravity = 0.82 and viscosity = 2.5 cSt.

(a) Calculate the total station head for the pumps
(b) If liquid is transferred at the rate of 2 500 gal/min, calculate the friction losses in the suction and discharge piping system.
(c) If one pump is operated while the other is on stand by, determine the pump (Q, H) requirement for above product movement
(d) Develop a system head curve for the piping system.
(e) What size electric motor (95% efficiency) will be required to pump at the above rate, assuming the pump has an efficiency of 80%.

# 8

# Pump Station Design

In this chapter we will look at some of the significant items in a pump station, that pertain to pumps and pipeline hydraulics. We will analyze the various pressures on both the suction and discharge side of the pump and how pressure control is implemented using a control valve downstream of the pumps. The amount of pump throttle pressure due to mismatch between pump head and system curve head will be analyzed and amount of horsepower wasted will be calculated. The use of VSD pumps to eliminate throttle pressure by providing just enough pressure for the specified flow rate will be explained. We will also perform a simple economic comparison between a VSD pump installation and a constant speed pump with a control valve.

## 8.1  Suction pressure and discharge pressure

A typical piping layout within a pump station is as shown in Figure 8.1 The pipeline enters the station boundary at point A, where the station block valve MOV-101 is located. The pipeline leaves the station boundary on the discharge side of the pump station at point B, where the station block valve MOV-102 is located.

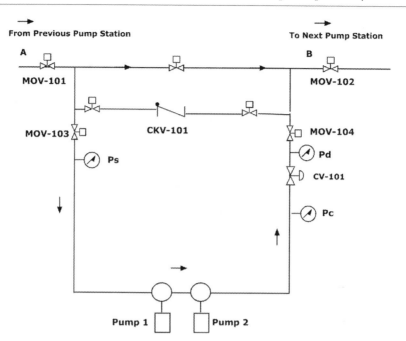

**Figure 8.1 Typical Pump Station Layout**

Station bypass valves designated as MOV-103 and MOV-104 are used for bypassing the pump station in the event of pump station maintenance or other reasons where the pump station must be isolated from the pipeline. Along the main pipeline there is located a check valve, CKV-101 that prevents reverse flow thru the pipeline. This typical station layout shows two pumps configured in series. Each pump pumps the same flow rate and the total pressure generated is the sum of the pressures developed by each pump. On the suction side of the pump station the pressure is designated as $P_s$ while the discharge pressure on the pipeline side is designated as $P_d$. With constant speed motor driven pumps, there is always a control valve on the discharge side of the pump station shown as CV-101 in Figure 8.1. This control valve controls the pressure to the required value $P_d$ by creating a pressure drop across it between the pump discharge pressure $P_e$ and the station discharge pressure $P_d$. Since the pressure within the case of the second pump represents the sum of the suction pressure and the total pressure generated by both pumps, it is referred to as the case pressure $P_c$.

If the pump is driven by a variable speed drive (VSD) motor or an engine, the control valve is not needed as the pump may be slowed down or speeded up as required to generate the exact pressure $P_d$. In such a situation, the case pressure will equal the station discharge pressure $P_d$.

In addition to the valves shown, there will be additional valves on the suction and discharge of the pumps. Also, not shown in the figure, is a check valve located immediately after the pump discharge that prevents reverse flow through the pumps.

## 8.2   Control Pressure and Throttle Pressure

Mathematically, if $\Delta P_1$ and $\Delta P_2$ represent the differential head produced by Pump1 and Pump2 in series, we can write

$$P_c = P_s + \Delta P_1 + \Delta P_2 \tag{8.1}$$

where

$P_c$    - Case pressure in Pump2 or upstream pressure at control valve

The pressure throttled across the control valve is defined as

$$P_{thr} = P_c - P_d \tag{8.2}$$

where

$P_{thr}$   - Control valve throttle pressure
$P_d$    - Pump station discharge pressure

The throttle pressure represents the mismatch that exists between the pump and the system pressure requirements at a particular flow rate. $P_d$ is the pressure at the pump station discharge needed to transport liquid to the next pump station or delivery terminus based on pipe length, diameter, elevation profile and liquid properties. The case pressure $P_c$ on the other hand is the available pressure due to the pumps. If the pumps were driven by variable speed motors, $P_c$ would exactly match $P_d$ and there would be no throttle pressure. In other words, the control valve would be wide open and obviously will not be needed in operation. The case pressure is also referred to as Control Pressure since it is the pressure upstream of the control valve. The control valve also functions as a means for protecting the discharge piping. The throttle pressure represents unused pressure developed by the pump and hence results in wasted HP and dollars. The objective should be to reduce the amount of throttle pressure in any pumping situation.

As mentioned earlier, VSD pumps can speed up or slow down to match pipe pressure requirements. In such situations, there is no control valve and therefore the throttle pressure is zero. With VSD pumps there is no HP wasted since the pump case pressure exactly matches the station discharge pressure.

A simplified pump station schematic with one pump is shown in Figure 8.2

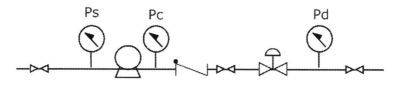

**Figure 8.2 Single Pump Schematic**

## Example Problem 8.1

A pump station installation at Corona consists of two pumps configured in series each developing 1 500 ft of head at 4 000 gal/min. Diesel fuel (Spgr = 0.85 and visc = 5.9 cSt) is pumped from Corona to a delivery terminal at Sunnymead, 75

miles away. The required discharge pressure at Corona based on pipe size, pipe length, liquid properties, elevation profile between Corona and Sunnymead and the required delivery pressure at Sunnymead has been calculated to be 1 050 psi. The pump station suction pressure is maintained at 50 psi to prevent pump cavitation. Assume combined efficiency of both pumps at Corona at the given flow rate to be 82%.

(a) Analyze the pump station pressures and determine the amount of throttle pressure and HP wasted.

(b) If electrical energy costs eight cents per kWh, estimate the dollars lost in control valve throttling

(c) What recommendation can you make to improve pipeline operation?

**Solution**

(a) The total pump pressure developed by two pumps in series is

$$\frac{(1\,500 + 1\,500) \times 0.85}{2.31} = 1\,104 \text{ psi, rounded off.}$$

adding the 50 psi suction pressure gives a control pressure upstream of the control valve of

$$1\,104 + 50 = 1\,154 \text{ psi}$$

This is also the case pressure in the second pump. Since the station discharge pressure is 1 050 psi, the control valve throttle pressure is

Throttle pressure = 1 154 - 1 050 = 104 psi

The HP wasted due to pump throttling can be calculated as

$$HP \text{ wasted} = \frac{gal/min \times psi}{1\,714 \times efficiency} \qquad \text{using Equation (5.18)}$$

or

$$HP \text{ wasted} = \frac{4\,000 \times 104}{1\,714 \times 0.82} = 296$$

(b) At 8 cents per kWh electric cost, above wasted HP is equivalent to

$$296 \times 0.746 \times 24 \times 350 \times 0.08 = \$148\,388 \text{ per year}$$

assuming 350 days of continuous operation per year. This is a substantial loss.

(c) If plans are to continue operating this pump station at the current flow rate on a continuous basis, it would be preferable to trim the pump impellers to eliminate the throttle pressure and hence wasted HP. A spare rotating element for the pump can easily be purchased and installed for about $50 000. This which represents almost a third of the dollars wasted in throttling. Therefore, the recommendation is to purchase and install a new trimmed rotating element for one of the pumps that will result in no throttling at 4 000 gal/min. The existing larger impeller may be stored as a spare for future use when increased flow rates are anticipated.

## 8.3   Variable speed pumps

It was mentioned in the preceding section that when pumps are driven by variable speed motors or gas turbine drives, the control valve is not needed and the pump throttle pressures will be zero. The variable speed pump will be able to speed up or slow down to match the pipeline pressure requirements at any flow rate, thereby not requiring the use of a control valve. Of course, depending upon the pump, there will be a minimum and maximum permissible pump speeds.

If there are two or more pumps in series configuration, one of the pumps may be driven by a variable speed drive (VSD) motor or an engine. With parallel pumps all pumps will have to be VSD pumps, since parallel configuration require matching heads at the same flow rate.

In the case of two pumps in series, similar to Example Problem 8.1, we could convert one of the two pumps to be driven by a variable speed motor. This pump can then slow down to the required speed that would develop just the right amount of head (at 4 000 gal/min) which when added to the 1 500 ft developed by the constant speed pump would provide exactly the total head required to match the pipeline system requirement of 1 050 psi.

**Example Problem 8.2**

Use the data in Problem 8.1 and consider one of the two pumps driven by a variable speed electric motor. Calculate the speed at which the VSD pump should be operated to match the pipeline requirements at 4 000 gal/min. Assume that the rated speed for the pump at 3 560 RPM produces a head of 1 500 ft at 4 000 gal/min.

## Solution

The pipeline pressure requirement is given as 1 050 psi. Assuming the VSD pump develops a head of H ft, the total head produced by the two pumps in series is (H + 1 500) ft. Since the pump suction pressure is 50 psi, the total pressure on the discharge side of the pumps is

(H + 1 500) x 0.85 / 2.31 + 50 psi

This must exactly be equal to the discharge pressure requirement of 1 050 psi at the 4 000 gal/min flow rate.

Therefore, we can state that

1 050 = 50 + (H + 1 500) x 0.85/2.31

From which, we get

H = 1 218 ft

This is the head that the VSD pump must develop, compared to 1 500 ft for the constant speed pump.

At the rated speed of 3 560 RPM this pump develops 1 500 ft at 4 000 gal/min. In order to reduce the head to 1 218 ft, the VSD pump has to be slowed down to some speed N RPM. We will calculate this speed using Affinity Laws,

Since head is proportional to square of the speed,

Speed ratio = N / 3 560 = (1 218 / 1 500)$^{1/2}$ using Equation (7.8)

Solving for N we get

N = 3 208 RPM approximately.

By operating the second pump (as a VSD pump) at 3 208 RPM in series with the first pump (constant speed) the pump station will produce exactly 1 050 psi required to pump the diesel fuel at 4 000 gal/min. Therefore, no pump throttling will be required and no pump BHP will be wasted.

### 8.3.1 VSD Pump versus Control Valve

In a single pump station pipeline with one pump used to provide the pressure required to pump the liquid, a control valve is used to regulate the pressure for a given flow rate.

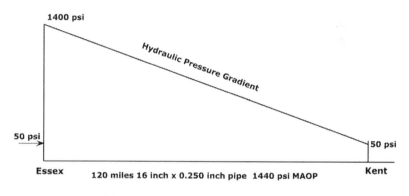

**Figure 8.3 Essex to Kent Pipeline**

Suppose the pipeline from Essex pump station to the Kent delivery terminal is 120 mi long and is constructed of a 16 in diameter and 0.250 in wall thickness pipe with an MAOP of 1 440 psi. The pipeline is designed to operate at 1 400 psi, pumping 4 000 bbl/hr of liquid (specific gravity - 0.89 and viscosity - 30 cSt at 60°F) on a continuous basis. The delivery pressure required at Kent is 50 psi. If the pump suction pressure at Essex is 50 psi, the required pump differential pressure is 1 350 psi. Unless the single pump unit at Essex was specifically selected for this application, chances are that the Essex pump H-Q curve may indicate the following

$$Q = 2\ 800 \text{ gal.min} \quad H = 3\ 800 \text{ ft}$$

The flow rate of 2 800 gal/min is the equivalent of 4 000 bbl/hr. Converting the head available 3 800 ft into psi, the pump pressure developed is

3 800 x 0.89 / 2.31 = 1 464 psi

On the other hand, if the pump head at Q = 2 800 gal/min were only 3 200 ft, the pressure developed would only be

3 200 x 0.89 / 2.31 = 1 233 psi

which is inadequate for our application that requires a pressure of 1 350 psi to pump the liquid at 4 000 bbl/hr.

Assuming the first case where the pump is able to develop 3 800 ft that corresponds to 1 464 psi, we can see that this pressure combined with the 50 psi pump suction pressure would produce a pump discharge pressure (actually the pump case pressure) of

1 464 + 50 = 1 514 psi

As mentioned earlier, the pipeline pressure limit (MAOP) is 1 440 psi, as a result we will over pressure the pipeline by 74 psi. Obviously, this cannot be tolerated and we must have some means to control the pump pressure to the required 1 400 psi pump station discharge we mentioned earlier. A control valve located downstream of the pump discharge will be used to reduce the discharge pressure to the required pressure of 1 400 psi. This is depicted by the modified system curve (2) in Figure 8.4

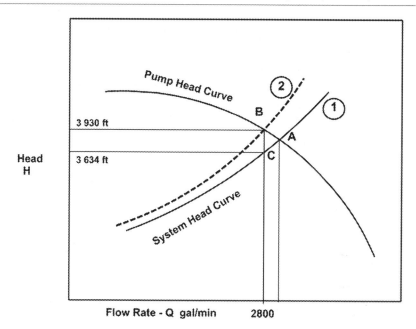

**Figure 8.4 System Curve and Control Valve**

The system head curve (1) in Figure 8.4 represents the pressure versus flow rate variation for our pipeline from Essex to Kent. At Q = 2 800 gal/min (4 000 bbl/hr) flow rate, point C on the pipeline system head curve shows the desired operating point that requires a pipeline discharge pressure of 1 400 psi at Essex. Since we have superimposed the pump H-Q curve and pressures are in ft of liquid head, the pressure at C is

1 400 x 2.31 / 0.89 = 3 634 ft

Since the pump suction pressure is 50 psi, we have plotted the pump curve in Figure 8.4 to include this suction head as well. The pump H-Q curve shows that at a flow rate of 2 800 gal/min, the differential head generated is 3 800 ft. Adding the 50 psi suction pressure point B on the H-Q curve corresponds to

3 800 + 50 x 2.31 / 0.89 = 3 930 ft

We can see from the point of intersection of the pump H-Q curve and the pipeline system curve, the operating point will be at point A, which is at a higher flow rate than 2 800 gal/min. Also, the pump discharge pressure at A will be at a higher flow rate than that at C. Unless the pipe MAOP is greater than the pressure at A, we cannot allow this pump to operate without pressure control. Thus utilizing the control valve on the pump discharge we will move the operating point from A to B on the pump curve corresponding to 2 800 gal/min and a higher head of 3 930 ft as calculated above. The control valve will then cause the pressure drop equivalent to the head difference BC calculated as follows

Control valve pressure drop, BC = 3930 - 3634 = 296 ft

This pressure drop across the control valve is also called the pump throttle pressure

Throttle pressure = 296 x 0.89 / 2.31 = 114 psi

Thus we conclude that due to the slightly oversized pump and the pipe MAOP limit, we have to utilize a control valve on the pump discharge to limit the pipeline discharge pressure to 1 400 psi at the required flow rate of 4 000 bbl/hr (2 800 gal/min).

A dashed hypothetical system head curve designated as (2), passing through the point B on the pump curve is the artificial system head curve due to the restriction imposed by the control valve.

It must be pointed out that as far as the pump is considered, the suction pressure, case pressure, discharge pressure and throttle pressure are as follows

| | |
|---|---|
| Pump suction pressure | - 50 psi |
| Pump case pressure | - 1 514 psi |
| Pump station discharge pressure | - 1 400 psi |
| Pump throttle pressure | - 114 psi |

The above analysis applies to a pump driven by a constant speed electric motor. If we had a VSD pump that can vary the pump speed from 60% to 100% rated speed and if the rated speed were 3 560 RPM, the pump speed range will be

3 560 x 0.60 = 2 136 RPM to 3 560 RPM

Since the pump speed can be varied from 2 136 RPM to 3 560 RPM, the pump head curve will correspondingly vary according to the centrifugal pump Affinity Laws. Therefore we can find some speed (less than 3 560 RPM) at which the pump will generate the required head corresponding to point C in Figure 8.4

Point C represents a pump differential head of

(1 400 - 50) x 2.31 / 0.89 = 3 504 ft

where 50 psi represents the pump suction pressure. If the given H-Q curve is based on the rated speed of 3 560 RPM, using Affinity Laws we will calculate the approximate speed required for point C as

3 560 (3 504 / 3 800) $^{1\!/\!2}$ = 3 419 RPM

If we had the full H-Q curve data at 3 560 RPM we can generate the revised H-Q curve at the desired speed of 3 419 RPM. Actually, we may have to adjust the speed a little to get the required head corresponding to point C.

From the foregoing analysis we can conclude that use of VSD pumps can provide the right amount of pressure required for a given flow rate pump thus avoiding pump throttle pressures (and hence wasted HP), common with constant speed motor driven pumps with control valves. However, VSD pumps are expensive to install and operate compared to the use of a control valve. A typical control valve installation may cost $100 000, whereas a VSD may require $300 000 to 500 000 incremental cost compared to the constant speed motor driven pump. We will have to factor in the increased operating cost of the VSD pump compared to the dollars lost in wasted HP from control valve throttling.

Let us illustrate this function by considering the same Essex to Kent Pipeline. In the first case, we will assume a constant speed pump at

Essex with a control valve. The pump and control valve installation will cost approximately

$ (X + 100 000)

where x represents the cost of the constant speed motor driven pump. The annual maintenance cost for the control valve is estimated at $10 000. Due to pump throttling. Let us assume an average pressure drop in the control valve to be 100 psi over a 12 month period at an average pipeline flow rate of 4 000 bbl/hr. This is equivalent to a wasted HP 0f

$$\frac{2\ 800 \times 100}{1\ 714 \times 0.8} = 204.2 \text{ HP}$$

Assuming 80% pump efficiency, at an average electrical cost of 10 cents/ kWh and 350 days operation per year, the throttled pressure represents a loss of

$ (204.2 x 0.746 x 24 x 350 x 0.01) = 127 960 per year.

If we now replaced the pump at Essex with a VSD pump consisting of a variable frequency drive (VFD) electric motor with the same pump without the control valve, the capital cost will be approximately

$ (X + 500 000)

where $500 000 has been added to account for the additional switchgear and the VFD motor. The annual maintenance cost for this installation may be as high as $50 000.

Comparing the two alternatives, we can summarize as follows,

assuming x = 250 000 for the pump motor.

Case A - Constant speed motor driven pump with control valve

Capital cost        - $350 000
Annual cost        - $137 960
(Includes HP lost due to control valve)

Case B - VFD motor driven pump, no control valve

Capital cost        - $750 000
Annual cost        - $ 50 000

It is clear that the VSD pump has higher initial cost but lower annual cost compared to the constant speed motor/control valve combination.

In this example, we can perform an economic analysis and compare Case A with Case B considering discounted cash flow. If interest rate is 8% per year the present value of Case A for a 15 year project life is

$$PV_A = 350\ 000 + PV(137\ 960,15,8)$$

Where $PV(A, N, i)$ represents the present value of a series of cash flows of $A, each for N years at an annual interest rate of $i$ percent.

Therefore

$$PV_A = 350\ 000 + 1\ 180\ 866$$
$$= \$\ 1\ 530\ 866$$

Similarly,

$$PV_B = 750\ 000 + PV(50\ 000, 15, 8)$$
$$= 750\ 000 + 427\ 974$$
$$= \$1\ 177\ 974$$

It can be seen from the above that the VSD case will be more economical alternative, since it has a lower present value of investment.

The advantages of VSD pumps include operational flexibility and lower power requirements since no HP is wasted as with a control valve. The disadvantage of VSD pumps include higher capital and operating cost compared to a constant speed pump with a control valve.

## 8.4   Summary

This chapter covered the main components found in a typical pump station as it relates to pumps and pipeline hydraulics. The concept of pump suction pressure, case pressure, pump station discharge pressure and throttle pressure were introduced and calculations illustrated using examples. The impact of pump throttle pressure and how the control valve is used to protect the discharge piping as well as provide the necessary pressure for a particular flow rate were explained. We found that the throttle pressure contributes to energy loss and therefore wasted dollars. The advantage of using VSD pumps to provide just enough pressure for the specified flow, thereby eliminating throttle pressures were analyzed. Also illustrated was a comparison of using a VSD pump versus the control valve.

## 8.5   Problems

8.5.1   The origin pump station at Harvard on a 50 mile pipeline consists of two pumps in series each developing 1 450 ft of head at 3 000 bbl/hr. Diesel fuel (Specific gravity of 0.85 and viscosity of 5.9 cSt) is pumped from Harvard to a delivery terminal at Banning. The required discharge pressure at Harvard has been calculated to be 995 psi. The pump station suction pressure is maintained at 50 psi. Use a combined pump efficiency at the given flow rate to be 84%.

   (a) Analyze the pump station pressures and determine the amount of throttle pressure and HP wasted.
   (b) If electrical energy costs 10 cents/kWh, estimate the dollars lost in control valve throttling

8.5.2   If a Variable Speed Drive pump is used in problem above, determine the speed at which the pump should be run to maintain the flow rate of 3 000 bbl/hr. The rated pump speed is 3560 RPM.

8.5.3  A water pipeline consists of three pump stations, each having one constant speed motor driven pump of 2 000 HP. The analysis of current operations show that the pipeline has been operating at a fairly constant flow rate of 3 000 gal/min and the pump throttle pressures are as follows:

Pump station 2 - 75 psi
Pump station 3 - 85 psi

Pump station 1 has zero throttle pressure. If the pump impellers are trimmed to the correct size so as to eliminate wasted HP from pump throttle it is estimated to cost $50 000 per pump station. Retrofitting each pump station with VFD motor driven pumps is estimated to cost $500 000 per pump station. The operating cost for the VFD units is $50 000 per site.

(a) Determine the economics of installing trimmed impellers and the payment period based on 8% interest rate. Assume 80% pump efficiency for HP calculations and electric cost of 10 cents/kWh.

(b) If it was decided not to trim the pump impellers, compare the VFD motor option with the constant speed pump option. Consider control valve annual operating cost of $5 000 per site.

8.5.4  A refined products pipeline is used for batched operations with gasoline, kerosene and diesel fuel. The pipeline is 80 miles long and is 14 in diameter, 0.250 in wall thickness. The MAOP of the pipeline is limited to 1 200 psi.

(a) Determine the maximum throughput possible with one pump station for each product considering each product pumped alone with no batching. Assume flat elevation profile with 50 psi pipe delivery pressure. For liquid properties use the following:

| Product | Specific Gravity | Viscosity cSt |
|---------|------------------|---------------|
| Gasoline | 0.74 | 0.6 |
| Kerosene | 0.82 | 2.0 |
| Diesel | 0.85 | 5.5 |

(b) Investigate the use of VSD pumps to provide the needed operational control and flexibility when this pipeline is operated in a batched mode. Pick a typical pump curve for the origin pump station to satisfy the needs of maximizing gasoline throughput first. Having based pump requirements for gasoline, determine the speed variants required for the other two products without exceeding MAOP.

(c) Estimate the annual power cost for each product based on 10 cents/ kWh.

(d) How would the operating scenario change if VSD pumps were not used? Estimate the cost of HP wasted in pump throttling with the control valve.

# 9

# Thermal Hydraulics

Thermal hydraulics takes into account the temperature variation of a liquid as it flows through the pipeline. This is in contrast with isothermal hydraulics, where there is no significant temperature variation in the liquid. In prior chapters, we concentrated on water pipelines, refined petroleum products (gasoline, diesel etc.) and other light crude oil pipelines where the liquid temperature was close to ambient temperatures. In many cases where heavy crude oil and other liquids of high viscosity have to be pumped, the liquid is heated to some temperature (such as 150 °F to 180 °F) prior to being pumped through the pipeline. In this chapter we will explore how calculations are performed in thermal hydraulics.

## 9.1   Temperature Dependent Flow

In the preceding chapters we concentrated on steady state liquid flow in pipelines without paying much attention to temperature variations along the pipeline. We assumed the liquid entered the pipeline inlet at some temperature such as 70°F. The liquid properties such as specific gravity and viscosity at the inlet temperature were used to calculate the Reynold's number and friction factor and finally the pressure drop due to friction. Similarly, we also used the specific gravity at inlet temperature to calculate the elevation head based on the pipeline topography. In all cases the liquid properties were considered at some constant flowing temperature. These calculations are therefore based on isothermal (constant temperature) flow.

The above may be valid in most cases where the liquid transported such as water, gasoline, diesel or light crude oil is at ambient temperatures. As the liquid flows through the pipeline, heat may be transferred to or from the liquid to the surrounding soil (buried pipeline) or the ambient air (above ground pipeline). Significant changes in liquid temperatures due to heat transfer with surroundings will affect liquid properties such as specific gravity and viscosity. This in turn will affect pressure drop calculations. So far we have ignored this heat transfer effect assuming minimal temperature variations along the pipeline. However, there are instances when the liquid has to be heated to a much higher temperature than ambient conditions to reduce the viscosity and make it flow easily. Pumping a higher viscosity liquid that is heated will also require less pump horsepower.

For example, a high viscosity crude oil (200 cSt to 800 cSt or more at 60 °F) may be heated to 160°F temperature before it is pumped into the pipeline. This high temperature liquid loses heat to the surrounding soil as it flows through the pipeline by conduction of heat from the interior of the pipe, to the soil through the pipe wall. The ambient soil temperature may be 40°F to 50°F during the winter and 60°F to 80°F during the summer. Therefore, considerable temperature difference exists between the hot liquid in the pipe and the surrounding soil.

The temperature difference of about 120°F in winter and 100°F during summer, will cause significant heat transfer between the crude oil and surrounding soil. This will result in temperature drop of the liquid and variation in liquid specific gravity and viscosity as it flows through the pipeline. Therefore, in such instances, we will be wrong in assuming a constant flowing temperature to calculate pressure drop as we do in isothermal flow. Such heated liquid pipeline may be bare or insulated. In this chapter we will study the effect of temperature variation and friction loss along the pipeline known as Thermal Hydraulics.

To illustrate further, consider a 20 inch buried pipeline transporting 8 000 bbl/hr of a heavy crude oil that enters the pipeline at an inlet temperature of 160°F. Assume that the liquid temperature has dropped to 124°F at a location 50 miles from the pipeline inlet. Suppose, the crude oil properties at 160°F inlet conditions and at 124°F at milepost 50 are as follows:

| Temperature (°F) | Specific Gravity | Viscosity (cSt) |
|---|---|---|
| 160 | 0.9179 | 40.55 |
| 124 | 0.9306 | 103.69 |

Using the 160°F inlet temperature we calculate the frictional pressure drop at inlet conditions to be 7.6 psi/mi. At milepost 50 temperature conditions, using the given liquid properties we find that the frictional pressure drop has increased to 23.97 psi/mi. Thus, in a 50 mile section of pipe the pressure drop due to friction varies from 7.6 psi/mi to 23.97 psi/mi as shown in Figure 9.1.

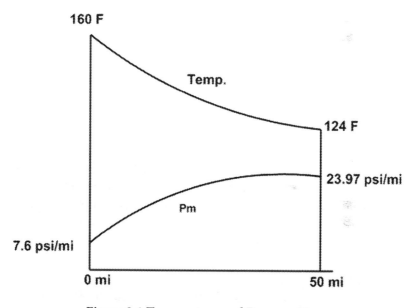

**Figure 9.1 Temperature and Pressure Drop**

We could use an average value of the pressure drop per mile to calculate the total frictional pressure drop in the first 50 miles of the pipe. However, this will be a very rough estimate.

A better approach would be to sub-divide the 50 mile section of the pipeline into smaller segments 5 or 10 mile each and compute the pressure drop per mile for each segment. We will then add the individual pressure drops for each 5 or 10 mile segment to get the total friction drop in the 50 mile length of pipeline. Of course, this assumes that we have available to us the temperature of the liquid at 5 or 10 mile increments up to milepost 50. The liquid properties and pressure drop due to friction can then be calculated at the boundaries of each 5 or 10 mile segment as illustrated in Figure 9.2.

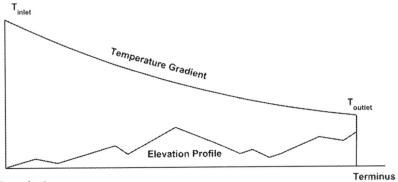

**Figure 9.2 Thermal Temperature Gradient**

How do we get the temperature variation along the pipeline so we can calculate the liquid properties at each temperature and then calculate the pressure drop due to friction? This represents the most complicated portion of thermal hydraulic analysis. Several approaches have been put forth for calculating the temperature variation in a pipeline transporting heated liquid. We must consider soil temperatures along the pipeline, thermal conductivity of pipe material, pipe insulation if any, thermal conductivity of soil and pipe burial depth.

We will present in this Chapter a simplified approach to calculating the temperature profile in a buried pipeline. The method and formulas used were developed originally for the Trans Alaskan Pipeline System (TAPS). These have been found to be quite accurate over the range of temperatures and pressures encountered in heated liquid pipelines today.

The hydraulic gradient showing the pressure profile in a batched liquid pipeline as shown in Figure 9.3. Note the curved shape of the gradient, compared to the straight line gradient in isothermal flow (such as Figures 6.2 or 8.3)

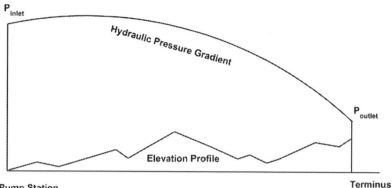

**Figure 9.3 Thermal Hydraulic Pressure Gradient**

An example will illustrate the use of these formulas. More accurate methods include using a computer software program that will sub-divide the pipeline into small segments and compute the heat balance and pressure drop calculations, to develop the pressure and temperature profile for the entire pipeline. One such commercially available program is LIQTHERM developed by SYSTEK (www.systek.us).

## 9.2 Formulas for Thermal Hydraulics

### 9.2.1 Thermal conductivity

Thermal conductivity is the property used in heat conduction through a solid. In English units it is measured in Btu/hr/ft/ °F. In SI units thermal conductivity is expressed in W/m/ °C.

For heat transfer through a solid of area A and thickness dx, with a temperature difference dT, the formula in English units is:

$$H = K (A)(dT/dx) \qquad (9.1)$$

where

H     - Heat flux perpendicular to the surface area, Btu/hr
K     - Thermal conductivity of solid, Btu/hr/ft/ °F.
A     - Area of heat flux, $ft^2$

dx    - Thickness of solid, ft
dT    - Temperature difference across the solid, °F

The term dT/dx represents the temperature gradient in °F /ft.

Equation (9.1) is also known as the Fourier heat conduction formula.

It can be seen from Equation (9.1) that the thermal conductivity of a material is numerically equal to the amount of heat transferred across a unit area of the solid material with unit thickness, when the temperature difference between the two faces of the solid is maintained at one degree.

The thermal conductivity for steel pipe and soil are as follows:

K for steel pipe = 29 Btu/hr/ft/ °F
K for soil = 0.2 to 0.8 Btu/hr/ft/ °F.

Sometimes heated liquid pipelines are insulated on the outside with an insulating material. The K value for insulation may range from 0.01 to 0.05 Btu/hr/ft/ °F

In SI units, Equation (9.1) becomes

$$H = K\,(A)(dT/dx) \quad (9.2)$$

where

H    - Heat flux, W
K    - Thermal conductivity of solid, W/m /°C.
A    - Area of heat flux, $m^2$
dx    - Thickness of solid, m
dT    - Temperature difference across the solid, °C

In SI units, the thermal conductivity for steel pipe, soil and insulation are as follows:

K for steel pipe     = 50.19 W/m /°C
K for soil           = 0.35 W/m /°C to 1.4 W/m /°C
K for insulation     = 0.02 to 0.09 W/m /°C

As an example, heat transfer occurs across a flat steel plate of thickness 8 inch and area 100 ft$^2$ where the temperature difference across the plate thickness is 20 ° F. From Equation (9.1)

Heat transfer = 29 x (100)x 20 x 12 / 8 = 87 000 Btu/hr

### 9.2.2     Overall Heat Transfer Coefficient

The overall heat transfer coefficient is also used in heat flux calculations. Equation (9.1) for heat flux can be written in terms of overall heat transfer coefficient as follows:

$$H = U (A)(dT) \tag{9.3}$$

where

U - Overall heat transfer coefficient, Btu/hr/ft$^2$/ °F

Other symbols in Equation (9.3) are the same as in Equation (9.1)

In SI units, Equation (9.3) becomes

$$H = U (A)(dT) \tag{9.4}$$

where

U - Overall heat transfer coefficient, W/m$^2$/ °C

Other symbols in Equation (9.4) are the same as in Equation (9.2)

The value of U may range from 0.3 to 0.6 Btu/hr/ft$^2$/°F in English units and 1.7 to 3.4 W/m$^2$/°C in SI units.

When analyzing heat transfer between the liquid in a buried pipeline and the outside soil, we consider flow of heat through the pipe wall and pipe insulation, (if any) to the soil. If U represents the overall heat transfer coefficient, we can write from Equation (9.3)

$$H = U (A)(T_L - T_S) \qquad\qquad (9.5)$$

where

A     - Area of pipe under consideration, $ft^2$
$T_L$     - Liquid temperature, °F
$T_S$     - Soil temperature, °F
U     - Overall heat transfer coefficient, $Btu/hr/ft^2/°F$

Since we are dealing with temperature variation along the pipeline length, we must consider a small section of pipeline at a time, when applying above Equation (9.5) for heat transfer.

For example consider a 100 ft length of 16 inch pipe carrying a heated liquid at 150°F. If the outside soil temperature is 70°F and the overall heat transfer coefficient

$$U = 0.5 \; Btu/hr/ft^2/ °F$$

We can calculate the heat transfer using Equation (9.5) as follows:

$$H = 0.5 \; A \; (150 - 70)$$

where A is the area through which heat flux occurs.

$$A = \pi \times (16/12) \times 100 = 419 \; ft^2$$

Therefore

$$H = 0.5 \times 419 \times 80 = 16\,760 \; Btu/hr$$

### 9.2.3    Heat Balance

The pipeline is sub-divided and for each segment, the heat content balance is computed as follows:

Hin - DeltaH + Hw = Hout                                                                         (9.6)

where

Hin          - Heat content entering line segment, Btu/hr
DeltaH       - Heat transferred from line segment to surrounding
               medium(soil or air), Btu/hr
Hw           - Heat content from frictional work, Btu/hr
Hout         - Heat content leaving line segment, Btu/hr

In the above we have included the effect of frictional heating in the term Hw. With viscous liquids the effect of friction is to create additional heat which would raise the liquid temperature. Therefore in thermal hydraulic analysis frictional heating is included to improve the calculation accuracy.

In SI units, Equation (9.6) will be the same, with each term expressed in Watts instead of Btu/hr.

The heat balance Equation (9.6) forms the basis for computing the outlet temperature of the liquid in a segment starting with its inlet temperature and taking into account the heat loss (or gain) with the surroundings and accounting for frictional heating. In the following sections we will formulate the method of calculating each term in Equation (9.6)

### 9.2.4    Logarithmic Mean Temperature Difference (LMTD)

In heat transfer calculations, due to varying temperatures it is customary to use a slightly different concept of temperature difference called Logarithmic Mean Temperature Difference (LMTD). The LMTD between the liquid in the pipeline and the surrounding medium is calculated as follows:

Consider a pipeline segment of length $\Delta x$ with liquid temperatures $T_1$ at the upstream end and $T_2$ at the downstream end of the segment. If $T_s$ represents the average soil temperature (or ambient air temperature, if above ground pipeline) surrounding this pipe segment, the Logarithmic Mean Temperature of the pipe ($T_m$) segment is calculated as follows:

$$T_m - T_S = \frac{(T_1 - T_S) - (T_2 - T_S)}{Log_e \left[(T_1 - T_S)/(T_2 - T_S)\right]} \tag{9.7}$$

where

$T_m$    - Logarithmic mean temperature of pipe segment, °F
$T_1$    - Temperature of liquid entering pipe segment, °F
$T_2$    - Temperature of liquid leaving pipe segment, °F
$T_S$    - Sink temperature(soil or surrounding medium), °F

In SI units, Equation (9.7) will be the same, with all temperatures expressed in °C instead of °F.

For example, if the average soil temperature is 60°F and the temperature of a pipe segment upstream and downstream are 160 °F and 150 °F respectively, the Logarithmic Mean Temperature of the pipe segment is:

$$T_m = 60 + \frac{(160 - 60) - (150 - 60)}{Log_e \left[(160 - 60)/(150 - 60)\right]} = 60 + 94.88 = 154.88°F$$

Thus we calculated the Logarithmic Mean Temperature of the pipe segment to be 154.88°F. If we had used a simple arithmetic average we would get the following for mean temperature of pipe segment

Arithmetic Mean temperature = (160+150)/2 = 155°F

This is not too far off from the Logarithmic Mean Temperature $T_m$ calculated above. It can be seen that the Logarithmic Mean Temperature approach gives a slightly more accurate representation of the average liquid temperature in the pipe segment. Note that the use of natural Logarithm in Equation (9.7) signifies an exponential decay of the liquid temperature in the pipeline segment. In this example, the LMTD for the pipe segment is

LMTD = 154.88 - 60 = 94.88°F

If we assume an overall heat transfer coefficient U = 0.5 Btu/hr/ft²/°F, we can estimate the heat flux from this pipe segment to the surrounding soil using Equation (9.4) as follows:

Heat flux = 0.5 x 1 x 94.88 = 47.44 Btu/hr per ft² of pipe area.

### 9.2.5   Heat Entering and Leaving Pipe Segment

The heat content of the liquid entering and leaving a pipe segment is calculated using the mass flow rate of the liquid, its specific heat and the temperatures at the inlet and outlet of the segment. The heat content of the liquid entering the pipe segment is calculated from

$$H_{in} = w(C_{pi})(T_1) \tag{9.8}$$

The heat content of the liquid leaving the pipe segment is calculated from

$$H_{out} = w(C_{po})(T_2) \tag{9.9}$$

where

$H_{in}$    - Heat content of liquid entering pipe segment, Btu/hr
$H_{out}$    - Heat content of liquid leaving pipe segment, Btu/hr
$C_{pi}$    - Specific heat of liquid at inlet, Btu/lb/°F
$C_{po}$    - Specific heat of liquid at outlet, Btu/lb/°F
w      - Liquid flow rate, lb/hr
$T_1$      - Temperature of liquid entering pipe segment, °F
$T_2$      - Temperature of liquid leaving pipe segment, °F

The specific heat $C_p$ of most liquid range between 0.4 to 0.5 Btu/lb/ °F (0.84 to 2.09 kJ/kg/ °C) and increases with liquid temperature. For petroleum fluids $C_p$ can be calculated if the specific gravity or API gravity and temperatures are known

In SI units Equations (9.8) and (9.9) become

$$H_{in} = w(C_{pi})(T_1) \tag{9.10}$$

$$H_{out} = w(C_{po})(T_2) \hspace{4cm} (9.11)$$

where

| | |
|---|---|
| $H_{in}$ | - Heat content of liquid entering pipe segment, J/s (W) |
| $H_{out}$ | - Heat content of liquid leaving pipe segment, J/s (W) |
| $C_{pi}$ | - Specific heat of liquid at inlet, kJ/kg/ °C |
| $C_{po}$ | - Specific heat of liquid at outlet, kJ/kg/ °C |
| w | - Liquid flow rate, kg/s |
| $T_1$ | - Temperature of liquid entering pipe segment, °C |
| $T_2$ | - Temperature of liquid leaving pipe segment, °C |

## 9.2.6   Heat Transfer - Buried Pipeline

Consider a buried pipeline, with insulation, that transports a heated liquid. If the pipeline is divided into segments of length L we can calculate the heat transfer between the liquid and the surrounding medium using the following equations:

In English units,

$$H_b = 6.28 \ (L) \ (T_m - T_s) \ / \ (Parm1 + Parm2) \hspace{1.5cm} (9.12)$$
$$Parm1 = (1/K_{ins}) \ Log_e(R_i \ /R_p) \hspace{2.6cm} (9.13)$$
$$Parm2 = (1/K_s) \ Log_e[2S/D + ((2S/D)^2 - 1)^{1/2}] \hspace{1cm} (9.14)$$

where

| | |
|---|---|
| $H_b$ | - Heat transfer, Btu/hr |
| $T_m$ | - Log mean temperature of pipe segment, °F |
| $T_s$ | - Ambient soil temperature, °F |
| L | - Pipe segment length, ft |
| $R_i$ | - Pipe insulation outer radius, ft |
| $R_p$ | - Pipe wall outer radius, ft |
| $K_{ins}$ | - Thermal conductivity of insulation, Btu/hr/ft/ °F |
| $K_s$ | - Thermal conductivity of soil, Btu/hr/ft/ °F |
| S | - Depth of cover (pipe burial depth) to pipe centerline, ft |
| D | - Pipe outside diameter, ft |

Parm1 and Parm2 are intermediate values that depend on parameters indicated.

In SI units, Equations (9.12), (9.13) and (9.14) become:

$$H_b = 6.28 \ (L) \ (T_m - T_s) \ / \ (Parm1 + Parm2) \tag{9.15}$$
$$Parm1 = (1/K_{ins}) \ Log_e(R_i / R_p) \tag{9.16}$$
$$Parm2 = (1/K_s) \ Log_e[2S/D + ((2S/D)^2 - 1)^{1/2}] \tag{9.17}$$

where

$H_b$ - Heat transfer, W
$T_m$ - Log mean temperature of pipe segment, °C
$T_s$ - Ambient soil temperature, °C
L - Pipe segment length, m
$R_i$ - Pipe insulation outer radius, mm
$R_p$ - Pipe wall outer radius, mm
$K_{ins}$ - Thermal conductivity of insulation, W/m/ °C
$K_s$ - Thermal conductivity of soil, W/m/ °C
S - Depth of cover (pipe burial depth) to pipe centerline, mm
D - Pipe outside diameter, mm

### 9.2.7   Heat Transfer - Above Ground Pipeline

Similar to the buried pipeline discussed in the earlier section, an above ground insulated pipeline is used to transports a heated liquid. If the pipeline is divided into segments of length L we can calculate the heat transfer between the liquid and the ambient air using the following equations.

In English units

$$H_a = 6.28 \ (L) \ (T_m - T_s) \ / \ (Parm1 + Parm3) \tag{9.18}$$
$$Parm3 = 1.25/[R_i \ (4.8 + 0.008(T_m - T_s))] \tag{9.19}$$
$$Parm1 = (1/K_{ins}) \ Log_e(R_i / R_p) \tag{9.20}$$

where

H$_a$  - Heat transfer, Btu/hr
T$_m$  - Log mean temperature of pipe segment, °F
T$_s$  - Ambient soil temperature, °F
L    - Pipe segment length, ft
R$_i$  - Pipe insulation outer radius, ft
R$_p$  - Pipe wall outer radius, ft
K$_{ins}$ - Thermal conductivity of insulation, Btu/hr/ft/ °F
K$_s$  - Thermal conductivity of soil, Btu/hr/ft/ °F
S    - Depth of cover (pipe burial depth) to pipe centerline, ft
D   - Pipe outside diameter, ft

In SI units, Equations (9.18), (9.19) and (9.20) become

$$H_a = 6.28 \, (L) \, (T_m - T_s) \, / \, (Parm1 + Parm3) \qquad (9.21)$$
$$Parm3 = 1.25/[R_i \, (4.8 + 0.008(T_m - T_s))] \qquad (9.22)$$
$$Parm1 = (1/K_{ins}) \, Log_e(R_i \, /R_p) \qquad (9.23)$$

where

H$_a$  - Heat transfer, W
T$_m$  - Log mean temperature of pipe segment, °C
T$_s$  - Ambient soil temperature, °C
L    - Pipe segment length, m
R$_i$  - Pipe insulation outer radius, mm
R$_p$  - Pipe wall outer radius, mm
K$_{ins}$ - Thermal conductivity of insulation, W/m/ °C
K$_s$  - Thermal conductivity of soil, W/m/ °C
S    - Depth of cover (pipe burial depth) to pipe centerline, mm
D   - Pipe outside diameter, mm

## 9.2.8   Frictional Heating

The frictional pressure drop causes heating of the liquid. The heat gained by the liquid due to friction is calculated using the following equations:

$$H_w = 2\,545\,(HHP) \tag{9.24}$$
$$HHP = (1.7664 \times 10^{-4})\,(Q)(Sg)\,(h_f)(L_m) \tag{9.25}$$

where

$H_w$     - Frictional Heat gained, Btu/hr
HHP - Hydraulic horsepower required for pipe friction
Q      - Liquid flow rate, bbl/hr
Sg     - Liquid specific gravity
$h_f$      - Frictional head loss, ft/mi
$L_m$     - Pipe segment length, mi

In SI units, Equations (9.24) and (9.25) become

$$H_w = 1\,000\,(Power) \tag{9.26}$$
$$Power = (0.00272)\,(Q)(Sg)((h_f)(L_m) \tag{9.27}$$

where

$H_w$     - Frictional Heat gained, W
Power - Power required for pipe friction, kW
Q      - Liquid flow rate, m³/hr
Sg     - Liquid specific gravity
$h_f$      - Friction loss, m/km
$L_m$     - Pipe segment length, km

## 9.2.9    Pipe Segment Outlet Temperature

Using the formulas developed in the preceding sections and referring to the heat balance Equation (9.6), we can now calculate the temperature of the liquid at the outlet of the pipe segment as follows:

For buried pipe:

$$T_2 = (1/wC_p)[2\,545\,(HHP) - H_b + (wC_p)T_1] \tag{9.28}$$

For above ground pipe:

$$T_2 = (1/wC_p)[2\ 545\ (HHP) - H_a + (wC_p)T_1] \qquad (9.29)$$

where

$H_b$    - Heat transfer for buried pipe, Btu/hr from Equation (9.12)
$H_a$    - Heat transfer for above ground pipe, Btu/hr from Equation
     (9.18)
$C_p$    - Average specific heat of liquid in pipe segment

For simplicity, we have used the average specific heat above for the pipe segment based on $C_{pi}$ and $C_{po}$ discussed earlier in Equations (9.8) and (9.9).

In SI units, Equations (9.28) and (9.29) can be expressed as

For buried pipe:

$$T_2 = (1/wC_p)[1\ 000\ (Power) - H_b + (wC_p)T_1] \qquad (9.30)$$

For above ground pipe:

$$T_2 = (1/wC_p)[1\ 000\ (Power) - H_a + (wC_p)T_1] \qquad (9.31)$$

where

$H_b$    - Heat transfer for buried pipe, W
$H_a$    - Heat transfer for above ground pipe, W
Power - Frictional Power defined in Equation (9.27), kW

## 9.2.10   Liquid Heating due to Pump Inefficiency

Since a centrifugal pump is not 100% efficient, the difference between the hydraulic horsepower and the brake horsepower represents power lost. Most of this power lost is converted to heating the liquid being pumped. The temperature rise of the liquid due to pump inefficiency may be calculated from the following equation

$$\Delta T = (H / 778 \, C_p) (1 / E - 1) \qquad (9.32)$$

where

$\Delta T$    - Temperature rise, °F
H    - Pump head, ft
$C_p$    - Specific heat of liquid, Btu/lb/ °F
E    - Pump efficiency as a decimal value, less than 1.0

When considering thermal hydraulics the above temperature rises as the liquid moves through a pump station, should be included in the temperature profile calculation. For example, if the liquid temperature has dropped to 120°F at the suction side of a pump station and the temperature rise due to pump inefficiency causes 3°F rise, the liquid temperature at the pump discharge will be 123°F.

**Example Problem 9.1**

Calculate the temperature rise of a liquid (specific heat = 0.45 Btu/ lb/ °F) as it flows through a pump, due to pump inefficiency.

Pump head = 2 450 ft and Pump efficiency = 75%

**Solution**

From Equation (9.32), the temperature rise is

$$\Delta T = (2\,450 / (778 \times 0.45)) (1 / 0.75 - 1) = 2.33°F$$

We will now use the equations discussed in this chapter to calculate the thermal hydraulic temperature profile of a crude oil pipeline.

**Example Problem 9.2**

A 16 inch, 0.250 inch wall thickness, 50 mile long buried pipeline transports 4 000 bbl/hr of heavy crude oil that enters the pipeline at 160°F. The crude oil has a specific gravity and viscosity as follows:

| Temperature (°F) | 100 | 140 |
|---|---|---|
| Specific gravity | 0.967 | 0.953 |
| Viscosity (cSt) | 2 277 | 348 |

Assume a pipe burial depth of 36 inches to the top of pipe and 1.5 inch insulation thickness with a thermal conductivity (K value) of 0.02 Btu/hr/ft/°F. Also assume a uniform soil temperature of 60°F with a K value of 0.5 Btu/hr/ft/°F. Using the heat balance equation calculate the outlet temperature of the crude oil at the end of the first mile segment. Assume average specific heat of 0.45 for the crude oil.

## Solution

First calculate the heat transfer for buried pipe using Equations (9.12) through (9.14)

$Parm1 = (1/0.02) \, Log_e \, (9.5 / 8) = 8.5925$

$Parm2 = (1/0.5) \, Log_e [2x \, 44 / 16 + ((2 \times 44/ 16)^2 - 1)^{1/2}] = 4.7791$

$H_b = 6.28 \, (5 \, 280) \, (T_m - 60) / (8.5925 + 4.7791)$

or

$$H_b = 2 \, 479.76 \, (T_m - 60) \, Btu/hr \tag{9.33}$$

The Log Mean Temperature $T_m$ of this one mile pipe segment has to be approximated first, since it depends on the inlet temperature, soil temperature and the unknown liquid temperature at the outlet of the one mile segment.

As a first approximation, assume outlet temperature at the end the first one mile segment to be $T_2 = 150$. Calculate $T_m$ using Equation (9.7)

$$T_m = 60 + \frac{(160 - 60) - (150 - 60)}{Log_e \, [(160 - 60)/(150 - 60)]} \tag{9.34}$$

Or

$T_m = 154.91°F$

Therefore, Hb from above Equation (9.32) becomes

$H_b$ = 2 479.76 (154.91- 60) = 235 354 Btu/hr

The frictional heating component $H_w$ will be calculated using Equation (9.24) and (9.25). The friction drop $h_f$ depends on specific gravity and viscosity at the calculated mean temperature $T_m$. Using the viscosity temperature relationship from Chapter 2, we calculate the specific gravity and viscosity at 154.91°F to be 0.9478 and 200.22 cSt respectively.

Reynold's number

$$R = \frac{92.24 \times (4\ 000 \times 24)}{15.5 \times 200.22} = 2\ 853$$

Using Colebrook-White equation, the friction factor is

f = 0.034

Friction drop $h_f$ will be calculated from Darcy-Weisbach Equation (3.26) as follows:

$h_f$ = 0.034 (5 280 x 12 / 15.5)($V^2$/ 64.4)

The velocity V is calculated using Equation (3.12) as follows

V = 0.2859 (4 000) / $(15.5)^2$ = 4.76 ft/s

Therefore, friction pressure drop is

$h_f$ = 0.034 (5 280 x 12 / 15.5)(4.76 x 4.76 / 64.4) = 48.89 ft

From Equation (9.25) the frictional HP is

HHP = (1.7664 x$10^{-4}$) x 4 000 x 0.9478 x 48.89 x 1.0 = 32.74

Therefore frictional heating from Equation (9.24) is

$H_w = 2\,545 \times 32.74 = 83\,323$ Btu/hr

The mass flow rate is

$w = 4\,000 \times 5.6146 \times 0.9478 \times 62.4 = 1.328 \times 10^6$ lb/hr

From Equation (9.30) the liquid temperature at the outlet of the one mile segment is

$T_2 = (1/(1.328 \times 10^6 \times 0.45))\,[83\,323 - 235\,354 + 1.328 \times 10^6 \times 0.45 \times 160]$
$T_2 = [-0.255 + 160] = 159.75\ °F$

This value of $T_2$ is used as a second approximation in Equation (9.34) to calculate a new value of $T_m$ and subsequently the next approximation for $T_2$. Calculations are repeated until successive values of $T_2$ are within close agreement. This is left as an exercise for the reader.

It can be seen from the foregoing, that manual calculation of temperatures and pressures along a heated oil pipeline is definitely a laborious process, that can be eased using programmable calculators and personal computers.

Thermal hydraulics is very complex and calculations require utilization of some type of computer program to generate quick results. Such a program can sub-divide the pipeline into short segments and calculate the temperatures, liquid properties and pressure drops as we have seen in the examples in this chapter. Several software packages are commercially available to perform thermal hydraulics. One such software is LIQTHERM developed by SYSTEK Technologies, Inc. (www. systek.us). For a sample output report from a liquid pipeline thermal hydraulics analysis using LIQTHERM software, refer to Appendix A.15.

## 9.3  Summary

We explored thermal effects of pipeline hydraulics in this chapter. To transport viscous liquids they have to be heated to a temperature, sometimes much higher than the ambient conditions. This temperature differential between the pumped liquid and the surrounding soil (buried pipeline) or ambient air (above ground pipeline) causes heat transfer to occur resulting in temperature variation of the liquid along the pipeline. Unlike isothermal flow, where the liquid temperature is uniform throughout the pipeline, heated pipeline hydraulics requires subdividing the pipeline into short segments and calculating pressure drops based on liquid properties at the average temperature of each segment. We illustrated this, using an example that showed how the temperature varies along the pipeline. The concept of LMTD was introduced for determining a more accurate average segment temperature. Also, a method to compute the heat transfer between the liquid and the surrounding medium was shown taking into account thermal conductivities of pipe, soil and insulation, if present. The heating of liquid due to friction was also quantified as was the liquid heating associated with pump inefficiency.

It was pointed out that unlike isothermal hydraulics, thermal hydraulics is a complex phenomenon that requires computer methods to correctly solve equations for temperature variation and pressure drop calculations.

## 9.4  Problems

9.4.1   An 8 inch nominal diameter pipe is used to move heavy crude oil from a heated storage tank to a refinery 12 miles away. The inlet temperature is 180°F and the crude oil has the following characteristics:

| Temperature (°F) | 100 | 180 |
|---|---|---|
| Specific gravity | 0.985 | 0.912 |
| Viscosity (cSt) | 4 000 | 25 |

If the outlet temperature cannot drop below 150 °F, what minimum flow rate needs to be maintained in the pipeline? Use MIT Equation and 0.002 inch roughness, 70°F soil temperature, 36 inch burial depth and 0.25 inch insulation with K = 0.02 Btu/hr/ft/ °F. K value for soil = 0.6 Btu/hr/ft/ °F. Pipeline pressures are limited to ANSI 600 rating (1 440 psi)

9.4.2    In the previous problem if insulation were absent, what minimum flow rate can be tolerated? Compare HP requirements in both cases. Assume 15 psi pump suction pressure.

9.4.3    In Problem 9.4.1 develop a set of data points for flow rate versus pressure required for plotting a system head curve. Assume 50 psi delivery pressure.

# 10

# Flow Measurement

In this chapter we will discuss the various methods and instruments used in the measurement of liquid that flows through a pipeline. The formulas used for calculating the liquid velocities, flow rates, etc from the pressure readings, their limitations and the degree of accuracy attainable will be covered for some of the more commonly used instruments. These days, considerable work is being done in this field of flow measurement to improve the accuracy of instruments, particularly when custody transfer of products is involved. For more detailed analysis of the various flow measurement devices, the reader is referred to the publications listed in the Reference section.

## 10.1 History

Measurement of flow of liquids and solids have been going on for centuries. In the early days, the Romans used some form of flow measurement to allocate water from the aqueducts to the houses in their cities. This was necessary to control the quantity of water used by the citizens and prevent waste. Similarly, it is reported that the Chinese used to measure salt water to brine pots that were used for salt production. In later years, a commodity had to be measured so that it may be properly allocated and the ultimate user charged for it appropriately. Today, gasoline is dispensed from meters at gas stations and the recipient is billed according to the volume of gasoline so measured. Water companies measure water consumed by a household using water meters, while natural gas for residential and industrial consumers are measured by gas meters. In all these instances, the objective is to ensure that the supplier gets paid for the commodity and the user of the commodity is billed for the product at the agreed upon price. In addition, in industrial processes that involve the use of liquids and gases to perform a specific function require accurate quantities to be dispensed so that the desired effect of the processes may be realized. In most cases involving consumers, certain regulatory or public agencies (for example, the department of weights and measures) periodically check flow measurement devices to ensure that they performing accurately and if necessary to calibrate them against a very accurate master device.

## 10.2 Flow Meters

Several types of instruments are available to measure the flow rate of a liquid in a pipeline. Some measure the velocity of flow, while others directly measure the volume flow rate or the mass flow rate.

The following flow meters are used in the pipeline industry

- Venturi meter
- Flow nozzle
- Orifice meter
- Flow tube
- Rotameter
- Turbine meter
- Vortex flowmeter
- Magnetic flowmeter
- Ultrasonic flowmeter
- Positive displacement meter
- Mass flowmeter
- Pitot tube

The first four items listed above are called *variable head meters* since the flow rate measured is based on the pressure drop due to a restriction in the meter, which varies with the flow rate. The last item, the Pitot tube is also called a velocity probe, since it actually measures the velocity of the liquid. From the measured velocity, we can calculate the flow rate using the conservation of mass equation, discussed in an earlier chapter.

Mass flow = (Flow rate) x (density)
= (Area) x (Velocity) x (density)

Or

$$M \quad = Q\rho \ = AV\rho \qquad (10.1)$$

Since the liquid density is practically constant, we can state that:

$$Q = AV \qquad (10.2)$$

where

A  - cross sectional area of flow
V  - velocity of flow
$\rho$  - liquid density

We will discuss the principle of operation and the formulas used for some of the more common meters described above. For a more detailed discussion on flow meters and flow measurement, the reader is referred to one of the fine texts listed in the Reference section of this book.


## 10.3 Venturi Meter

The venturi meter, also known as venturi tube, belongs to the category of variable head flowmeters. The principle of a venturi meter is depicted in Figure 10.1. This type of a venturi meter is also known as the Herschel type and it consists of a smooth gradual contraction from the main pipe size to the throat section, followed by a smooth, gradual enlargement from the throat section to the original pipe diameter.

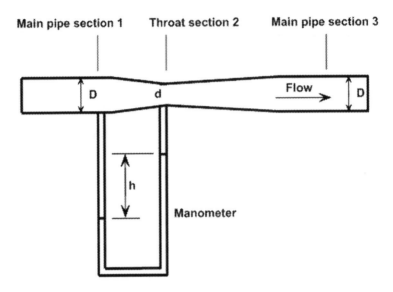

**Figure 10.1 Venturi meter**

The included angle from the main pipe to the throat section in the gradual contraction is generally in the range of 21° +/- 2°. Similarly, the gradual expansion from the throat to the main pipe section is limited to a range of 5° to 15° in this design of venturi meter. This construction results in minimum energy loss, causing the discharge coefficient (discussed later) to approach the value of 1.0. This type of a venturi meter is generally rough cast with a pipe diameter range of 4.0 inch to 48.0 inch. The throat diameter may vary quite a bit, but the ratio of the throat diameter to the main pipe diameter (d/D), also known as the *beta ratio*, represented by the symbol $\beta$, should range between 0.30 and 0.75.

The venturi meter consists of a main piece of pipe which decreases in size to a section called the throat, followed by a gradually increasing size back to the original pipe size. The liquid pressure at the main pipe section 1 is denoted by $P_1$ and that at the throat section 2 is represented by $P_2$. As the liquid flows through the narrow throat section, it accelerates to compensate for the reduction in area, since the volume flow rate is constant (Q = AV from Equation 10.2). As the velocity increases in the throat section, the pressure decreases (bernoulli's equation). As the flow continues past the throat, the gradual increase in the flow area results in reduction of flow velocity back to the level at section 1, correspondingly the liquid pressure will increase to some value higher than at the throat section.

We can calculate the flow rate using the Bernoulli's equation, introduced in Chapter 2, along with the Continuity equation, based on the Conservation of Mass as shown in Equation 10.1 earlier.

If we consider the Main pipe section 1 and the Throat section 2 as reference, and apply Bernoulli's equation for these two sections, we get

$$P_1/\gamma + V_1^2/(2g) + Z_1 = P_2/\gamma + V_2^2/(2g) + Z_2 + h_L \tag{10.3}$$

And from the Continuity equation

$$Q = A_1 V_1 = A_2 V_2 \tag{10.4}$$

Where

$\gamma$ is the specific weight of liquid and $h_L$ represents the pressure drop due to friction between section 1 and section 2.

Simplifying the above equations, yields the following:

$$V_1 = \sqrt{[2g(P_1 - P_2)/\gamma + (Z_1 - Z_2) - h_L] / [(A_1/A_2)^2 - 1]} \tag{10.5}$$

The elevation difference $Z_1 - Z_2$ is negligible even if the venturi meter is positioned vertically. We will also drop the friction loss term $h_L$ and include it in a coefficient C, called the Discharge Coefficient, and rewrite Equation 10.5 as follows:

$$V_1 = C \sqrt{[2g(P_1 - P_2)/\gamma] / [(A_1/A_2)^2 - 1]} \tag{10.6}$$

The above equation gives us the velocity of the liquid in the main pipe section 1. Similarly, the velocity in the throat section, $V_2$ can be calculated using Equation 10.4 as follows:

Velocity in throat,

$$V_2 = C \sqrt{[2g(P_1 - P_2)/\gamma] / [1- (A_2/A_1)^2]} \tag{10.7}$$

The volume flow rate Q can now be calculated using Equation 10.4 as follows

$$Q = A_1 V_1$$

Therefore,

$$Q = CA_1 \sqrt{[2g(P_1 - P_2)/\gamma] / [(A_1/A_2)^2 - 1]} \tag{10.8}$$

Since the beta ratio $\beta-d/D$ and $A_1/A_2 = (D/d)^2$

We can write the above equations in terms of the beta ratio $\beta$ as follows:

$$Q = CA_1 \sqrt{[2g(P_1 - P_2)/\gamma] / [(1/\beta)^4 - 1]} \tag{10.9}$$

It can be seen by examining Equations 10.5, 10.6 and 10.7 that the discharge coefficient C actually represents the ratio of the velocity of liquid through the venturi to the ideal velocity when the energy loss($h_L$) is zero. C is therefore be a number less than 1.0. The value of C depends on the Reynolds number in the main pipe section 1 and is shown graphically in Figure 10.2 Reynolds number greater than $2x10^5$ the value of C remains constant at 0.984

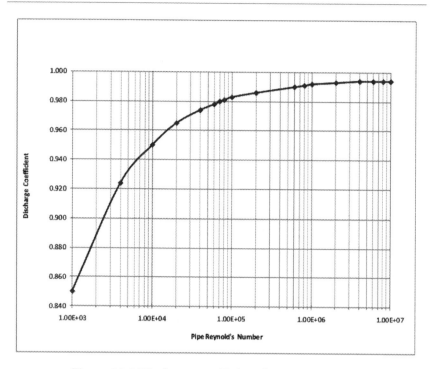

**Figure 10.2 Discharge coefficient for Venturi meter**

For smaller size piping, 2 inch to 10 inch size, venturi meters are machined and hence have a better surface finish than the larger, rough cast meters. These smaller venturi meters have a C value of 0.995 for Reynolds number greater than $2x10^5$.

## 10.4 Flow Nozzle

A typical *flow nozzle* is illustrated in Figure 10.3. It consists of the main pipe section 1, followed by a gradual reduction in flow area and a subsequent short cylindrical section, and finally an expansion to the main pipe section 3 as shown in the figure.

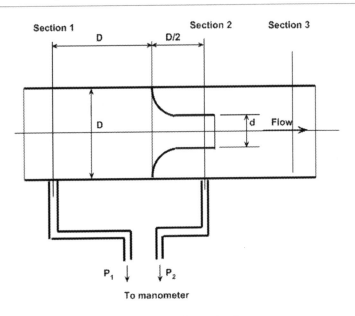

**Figure 10.3 Flow Nozzle**

The American Society of Mechanical Engineers (ASME) and the International Standards Organization (ISO) have defined the geometries of these flow nozzles and published equations to be used with the flow nozzles. Due to the smooth gradual contraction from the main pipe diameter D to the nozzle diameter d, the energy loss between the sections 1 and 2 is very small. We can apply the same Equations 10.6 through 10.8 for the venturi meter as well as the flow nozzle. The discharge coefficient C for the flow nozzle is found to be 0.99 or better, for Reynold's numbers above $10^6$. At lower Reynold's numbers, there is greater energy loss immediately following the nozzle throat, due to sudden expansion, and hence C values are lower.

Depending on the beta ratio and the Reynold's number, the discharge coefficient C can be calculated from the equation,

$$C = 0.9975 - 6.53 \sqrt{(\beta/R)} \tag{10.10}$$

where

$$\beta - d/D$$

and

R is the Reynold's number based on the pipe diameter D.

Compared to the venturi meter, the flow nozzle is more compact in size, since it does not require the length for gradual decrease in diameter at the throat, and the additional length for the smooth, gradual expansion from the throat to the main pipe size. However, there is more energy loss (and therefore, pressure head loss) in the flow nozzle, due to the sudden expansion from the nozzle diameter to the main pipe diameter. The latter causes greater turbulence and eddies compared to the gradual expansion in the venturi meter.

## 10.5  Orifice Meter

An *orifice meter* consists of a flate plate that has a sharp edged hole accurately machined in it and placed concentrically in a pipe as shown in Figure 10.4. As liquid flows through the pipe, the flow suddenly contracts as it approaches the orifice and then suddenly expands after the orifice back to the full pipe diameter. This forms a *vena contracta* or a throat immediately past the orifice. This reduction in flow pattern at the vena contracta causes increased velocity and hence lower pressure at the throat, similar to the venturi meter, discussed earlier.

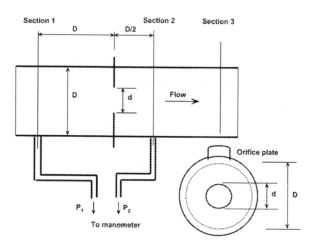

**Figure 10.4 Orifice meter**

The pressure difference between section 1, with the full flow and section 2 at the throat, can then be used to measure the liquid flow rate, using equations developed earlier for the venturi meter and the flow nozzle. Due to the sudden contraction at the orifice and the subsequent sudden expansion after the orifice, the coefficient of discharge C for the orifice meter is much lower than that of a venturi meter or a flow nozzle. In addition, depending on the pressure tap locations, section 1 and section 2, the value of C is different for orifices.

There are three possible pressure tap locations for an orifice meter as listed below:

### Table 10.1 - Pressure Taps for Orifice

| | Inlet pressure tap, P1 | Outlet pressure tap, P2 |
|---|---|---|
| 1 | One pipe diameter upstream from plate face of plate | One-half pipe diameter downstream of inlet |
| 2 | One pipe diameter upstream from plate | At vena contracta |
| 3 | Flange taps, one-inch upstream from plate outlet face of plate | Flange taps, one-inch downstream from |

Figure 10.5 shows the variation of C with the beta ratio d/D for various values of pipe Reynold's number.

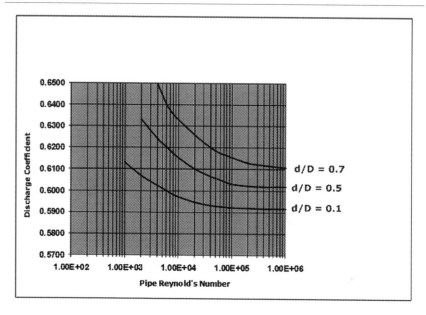

**Figure 10.5 Orifice Meter Discharge Coefficient**

Comparing the three types of flow meters discussed above, we can conclude that the orifice plate has the highest energy loss due to the sudden contraction followed by the sudden expansion. On the other hand, the venturi meter has a lower energy loss compared to flow nozzle due to the smooth, gradual reduction at the throat, followed by the smooth gradual expansion after the throat.

Flow tubes are proprietary varable head flow meters that are streamlined in design to cause the least amount of energy loss. The flow tube is also a variable head type flow meter, that is manufactured by different companies with their own special, proprietary designs.

## 10.6 Turbine meter

Turbine meters are used in a wide variety of applications. The food industry uses turbine meters for measurement of milk, cheese, cream, syrups, vegetable oils, etc. Turbine meters are also used in the oil industry.

Basically the turbine meter is a velocity measuring instrument. Liquid flows through a free turning rotor that is mounted co-axially inside the meter. Upstream and downstream of the meter, certain amount of straight piping lengths are required to ensure smooth velocities through the meter. The liquid striking the rotor causes it to rotate, the velocity of rotation being proportional to the flow rate. From the rotor speed measured and the flow area, we can compute the flow rate through the meter. The turbine meter must be calibrated since flow depends on the friction, turbulence and the manufacturing tolerance of the rotor parts.

For liquids with viscosity close to that of water the range of flow rates is 10:1. With higher and lower viscosity liquids it drops to 3:1. Density effect is similar to viscosity.

A turbine meter requires straight length of pipe before and after the meter. The straightening vanes located in the straight lengths of pipe helps eliminate swirl in flow. The upstream length of straight pipe has to be 10 D in length, where D is the pipe diameter. After the meter, the straight length of pipe is 5D long. Turbine meters may be two-section type or three-section types, depending upon the number of pieces the meter assembly is composed of. For liquid flow measurement, API Manual of Petroleum Measurement Standard must be followed. In addition, bypass piping must also be installed to isolate the meter for maintenance and repair. To maintain flow measurements on a continuous basis for custody transfer, an entire spare meter unit will be required on the bypass piping, when the main meter is taken out of service for testing and maintenance.

From the turbine meter reading at flowing temperature, the flow rate at some base temperature (such as 60°F) is calculated using the following equation:

$$Q_b = Q_f \times M_f \times F_t \times F_p \qquad (10.11)$$

where

$Q_b$     - Flow rate at base conditions, such as 60°F and 14.7 psi

$Q_f$     - Measured Flow rate at operating conditions, such as 80°F and 350 psi

$M_f$     - Meter factor for correcting meter reading, based on meter calibration data.

$F_t$  - Temperature correction factor for correcting from flowing temperature to the base temperature.

$F_p$  - Pressure correction factor for correcting from flowing Pressure to the base pressure.

## 10.7 Positive Displacement meter

The positive displacement meter, also known as PD meter is most commonly used in residential applications for measuring water and natural gas consumption. PD meters can measure small flows with good accuracy. These meters are used when consistently high accuracy is required under steady flow through a pipeline. PD meters are accurate within +/- 1% over flow range of 20: 1. They are suitable for batch operations, mixing and blending of liquids that are clean, without deposits and with non-corrosive liquids.

PD meters basically operate by measuring fixed quantities of the liquid in *containers* that are alternatively filled and emptied. The individual volume are then totaled and displayed. As the liquid flows through the meter, the force from the flowing stream causes the pressure drop in the meter, which is a function of the internal geometry. The PD meter does not require the straightening vanes as with a turbine meter and hence is a more compact design. However PD meters can be large and heavy compared to an equivalent turbine meter. Also, meter jamming can occur when stopped and restarted. This necessitates some form of bypass piping with valves to prevent pressure rise and damage to meter.

The accuracy of a PD meter depends on clearance between the moving and stationary components. Hence precision machined parts are required to maintain accuracy. They are not suitable for slurries or liquids with suspended particles that could jam the components and cause damage to the meter. PD meters include several types, such as reciprocating piston type, rotating disk, rotary piston, sliding vane and rotary vane type.

Many modern petroleum installations employ PD meters for crude oil and refined products. These have to be calibrated periodically, to ensure accurate flow measurement. Most PD meters used in the oil industry are tested at regular intervals, such as daily or weekly, using a master meter, called a *meter prover* system that is installed as a fixed unit along with the PD meter.

### Example problem 10.1

A venturi meter is used for measuring the flow rate of water at 70 °F. The flow enters a 6.625 inch pipe, 0.250 inch wall thickness and the throat diameter is 2.4 inch. The venturi is of the Herschel type and is rough cast with a mercury manometer. The manometer reading shows a pressure difference of 12.2 inch of mercury. Calculate the flow velocity in the pipe and the volume flow rate. Use specific gravity of 13.54 for mercury.

### Solution:

At 70 °F, the specific weight and viscosity are:

$\gamma = 62.3$ lb/ft$^3$ and $v = 1.05$ x $10^{-5}$ ft$^2$/s

Assume first that the Reynold's number is greater than $2x\ 10^5$

Therefore, the discharge coefficient C will be 0.984 from Figure 10.2

The beta ratio $\beta = 2.4/(6.625-.5) = 0.3918$

and therefore $\beta$ is between 0.3 and 0.75.

$A_1/A_2 = (1/0.3918)^2 = 6.5144$

The manometer reading will give us the pressure difference, by equating the pressures at the two points in the manometer.

$P1 + \gamma_w(y + h) = P2 + \gamma_w y + \gamma_m h$

where y is depth of the higher mercury level below the center line of the venture and $\gamma_w$ and $\gamma_m$ are the specific weights of water and mercury respectively.

Simplifying the above, we get:

$$(P1 - P2)/\gamma_w = (\gamma_m/\gamma_w - 1)\, h$$
$$= (13.54*62.4/62.3 - 1)x\ 12.2/12 = 12.77\ ft$$

And using Equation 10.6, we get

$$V1 = 0.984\ \sqrt{[(64.4x12.77)/(6.5144^2 - 1)]}$$
$$= 4.38\ ft/s$$

Now we will check the value of Reynolds number:

$$R = 4.38\ x\ (6.125/12)/\ 1.05\ x\ 10^{-5}$$
$$= 2.13\ x10^5$$

Since R is greater than $2\ x10^5$ our assumption for C is correct.

The volume flow rate is

$$Q = A_1 V_1$$

Flow rate $= 0.7854(6.125/12)^2\ x\ 4.38 = 0.8962\ ft^3/s$

Or

Flow rate $= 0.8962\ x\ (1728/231)\ x\ 60 = 402.25\ gal/min$

## Example problem 10.2

An orifice meter is used to measure the flow rate of kerosene in a 2 inch schedule 40 pipe at 70°F. The orifice is 1.0 inch diameter. Calculate the volume flow rate if the pressure difference across the orifice is 0.54 psi. Specific gravity of kerosene = 0.815. Viscosity = 2.14 x 10$^{-5}$ ft$^2$/s

**Solution:**

Since the flow rate depends on the discharge coefficient C, which in turn depends on the beta ratio and the Reynold's number, we will have to solve this problem by trial and error.

The 2 inch schedule 40 pipe is 2.375 inch outside diameter and 0.154 inch wall thickness.

The beta ratio

$$\beta = 1.0/(2.375 - 2 \times 0.154) = 0.4838$$

First, we assume a value for the discharge coefficient C (say 0.61) and calculate the flow rate from Equation 10.6. We will then calculate the Reynold's number and get a more correct value of C from Figure 10.5. Repeating the method a couple of times will yield a more accurate value of C and finally the flow rate.

The density of kerosene = $0.815 \times 62.3 = 50.77$ lb/ft$^3$

Ratio of areas

$$A_1/A_2 = (1/0.4838)^2 = 4.2717$$

Using Equation 10.6, we get

$$V1 = 0.61 \text{ sqrt } [(64.4 \times 0.54 \times 144/(50.77)/(4.2717^2 - 1)] = 1.4588 \text{ ft/s}$$

Reynold's number = $1.4588 \times (2.067/12)/ (2.14 \times 10^{-5)}$
                  = 11 742

Using this value of Reynold's number, we get from Figure 10.5, the discharge coefficient C = 0.612

Since we earlier assumed C = 0.61 we were not too far off.

Recalculating based on the new value of C:

V1 = 0.612 sqrt [(64.4x0.54x144/(50.77)/(4.2717² - 1)] = 1.46 ft/s

Reynold's number = 1.46 x(2.067/12)/ (2.14 x 10⁻⁵)
$$= 11\ 751$$

This is quite close to the previous value of the Reynold's number.

Therefore we will use the last calculated value for velocity

The volume flow rate is

$Q = A_1 V_1$

Flow rate = 0.7854(2.067/12)²x 1.46 = 0.034 ft³/s

Or

Flow rate = 0.034 x (1728/231)x60 = 15.27 gal/min

## 10.8 Summary

In this chapter we discussed the more commonly used devices for the measurement of liquid pipeline flow rates. The variable head flow meters, such as the venturi tube, flow nozzle, orifice meter and the equations for calculating the velocities and flow rate from the pressure drop were explained. The importance of the discharge coefficient and how it varies with the Reynold's number and the beta ratio were also discussed. A trial and error method for calculating the flow rate through an orifice meter was illustrated using an example. For a more detailed description and analysis of flow meters, the reader should refer to any of the standard texts used in the industry. Some of these are listed in the Reference section.

## 10.9 Problems

10.9.1   Water flow through a 4 inch internal diameter pipeline is measured using a venturi tube. The flow rate expected is 300 gal/min. Specific weight of water is 62.4 lb/ft³ and viscosity is 1.2 x 10⁻⁵ ft²/s

    (a) Calculate the upstream Reynold's number
    (b) If the pressure differential across the meter is 10 psi. Determine the dimensions of the meter.
    (c) Calculate the throat Reynold's number.

10.9.2   A 12 inch venturi with a 7 inch throat is used for measuring flow of diesel at 60°F. The differential pressure reads 70 inches of water. Calculate the volume and mass flow rates. Specific gravity of diesel is 0.85 and viscosity is 5 cSt.

10.9.3   An orifice meter with a beta ratio of 0.70 is used for measuring crude oil (0.895 specific gravity and 15 cSt viscosity). The pipeline is 10.75 inch diameter, 0.250 wall thickness and the line pressure is 250 psig. With pipe taps the differential pressure is 200 inches of water. Calculate the flow rate of crude oil.

10.9.4   Water flows through a venturi tube 300 mm main pipe diameter and 150 mm throat diameter respectively at 150 m³/hr. The differential manometer deflection is 106 cm. The specific gravity of the manometer liquid is 1.3. Calculate the discharge coefficient of the meter.

# 11

# Unsteady Flow in Pipelines

Throughout the last ten chapters we have addressed steady flow of liquids in pipelines. This means that the pipeline flow rates, velocities and pressures at any point along the pipeline do not change with time. In other words, *flow is not time dependant*. In realty this may not be true. Many factors affect pipeline operation. Even though volumes may be taken out of storage tanks at a constant rate and pump stations pump the liquid at a uniform rate there will always be some environmental or mechanical factors that could result in changes to the steady state flow in the pipeline. In this chapter we will briefly discuss unsteady flow and pipeline transients. A detailed analysis unsteady flow and pipeline transient pressures is beyond the scope of this book. The reader is advised to refer to more specialized texts on unsteady flow, listed in the Reference section.

## 11.1 Steady versus Unsteady Flow

In steady state flow, we assume the hydraulic gradient, representing the variation in pipeline pressures along the pipeline, to be fixed as long as the same liquid is being pumped at the same flow rate. Also, in steady state flow, we assume that liquids are incompressible. In reality, liquids *are* compressible, to some extent. The compressibility effect can cause *transients* or *surges* in otherwise steady state flow. If during the course of operation, different liquids enter the pipeline at different times, such as in a batched pipeline, the pressures, velocities and flow rates will vary with time, resulting in changes to the hydraulic pressure gradient.

Unsteady flow occurs when the flow rate, velocity and pressure at a point in a pipeline change with time. Examples of such pressure transients include the effect of opening and closing of valves, starting and stopping a pump, or starting and stopping pipeline injections or deliveries.

A basic example, of unsteady flow would be a situation when a downstream valve on a pipeline is suddenly closed as shown in Figure 11.1

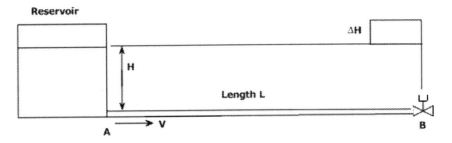

Figure 11.1 Transient with sudden valve closure

## 11.2 Transient Flow due to Valve Closure

To illustrate the effect of unsteady flow and transient pressure, we will consider a simple pipeline system from a storage tank at A, to a valve at the end of the pipe at B, as shown in Figure 11.1. The valve is located at a distance of L from the tank. The pipeline is assumed to be horizontal, with no elevation changes and uniform diameter D. In order to simplify the problem, we will further assume that the friction in the pipe is negligible. Later we will consider friction and its effect on transient flow. Since friction is ignored, the steady state hydraulic gradient line is horizontal.

Let us assume that initially, at time t = 0, the valve at B is fully open, steady state flow Q exists from A to B and the liquid velocity throughout the pipe is V. The tank head is assumed to be H. The velocity is assumed to be positive in the downstream direction from A to B. If nothing changed and the valve at B was left open long enough, the tank head H will be dissipated by flow through the valve.

When we suddenly close the valve at B, we introduce a transient or surge into the pipeline. As the valve closes, pressure, velocity and flow variations occur both upstream and downstream of the valve. Since we are only interested in the pipe section from A to B, we will ignore the effect downstream of the valve at B.

Closing the valve suddenly causes the velocity of flow at B to become zero instantaneously. This sudden drop in velocity at the valve, causes the pressure head at B to increase suddenly by some value $\Delta H$. This pressure rise is equal to the change in momentum of the liquid from a velocity of V to zero.

If the velocity changes from V to zero,

$$\Delta V = 0 - V = -V$$

The negative sign indicates a decrease in velocity

The magnitude of $\Delta H$ is given by

$$\Delta H = aV / g \tag{11.1}$$

where

| | |
|---|---|
| a | - Velocity of propagation of pressure wave, ft/s |
| V | - Velocity of liquid flow, ft/s |
| g | - Acceleration due to gravity, ft/s$^2$ |

If instead of closing the valve fully, we close it such that the velocity drops from V to $V_1$

$$\Delta V = V_1 - V$$

the corresponding head rise at the valve is

$$\Delta H = a (V_1 - V) / g \tag{11.2}$$

The pressure head increase ΔH at B causes a slight expansion of the pipe and also increase in liquid density, that depends upon the bulk modulus of the liquid, pipe size and pipe material. For most pipelines, this pipe expansion and increase in liquid density (decrease in liquid volume) will be insignificant. Depending upon liquid properties and pipe, the pressure increase propagates upstream from the valve at the speed of sound (wave speed), a, generally in the range of 2 000 ft/s to 5 000 ft/s. The wave traveling at the speed of sound reaches the tank in time L/a. At this point in time, the velocity of liquid throughout the pipe has become zero and correspondingly, the pressure everywhere has reached H + ΔH. The consequence of this is that the liquid has compressed somewhat and the pipe has stretched as well.

The above scenario is an unstable condition since the tank head is H, but the pipeline head is higher (H + ΔH). Due to this pressure differential, the liquid starts to flow *from* the pipeline *into* the tank with velocity -V. The velocity is thus changed from zero to -V which causes the pressure to drop from H + ΔH to H. Therefore, a negative pressure wave travels towards the valve B. Since friction is negligible, the magnitude of this reverse flow velocity is exactly equal to the original velocity V.

At time t = 2L/a, the pressure in the pipeline returns to the steady state value H, but reverse flow continues. However, since the valve is closed, no flow can take place upstream of the valve. Consequently, the pressure head drops by the amount ΔH forcing the reverse flow velocity to zero. This causes the pipe to shrink and the liquid expands.

At time t = 3L/a, this transient has reached the tank and the flow velocity is now zero throughout the pipe. The pipe pressure, however has dropped to a value ΔH below that of the tank. This causes flow from A to B equal to the original steady state flow. During this process the pipe pressure returns to the original steady state value.

Finally at time t = 4L/a, the pressure wave from A has reached the valve at B and the flow velocity reaches the original steady state value V. The total time elapsed is 4L/a and is defined as one wave cycle. Also known as the theoretical period of the pipeline. As the valve is completely closed, the preceding sequence of events starts again at t = 4 L/a. Since we assumed a frictionless system, the process continues indefinitely and the conditions are repeated at an interval of 4L/a.

However, in reality, due to pipe friction, the transient pressure waves are dissipated over a definite period of time and the liquid becomes stationary with no flow anywhere and pressure head equal to that of the tank.

Figure 11.2 shows a plot of the pressures at the valve at various times, beginning with the valve closure at time t = 0 to time t = 12L/a (three wave periods).

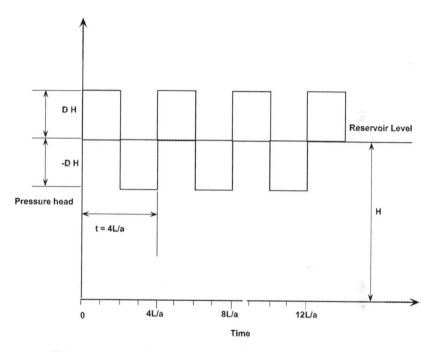

**Figure 11.2 Pressure Variation at Valve - Friction Neglected**

When friction losses are taken into consideration, the variation of pressure at the valve with time will be as shown in Figure 11.3

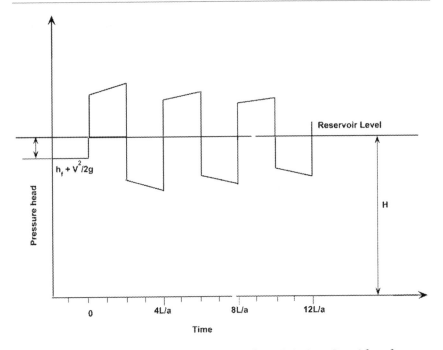

**Figure 11.3 Pressure Variation at Valve - Friction Considered**

It can be seen from Figure 11.3, due to pipe friction, the transient pressure waves will die down over a definite period of time and the liquid will reach equilibrium with zero flow throughout the pipeline with the pressure head exactly equal to the initial tank head.

As mentioned before, examples of unsteady flow include filling a pipeline, power failure and pump shut down and opening and closing of valves. In summary we can say that unsteady flow conditions occur when the pipeline flow rate continually varies with time or variations in flow between two steady state conditions occur. Turning on or shutting down a pump, changing a valve setting etc all cause unsteady flow. Unsteady flow conditions may be referred to as transient flow. Any change from a regular steady state flow condition can cause transients. Slow transients are called surges.

Depending on the magnitude of the transient and the rate at which the transient occurs, pipeline pressures may exceed steady state pressures and sometimes violate allowable maximum pressure in a pipeline. Design codes and federal regulations dictate that the pressure surge cannot exceed 110% of the Maximum Allowable Operating

Pressure (MAOP). Therefore, for a pipeline with an MAOP of 1 400 psi, during transient situations, the highest pressure reached anywhere in the pipeline cannot exceed 1 400 + 140 = 1 540 psi.

To control surge pressures and avoid pipeline overpressure, relief valves may be used, that are set to relieve the pipe pressure by allowing a certain amount of flow to a relief tank.

## 11.3 Wave Speed in Pipeline

We have seen from Equation (11.1) that the instantaneous head rise due to sudden stoppage of flow is

$$\Delta H = aV / g$$

The above equation can be derived easily considering a control element of flow in the tank - pipe example discussed in Section 11.2. However, we will not attempt to derive the above equation here. The reader is referred to one of the fine books on unsteady flow listed in the Reference section. At this point we will assume that the above equation is correct and proceed to apply it to some practical situations.

Since the transient head rise $\Delta H$ is directly dependent on the wave speed a from Equation (11.1), we must accurately determine its value for each pipeline.

It has been found that the wave speed in a liquid depends on the density and bulk modulus of the liquid. It depends also on the elasticity of the pipe material, pipe diameter and wall thickness. Also, if the liquid contains free air or gas, the wave speed will be affected by these.

An approximate value of the wave speed is calculated from

$$a = (K/\rho)^{1/2}$$

where

a      - Wave speed, ft/s
K      - Bulk modulus of liquid, psi
$\rho$      - Density of liquid, slugs/ft$^3$

However, the above value of wave speed does not take into account the the elastic nature of the pipe. Hence, the equation will be modified to account for the pipe material and its elastic properties, in the following discussion.

Consider an extreme case of a pipeline system where the pipe is considered rigid and liquid incompressible. Incompressible liquid means infinite wave speed, which is impossible. Since no pipe is totally rigid, the infinite wave speed situation is not possible.

According to Wylie and Streeter (Reference 14), a general equation for calculating the wave speed in terms of liquid properties, pipe size and elasticity of pipe material is as follows:

$$a = \frac{(K/\rho)^{1/2}}{[1+ C\,(K/E)\,(D/t)]^{1/2}} \qquad\qquad (11.3)$$

where

| | |
|---|---|
| a | - Wave speed, ft/s |
| K | - Bulk modulus of liquid, psi |
| $\rho$ | - Density of liquid, slugs/ft$^3$ |
| C | - Restraint factor, dimensionless |
| D | - Pipe outside diameter, in |
| t | - Pipe wall thickness, in |
| E | - Young's modulus of pipe material, psi |

An equivalent formula in SI units is

$$a = \frac{(K/\rho)^{1/2}}{[1+ C\,(K/E)\,(D/t)]^{1/2}} \qquad\qquad (11.4)$$

where

| | |
|---|---|
| a | - Wave speed, m/s |
| K | - Bulk modulus of liquid, kPa |
| $\rho$ | - Density of liquid, kg/m$^3$ |
| C | - Restraint factor, dimensionless |
| D | - Pipe outside diameter, mm |

t    - Pipe wall thickness, mm
E    - Young's modulus of pipe material, kPa

The Young's modulus is also referred to as the modulus of elasticity of the pipe material.

The restraint factor C depends on the type of pipe condition. The following three cases are possible:

Case1: Pipe is anchored at the upstream end only.
Case2: Pipe is anchored against any axial movements
Case3: Each pipe section anchored with expansion joints

The calculated value of the wave speed depends on the restraint factor C which in turn depends on the type of pipe support described above. It is found that the restraint type has less than 10% effect on the magnitude of wave speed.
The restraint factor C is defined in terms of the Poisson's ratio $\mu$ of the pipe material as (See Reference 14)

$$C = 1 - 0.5\mu \qquad \text{for Case 1} \qquad\qquad (11.5)$$

$$C = 1 - \mu^2 \qquad \text{for Case 2} \qquad\qquad (11.6)$$

$$C = 1.0 \qquad \text{for Case 3} \qquad\qquad (11.7)$$

Where $\mu$ = Poisson's ratio for pipe material, usually in the range of 0.20 to 0.45 and for steel pipe $\mu = 0.30$

It must be noted that the above values of the restraint factor C are applicable to thin-walled elastic pipes, with pipe diameter to thickness ratio (D/t) greater than 25. For thick-walled pipes with D/t ratio less than 25 the following C values may be used (Reference 14)

Case1

$$C = \frac{2t}{D}(1 + \mu) + \frac{D}{D+t}\frac{(1 - \mu)}{2}$$  (11.8)

Case2

$$C = \frac{2t}{D}(1 + \mu) + \frac{D(1 - \mu^2)}{D+t}$$  (11.9

Case3

$$C = \frac{2t}{D}(1 + \mu) + \frac{D}{D+t}$$  (11.10)

## Example Problem 11.1

Water flows through a steel pipe at velocity of 6 ft/s. A valve at the end of the pipe is partially closed and the velocity instantly reduces to 4 ft/s. Estimate the surge pressure rise at the valve. Use 3 000 ft/s wave speed for water in the pipe.

From Equation 11.1

Pressure rise = $aV/g$ = 3 000 x (6 - 4) / 32.2 = 186.33 ft

## Example Problem 11.2

A steel pipe carries water at 20 °C. The pipe diameter is 600 mm and 6 mm wall thickness. Assume Young's modulus E = 207 GPa and Poissons ratio = 0.3. Use 1 000 kg/m³ density of water and bulk modulus K = 2.2 GPa. Calculate the wave speed through this pipeline considering the three pipe constraints.

**Solution**

The D/t ratio is

$600 / 6 = 100$

Therefore thin-walled pipe values for restraint factor C apply.

Considering the pipe to be completely rigid, the wave speed is

$a = (K/\rho)^{\frac{1}{2}} = (2.2 \times 10^9 / 1\,000)^{\frac{1}{2}} = 1\,483$ m/s

(a) Pipe anchored at upstream end only, from Equation 11.5

$C = 1 - 0.5\mu = 1 - 0.5 \times 0.3$

or

$C = 0.85$

Therefore,

$$a \quad = \frac{(K/\rho)^{1/2}}{(1+ CKD/Et)^{\frac{1}{2}}}$$

$$= \frac{1\,483}{[1+ 0.85 \times 2.2 \times 600 / (207 \times 6)]^{1/2}}$$

$$a \quad = \frac{1\,483}{1.9034} \quad = 1\,075 \text{ m/s}$$

(b) Pipe anchored against any axial movement, from Equation 11.6

$C = 1 - \mu^2 = 1 - (0.3)^2$
$C = 0.91$

$$a = \frac{1\,483}{[1+ 0.91 \times 2.2 \times 600 / (207 \times 6)]^{1/2}}$$

$$a = 1\,057 \text{ m/s}$$

(c) Each pipe section anchored with expansion joints, from Equation 11.7

$$C = 1$$

and

$$a = \frac{1\,483}{[1+ 1.0 \times 2.2 \times 600 / (207 \times 6)]^{1/2}}$$

$$a = 1\,033 \text{ m/s}$$

## 11.4 Transients in Cross Country Pipelines

Cross country pipelines and other trunk pipelines transporting crude oil, refined products etc. are several hundred miles long and generally have multiple pump stations located along the pipeline. Average pump station spacing may be 30 to 60 miles or more. The pump stations are designed to provide the necessary pressure to overcome friction in pipeline between pump stations and any static head lift necessary due to pipeline elevation difference.

The study of transient pressures in long distance pipelines, called *surge analysis* is fairly complex due to the fact that the frictional pressure drop in the pipelines are large compared to the instantaneous pressure surge due to sudden stoppage of flow.

A rigorous mathematical analysis of the pressure versus flow variations due to transient conditions have found to result in partial differential equations that need to be solved with specified boundary conditions. Such solutions are difficult, if not impossible using manual methods, even with a scientific calculator. However, over the years engineers and scientists have simplified these equations using finite difference methods and other mathematical approximations to result in the solution of simple simultaneous equations at every nodal point along the pipeline. Of course, more the pipeline is sub-divided, higher is the accuracy of calculation. Today, with high powered desktop computers we can model these transient pressures in pipelines using the Method of Characteristics that has proven to be quite accurate when compared to field tests. Several commercial computer programs are available to model transient pressures in pipelines due to valve closures, pump startup and shutdown and other upset conditions.

The *potential surge* is the instantaneous pressure rise due to sudden stoppage of flow, such as closing of valve at the end of the pipeline. This pressure was shown from Equation 11.1 to be

$$\Delta H = aV / g$$

The increase in storage capacity of a pipeline due to increase in pressure is termed *line packing*. Line packing occurs when the flow in a pipeline is changed by gradual or sudden closure of a valve.

Consider a long crude oil pipeline flowing at a steady flow rate of Q bbl/hr. Due to sudden closure of a valve downstream there is a pressure rise immediately upstream of the valve. A pressure wave with a amplitude of $\Delta H$ travels from the valve upstream towards the pump station. Oil flows downstream of the wave front, pressure rises gradually at the downstream end and the hydraulic grade line gradually reaches the horizontal level. During this time more oil gets packed between the valve and the pressure wave. This is refereed to as *line pack*. The pressure rise at the valve is made up of two components

(1) The potential surge pressure, $\Delta H$ from Equation 11.1 due to instantaneous closure of the valve
(2) The pressure rise due to line pack $\Delta P$

The pressure rise $\Delta P$ due to line pack, in a long oil pipeline may be several times the size of the potential surge $\Delta H$, depending on the length of the pipeline, its diameter etc.

Attenuation is the process by which the amplitude of the surge pressure wave is reduced as it propagates along the pipeline due to the reduction in the velocity differential across the wave front. This is because, even though the wave front has passed a certain location on the pipeline, the oil continues to flow towards the valve. This causes a reduction in the velocity differential across the wave front as the wave propagates upstream.

In addition, the amplitude of the surge is also decreased due to pipe friction. This reduction, however, is a smaller component than that caused by the velocity differential across the wave front.

For a detailed analysis of transient analysis in long pipelines, see Reference 14.

## 11.5 Summary

In this chapter, we introduced the concept of unsteady flow in pipelines. If the velocity, flow rate and pressure at any point in a pipeline change with time, unsteady flow is said to occur. Closing and opening valves, starting and stopping pumps and injecting volumes or changing products all contribute to some form of unsteady flow. Some unsteady flow situations can cause pressure surges or transients that when superimposed on the existing steady state pressure gradient may cause overpressure of a pipeline. Design codes for petroleum pipelines dictate that such transient overpressures cannot exceed 110% of the maximum allowable operating pressure in a pipeline.

We illustrated the concept of unsteady flow using a simple example of a tank connected to a pipe with a valve at the end. Sudden closure of the valve causes a pressure rise at the valve and generates a pressure wave that travels in the opposite direction of the flow at the speed of sound in the liquid. It was shown that flow rate and pressure variations occur throughout the pipeline due to the pressure wave traveling back and forth until friction attenuates and equilibrium is established after a sufficient period of time. The wave speed depends mainly on the liquid bulk modulus and density. It also depends upon the pipe size and the elastic modulus of the pipe material.

Transient pressures developed in long distance pipelines are generally modeled using computer programs based on the method of characteristics. This approach reduces the complex partial differential equations into simpler simultaneous equations which are solved sequentially from node to node, by sub dividing the pipeline into equal segments. For more detailed analysis of unsteady flow the reader is referred to one of the many publications listed in the Reference section.

## 11.6 Problems

11.6.1  Crude oil flows through a 16 inch diameter, 0.250 inch wall thickness steel pipeline, at flow rate of 150 000 bbl/day. A valve located at the end of the pipeline is suddenly closed. Determine the potential surge generated and the wave speed in the pipeline assuming the pipe is anchored throughout against axial movement. Bulk modulus of crude oil = 280 000 psi. Specific gravity = 0.89. Modulus of elasticity of steel = 30 x 10$^6$ psi. Poisson's ratio = 0.3

11.6.2  Compare the wave speed in a rubber pipeline 250 mm diameter, 6 mm wall thickness carrying water with a similar pipeline 60 mm thick.

E = 0.1 GPa and μ = 0.45. K = 2.2 GPa. ρ = 1 000 kg/m³

Assume pipe anchored at one end only.

11.6.3  A steel pipeline 12.75 inch diameter and 0.250 inch wall thickness is used to transport water between two storage tanks. Calculate the wave speeds and restraint factor C using thin-walled and thick-walled formulas.

# 12

# Pipeline Economics

In this chapter we will discuss cost of a pipeline and the various components that contribute to the economics of pipelines. This will include the major components of the initial capital costs and that of the recurring annual costs.

We will also examine how the transportation charge is established based on throughput rates, project life, interest rate and financing scenarios.

## 12.1 Economic Analysis

In any pipeline investment project, we must perform an economic analysis of the pipeline system to ensure that we have the right equipment and materials at the right cost to perform the necessary service and provide a profitable income for the venture. The previous chapters helped determine the pipe size, pipe material, pumping equipment etc necessary to transport a given volume of a product. In this chapter we will analyze the cost implications and how to decide on the economic pipe size and pumping equipment required to provide the optimum rate of return on our investment.

The major capital components of a pipeline system consists of the pipe, pump stations, storage tanks, valves, fittings, and meter stations. Once this capital is expended and the pipeline is installed, pump station and other facilities built, we will incur annual operating and maintenance cost for these facilities. Annual costs will also include General and Administrative (G&A) costs including payroll costs, rental and lease costs and other recurring costs necessary for the safe and efficient operation of the pipeline system. The revenue for this operation will be in the form of pipeline tariffs collected from companies that ship products through this pipeline. The capital necessary for building this pipeline system may be partly owner equity and partly borrowed money. There will be investment hurdles and Rate of Return (ROR) requirements imposed by equity owners and financial institutions that lend the capital for the project. Regulatory requirements will also dictate the maximum revenue that may be collected and the ROR that may be realized, as transportation services. An economic analysis must be performed for the project taking into account all of these factors and a reasonable project life of 20 to 25 years or more in some cases.

These concepts will be illustrated by examples in subsequent sections of this chapter. Before we discuss the details of each cost component, it will be instructive and beneficial to calculate the transportation tariff and cost of service using a simple example.

**Example Problem 12.1**

Consider a new pipeline that is being built for transporting crude oil from a tank farm to a refinery. For the first phase (10 years) it is estimated that shipping volumes will be 100 000 bbl/day. Calculations indicate that the 16 inch pipeline, 100 miles long with two pump stations will be required. The capital costs for all facilities is estimated to be $ 72 million. The annual operating cost including electric power, O&M, G&A etc. is estimated to be $5 million. The project is financed at a debt equity ratio 80/20. The interest rate on debt is 8% and the rate of return allowed by regulators is 12%. Consider a project life of 20 years and overall tax rate of 40%.

(a) What is the annual cost of service for this pipeline?
(b) Based on the fixed throughput rate of 100 000 bbl/day, and a load factor of 95%, what tariff rate can be charged within the regulatory guidelines?
(c) During the second phase, volumes are expected to increase by 20% (Years 11 through 20). Estimate the revised tariff rate for

the second phase assuming no capital cost changes to pump stations and other facilities. Use an increased annual operating cost of $7 million and same load factor as before.

## Solution

(a) The total capital cost of all facilities is $72 million. We start by calculating the debt capital and equity capital based on the given 80/20 debt equity ratio.

Debt capital      = 0.80 x 72     = $57.6 million
Equity capital    = 72 - 57.6    = $14.4 million

The debt capital of $57.6 million is borrowed from a bank or financial institution at the 8% annual interest rate. To retire this debt over the project life of 20 years, we must account for this interest payment in our annual costs.

In a similar manner, the equity investment of 14.4 million may earn 12% ROR. Since the tax rate is 40%, the annual cost component of the equity will be adjusted to compensate for the tax rate.

Interest cost per year   = 57.6 x 0.08           = $4.61 million

Equity cost per year    = 14.4 x 0.12 / (1 - 0.4) = $2.88 million

Assuming a straight line depreciation for 20 years, our yearly depreciation cost for $72 million capital is calculated as follows:

Depreciation per year   = 72 / 20               = $3.6 million

By adding annual interest expense, annual Equity cost and annual depreciation cost to the annual O&M cost, we get the total annual service cost to operate the pipeline.

Total cost of service = 4.61 + 2.88 + 3.60 + 5.00 = $16.09 million/yr

(b) Based on the total cost of service of the pipeline, we can now calculate the tariff rate to be charged, spread over a flow rate of 100 000 bbl/day and a load factor of 95%:

Tariff = $(16.09 \times 10^6)$ / $(365 \times 100\ 000 \times 0.95)$ or \$0.4640 /bbl

(c) During Phase 2, the new flow rate is 120 000 bbl/day and the total cost of service becomes:

Total cost of service = $4.61 + 2.88 + 3.60 + 7.00 = \$18.09$ million/yr

And the revised tariff becomes:

Tariff = $(18.09 \times 10^6)$ / $(365 \times 120\ 000 \times 0.95)$ or \$0.4348 /bbl

## 12.2 Capital Costs

The Capital cost of a pipeline project consists of the following major components:

- Pipeline
- Pump stations
- Tanks and manifold piping
- Valves, fittings, etc.
- Meter stations
- SCADA and Telecommunication
- Engineering and construction management
- Environmental and permitting
- Right of Way acquisition cost
- Other project costs such as Allowance for Funds Used During Construction (AFUDC) and contingency

### 12.2.1 Pipeline Costs

The capital cost of pipeline consists of material and labor for installation. To estimate the material cost we will use the following method.

Pipe material cost = 10.68 (D - t) t x 2.64 x L x cost per ton

or

$$PMC = 28.1952 \, L \, (D-t) \, t \, (Cpt) \qquad\qquad (12.1)$$

where

PMC  - Pipe material cost, $
L       - Pipe length, mi
D       - Pipe outside diameter, in
t        - Pipe wall thickness, in
Cpt    - Pipe cost, $/ton

In SI units, Equation 12.1 can be written as

$$PMC = 0.02463 \, L \, (D-t) \, t \, (Cpt) \qquad\qquad (12.2)$$

where

PMC  - Pipe material cost, $
L       - Pipe length, km
D       - Pipe outside diameter, mm
t        - Pipe wall thickness, mm
Cpt    - Pipe cost, $/metric ton

Since pipe will be coated, wrapped and delivered to the site, we will have to increase the material cost by some factor to account for these items or add the actual cost of these items to the pipe material cost.

Pipe installation cost or labor cost is generally stated in $/ft or $/mile of pipe. It may also be stated based on an inch-diameter-mile of pipe. Construction contractors will estimate the labor cost of installing a given pipeline, based on detailed analysis of terrain, construction conditions, difficulty of access, and other factors. Historical data is available for estimating labor costs of various size pipelines. In this section we will use approximate methods that should be verified with contractors taking into account current labor rates, geographic and terrain issues. A good approach is to express the labor cost in terms of $/inch diameter per mile of pipe. Thus we can say that a particular 16 inch pipeline can be installed at a cost of $15 000 per inch-diameter-mile. Therefore for a 100 mile, 16 inch pipeline we can estimate the labor cost to be

Pipe Labor cost = $15 000 x 16 x 100 = $ 24 million

Based on $/ft cost this works out to be

$$\frac{24 \times 10^6}{100 \times 5\ 280} = \$45.50 \text{ per ft}$$

Table A.14 in Appendix A shows a summary of typical labor costs. It must be noted that these numbers are approximate and must be verified with contractor input for a specific geographic location and level of construction difficulty.

In addition to labor costs for installing straight pipe, there may be other construction costs such as road crossings, railroad crossings, river crossings, etc. These are generally estimated as a lump sum for each item and added to the total pipe installed costs. For example there may be ten highway and road crossings totaling $2 million and one major river crossing that may cost $500 000. For simplicity, however, we will ignore these items for the present.

### 12.2.2   Pump Station Costs

To estimate the pump station cost a detailed analysis would consist of preparing a material take-off from the pump station drawings and getting vendor quotes on major equipment such as pumps, drivers, switchgear, valves, instrumentation etc. and estimating the station labor costs.

An approximate cost for pump stations can be estimated using a value for cost in dollars per installed horsepower. This is an all-inclusive number considering all facilities associated with the pump station. For example, we can use an installed cost of $1 500 per HP and estimate that a pump station with 5 000 HP will cost

$1 500 x 5 000 = $7.5 million.

In the above, we used an all-inclusive number of $ 1 500 per installed HP. This figure takes into account all material and equipment cost and construction labor. Such values of installed cost per HP can be obtained from historical data on pump stations constructed in the recent past. Larger HP pump stations will have smaller dollar $/HP costs while smaller pump stations with less HP will have a higher $/HP cost, reflecting economies of scale.

### 12.2.3  Tanks and Manifold piping

Tanks and manifold piping can be estimated fairly accurately by detailed material take-offs from construction drawings and from vendor quotes.

Generally, tank vendors quote installed tank costs in $/bbl. Thus if we have a 50 000 bbl tank, it can be estimated at

50 000 x $10/bbl = $500 000

based on an installed cost of $10 /bbl

We would of course, increase the total tankage cost by a factor of 10-20% to account for other ancillary piping and equipment.

As with installed HP costs, the unit cost for tanks decreases with tank size. For example, a 300 000 bbl tank may be based on $6 or $8 per bbl compared to the $10/bbl cost for the smaller 50 000 bbl tank.

### 12.2.4  Valves and Fittings

This category of items may also be estimated as a percentage of the total pipe cost. However, if there are several mainline block valve locations that can be estimated as a lump sum cost, we can estimate the total cost of valves and fittings as follows:

A typical 16 inch mainline block valve installation may cost $100 000 per site including material and labor costs. If there are ten such installations spaced 10 miles apart on a pipeline, we would estimate cost of valves and fittings to be $1.0 million

## 12.2.5  Meter Stations

Meter stations may be estimated as a lump sum fixed price for a complete site. For example a 10 inch meter station with meter, valves, piping instrumentation may be priced at $250 000 per site including material and labor cost. If there are two such meter stations on the pipeline, we would estimate total meter costs at $500 000.

## 12.2.6  SCADA and Telecommunication System

This category covers costs associated with Supervisory Control and Data Acquisition (SCADA), telephone, microwave etc. SCADA system costs include the facilities for remote monitoring, operation and control of the pipeline from a central control center. Depending upon the length of the pipeline, number of pump stations, valve stations etc., the cost of these facilities may range from $2 million to $5 million or more. An estimate based on the total project cost may range from 2% to 5%.

## 12.2.7  Engineering and Construction Management

Engineering and construction management consists of preliminary and detail engineering design costs and personnel costs associated with management and inspection of the construction effort for pipelines, pump stations and other facilities. This category usually ranges from 15% to 20 % of total pipeline project costs.

## 12.2.8  Environmental and Permitting

In the past, environmental and permitting costs used to be a small percentage of the total pipeline system costs. In recent times, due to stricter environmental and regulatory requirements this category now includes items like environmental impacts report, environmental studies pertaining to the flora and fauna, fish and game, endangered species, sensitive area such as Native American burial sites and allowance for habitat mitigation. The latter cost include the acquisition of new acreage to compensate for areas disturbed by the pipeline route. This new acreage will then be allocated for parks, wildlife preserves, etc.

Permitting costs would include pipeline construction permits such as road crossings, railroad crossings, river and stream crossings and permitting for anti pollution devices for pump stations and tank farms.

Environmental and permitting costs maybe as high as 10% to 15% of total project costs.

### 12.2.9   Right of Way Acquisitions Cost

Right of Way (ROW) must be acquired for building a pipeline along private lands, farms, public roads and railroads. In addition to initial acquisition costs there may be annual lease costs that the pipeline company will have to pay railroads, agencies and private parties for pipeline easement and maintenance. The annual ROW costs would be considered an expense and would be included in the operating costs of the pipeline. For example the ROW acquisitions costs for a pipeline project may be $20 million, that would be included in the total capital costs of the pipeline project. In addition, annual lease payments for ROW acquired may be a total of $500 000 a year which would be included with other operating costs such as pipeline O&M, G&A costs etc

Historically, ROW costs have been in the range of 6% to 8% of total project costs for pipelines.

### 12.2.10  Other Project Costs

Other project costs would include Allowance for Funds Used During Construction (AFUDC), Legal and Regulatory costs and contingency costs. Contingency costs cover unforeseen circumstances and design changes including pipeline rerouting to bypass sensitive areas, pump stations and facilities modifications not originally anticipated at the start of the project. AFUDC and contingency costs will range between 15% and 20% of the total project cost.

## 12.3 Operating Costs

The annual operating cost of a pipeline consists mainly of the following:

- Pump station energy cost (electric or natural gas)
- Pump station equipment maintenance costs (equipment overhaul, repairs etc.)
- Pipeline maintenance cost including line rider, aerial patrol, pipe replacements, relocations etc.
- SCADA and Telecommunication costs
- Valve and meter station maintenance
- Tank farm operation and maintenance
- Utility costs - water, natural gas, etc.
- Ongoing environmental and permitting costs
- Right of Way lease costs
- Rentals and lease costs
- General & administrative costs including payroll.

In the above list, pump station costs include electrical energy and equipment maintenance costs, that can be substantial. Consider two pump stations of 5 000 HP each operating 24 hours a day, 350 days a year with two week shut down for maintenance. This can result in annual O&M costs of $6 to $7 million based on electricity costs of 8 to 10 cents per kWh. In addition to the power cost other components of O&M costs include annual maintenance and overhead which can range from $0.50 million to $1.0 million depending on the equipment involved.

## 12.4 Feasibility Studies and Economic Pipe Size

In many instances, we have to investigate the technical and economic feasibility of building a new pipeline system to provide transportation services for liquids from a storage facility to a refinery or from a refinery to a tank farm. Other types of studies may include technical and economic feasibility studies for expanding the capacity of an existing pipeline system to handle additional throughput volumes due to increased market demand or refinery expansion.

Grass roots pipeline projects where a brand new pipeline system needs to be designed from scratch, involve analysis of the best pipeline route, optimum pipe size and pumping equipment required to transport a given volume of liquid. In this section we will learn how an economic pipe size is determined for a pipeline system, based on an analysis of capital and operating costs.

Consider a project in which a 100 mile pipeline is to be built to transport 8 000 bbl/hr of refined products from a refinery to a storage facility. The question before us is: What pipe diameter and pump stations are the most optimum for handling this volume?

Let us assume that we selected a 16 inch diameter pipe to handle the designated volume and we calculated that this system needs two pump stations of 2 500 HP each. The total cost of this system of pipe and pump stations can be calculated and we will call this the 16 inch Cost Option. If we chose a 20 inch diameter pipeline, the design would require one 2 000 HP pump station. In the first case, more HP and less pipe will be required. The 16 inch pipeline system would require approximately 20% less pipe than the 20 inch option. However, the 16 inch option requires 2.5 times the HP required in the 20 inch case. Therefore, the annual pump station operating costs for the 16 inch system would be higher than the 20 inch case, since electric utility cost for the 5 000 HP pump stations will be higher than that for the 2 000 HP station required for the 20 inch system. Therefore, in order to determine the optimum pipe size required, we must analyze the capital costs and the annual operating costs to determine the scenario that gives us the least total cost, taking into account a reasonable project life. We would perform these calculations considering time value of money, and select the option that results in the lowest Present Value (PV) of investment.

Generally, in any situation we must evaluate at least three or four different pipe diameters and calculate the total capital costs and operating costs for each pipe size. As indicated in the previous paragraph, the optimum pipe size and pump station configuration will be the alternative that minimizes total investment, after taking into account the time value of money over the life of the project. We will illustrate this using an example.

**Example Problem 12.2**

A city is proposing to build a 24 mile long water pipeline to transport 14.4 million gal/day flow rate. There is static elevation head of 250 ft from the originating pump station to the delivery terminus. A minimum delivery pressure of 50 psi is required at the pipeline terminus.

The pipeline operating pressure must be limited to 1000 psi using steel pipe with a yield strength of 52 000 psi. Determine the optimum pipe diameter and the HP required for pumping this volume on a continuous basis, assuming 350 days operation, 24 hours a day. Electric costs for driving the pumps will be based on 8 cents/kWh. Interest rate on borrowed money is 8% per year. Use Hazen Williams equation for pressure drop with a C-factor of 100.

Assume $700/ton for pipe material cost and $20 000/inch-diameter-mile for pipeline construction cost. For pump stations, assume a total installed cost of $1 500 per HP. To account for items other than pipe and pump stations in the total cost use a 25% factor.

**Solution:**

First we have to bracket the pipe diameter range. If we consider 20 inch diameter pipe, 0.250 inch wall thickness, the average water velocity using Equation (3.11) will be:

$$V = 0.4085 \times 10\ 000 / 19.5^2 = 10.7 \text{ ft/s}$$

where 10 000 gal/min is the flow rate based on 14.4 million gal/day.

This is not a very high velocity. Therefore 20 inch pipe can be considered as one of the options. We will compare this with two other pipe sizes: 22 inch and 24 inch nominal diameter.

Initially, we will assume 0.500 in pipe wall thickness for 22 in and 24 in. Later we will calculate the actual required wall thickness for the given MAOP. Using ratios, the velocity in the 22 inch pipe will be approximately

10.7x(19.5/21)$^2$ or 9.2 ft/s

and the velocity in the 24 inch pipe will be

10.7x(19.5/23)$^2$ or 7.7 ft/s

Thus the selected pipe sizes 20 inch, 22 inch and 24 inch will result in a water velocity between 7.7 ft/s and 10.7 ft/s, which is an acceptable range of velocities in a pipe.

Next we need to choose a suitable wall thickness for each pipe size to limit the operating pressure to 1 000 psi.

Using the internal design pressure Equation (4.3), we calculate the pipe wall thickness required as follows

For 20 inch pipe, the wall thickness is

T = 1 000x20/(2x52 000x0.72) = 0.267 in

Similarly for the other two pipe sizes we calculate

For 22 inch pipe, the wall thickness is

T = 1 000x22/(2x52 000x0.72) = 0.294 in

And for 24 inch pipe, the wall thickness is

T = 1 000x24/(2x52 000x0.72) = 0.321 in

Using the closest commercially available pipe wall thicknesses, we choose the following three sizes:

20 inch, 0.281 inch wall thickness (MAOP = 1 052 psi)
22 inch, 0.312 inch wall thickness (MAOP = 1 061 psi)
24 inch, 0.344 inch wall thickness (MAOP = 1 072 psi)

The revised MAOP values for each pipe size, with the slightly higher than required minimum wall thickness, were calculated as shown within parentheses above. With the revised pipe wall thickness, the velocity calculated earlier will be corrected to

$$V_{20} = 10.81 \text{ ft/s}$$
$$V_{22} = 8.94 \text{ ft/s}$$

and

$$V_{24} = 7.52 \text{ ft/s}$$

Next, we will calculate the pressure drop due to friction in each pipe size at the given flow rate of 10 000 gal/min, using Hazen Williams equation with a C-factor of 100.

From Equation (3.36) for the 20 inch pipeline

$$10\ 000 \times 60 \times 24/42 = 0.1482 \times 100\ (20 - 2 \times 0.281)^{2.63}(Pm/1.0)^{0.54}$$

rearranging and solving for the pressure drop Pm we get

Pm = 63.94 psi/mi for the 20 inch pipe

Similarly, we get the following for the pressure drop in the 22 inch and 24 inch pipelines:

Pm = 40.25 psi/mi for the 22 inch pipe

and

Pm = 26.39 psi/mi for the 24 inch pipe

We can now calculate the total pressure required for each pipe size, taking into account the friction drop in the 24 mile pipeline and the elevation head of 250 ft along with a minimum delivery pressure of 50 psi at the pipeline terminus,.

Total pressure required at the origin pump station is:

(63.94x24) + 250x1.0/2.31 + 50 = 1 692. 79 psi for the 20 inch pipe

and

(40.25x24) + 250x1.0/2.31 + 50 = 1 124.23 psi for the 22 inch pipe

and

(26.39x24) + 250x1.0/2.31 + 50 = 791.59 psi for the 24 inch pipe

Since the MAOP of the pipeline is limited to 1 000 psi, it is clear that we would need 2 pump stations for the 20 inch and 22 inch pipeline cases while one pump station will suffice for the 24 inch pipeline case.

The total BHP required for each case will be calculated from the above total pressure and the flow rate of 10 000 gal/min using Equation (5.18) assuming a pump efficiency of 80%. We will also assume that the pumps require a minimum suction pressure of 50 psi.

BHP = 10 000 x (1 693 - 50) / (0.8x1 714)) = 11 983 for 20 inch
BHP = 10 000 x (1 124 - 50) / (0.8x1 714)) = 7 833 for 22 inch
BHP = 10 000 x (792 - 50) / (0.8x1 714)) = 5 412 for 24 inch

Increasing the BHP values above by 10% for installed HP and choosing the nearest motor size, we will use 14 000 HP for the 20 in pipeline system, 9 000 HP for the 22 inch system and 6 000 HP for the 24 inch pipeline system.

If we had factored in a 95% efficiency for the electric motor and picked the next nearest size motor we would have arrived at the same HP motors as above.

To calculate the capital cost of facilities, we will use $700 per ton for steel pipe, delivered to the construction site. The labor cost for installing the pipe will be based on $20 000 per inch-diameter-mile.

The installed cost for pump stations will be assumed at $1 500 /HP.

To account for other cost items discussed earlier in this chapter, we will add 25% to the sub-total of pipeline and pump station cost.

The summary of the estimated capital cost for the three pipe sizes are listed in the table below:

Table 12.1

| Capital Cost | 20-inch | 22-inch | 24-inch |
|---|---|---|---|
| million $ | | | |
| Pipeline | 2.62 | 3.21 | 3.85 |
| Pump stations | 21.00 | 13.50 | 9.00 |
| | | | |
| Other (25%) | 5.91 | 4.18 | 3.21 |
| | | | |
| | | | |
| Total | 29.53 | 20.88 | 16.07 |

Based on total capital costs alone, it can be seen that the 24-inch system is the best. However, we will have to look at the operating costs as well, before making a decision on the optimum pipe size.

Next, we calculate the operating cost for each scenario, using electrical energy costs for pumping. As discussed in an earlier section of this chapter, many other items enter into the calculation of annual operating costs, such as O&M, G & A costs, etc. For simplicity, we will increase the electrical cost of the pump stations by a factor to account for all other operating costs.

Using the BHP calculated at each pump station for the three cases and 8 cents/kWh for electricity cost, we find that the annual operating cost for 24 hr operation per day, 350 days per year the annual costs are:

11 983 x 0.746 x 24 x 350 x 0.08 = $ 6.0 million/yr for 20 inch
7 833 x 0.746 x 24 x 350 x 0.08 = $ 3.93 million/yr for 22 inch
5 412 x 0.746 x 24 x 350 x 0.08 = $ 2.71 million/yr for 24 inch

Strictly speaking the above costs will have to be increased to account for the demand charge for starting and stopping electric motors. The utility company may charge based on the kW rating of the motor. It will range from $4 to $6 per kW/month. Using an average demand charge of $5/kW/mo, we get the following demand charges for the pump station in a 12 month period

14 000 x 5 x 12     = $ 840 000
9 000 x 5 x 12      = $ 540 000
6 000 x 5 x 12      = $ 360 000

Adding the demand charges to the previously calculated electric power cost, we get the total annual electricity costs as

$6.84 million/yr for 20 in
$4.47 million/yr for 22 in
$3.07 million/yr for 24 in

Increasing above numbers by a 50% factor, to account for other operating costs, such as O&M, G&A, etc. we get the following for total annual costs for each scenario:

$ 10.26 million/yr for 20 inch
$ 6.7 million/yr for 22 inch
$ 4.6 million/yr for 24 inch

Next, we will use a project life of 20 years and interest rate of 8% to perform a discounted cash flow(DCF) analysis, to obtain the present value of these annual operating costs. Then the total capital cost calculated earlier listed in Table 12.1 will be added to the present values of the annual operating costs. The Present Value (PV) will then be obtained for each of the three scenarios.

PV of 20 inch system = $ 29.53 + present value of $10.26 million/yr at 8% for 20 years

Or

$PV_{20} = 29.53 + 100.74 = \$ 130.27$ million

Similarly for the 22 inch and 24 inch systems, we get

$PV_{22} = 20.88 + 65.78 = \$ 86.66$ million
$PV_{24} = 16.07 + 45.16 = \$ 61.23$ million

Thus based on the net present value of investment, we can conclude that the 24-inch pipeline system with one 6 000 HP pump station is the preferred choice.

In the preceding calculations for the sake of simplicity, we made several assumptions. We considered major cost components, such as pipeline and pump station costs and added a percentage of the sub-total to account for other costs. Also in calculating the PV of the annual costs we used constant numbers for each year. A more rigorous approach would require the annual costs be inflated by some percentage every year to account for inflation and cost of living adjustments. The Consumer Price Index (CPI) could be used in this regard. As far as capital costs go, we can get more accurate results if we perform a more detailed analysis of the cost of valves, meters and tanks instead of using a flat percentage of the pipeline and pump station costs. The objective in this chapter was to introduce the reader to the importance of economic analysis and to outline a simple approach to selecting the economical pipe size.

In addition, the earlier section on cost of services and tariff calculations provided an insight into how transportation companies finance a project and collect revenues for their services.

## 12.5 Summary

We reviewed the major cost components of a pipeline system consisting of pipe, pump station, etc. and illustrated methods of estimating the capital costs of these items. The annual costs such as electrical energy, O&M etc. were also identified and calculated for a typical pipeline. Using the capital cost and operating cost, the annual cost of service was calculated based on specified project life, interest cost etc. Thus we determined the transportation tariff that could be charged for shipments through the pipelines. Also a methodology of determining the optimum pipe size for a particular application using Present Value (PV) was explained. Considering three different pipe sizes, we determined the best option based on a comparison of PV of the three different cases.

# 12.6 Problems

12.6.1  Calculate the annual cost of service and transportation tariff to be charged for shipments through a refined products pipeline as follows. The pipeline is 90 km long, 400 mm diameter and 8 mm wall thickness and constructed of steel with a yield strength of 448 MPa. The terrain is essentially flat. It is used to transport gasoline(specific gravity = 0.74 and viscosity = 0.65 cSt at 60°F) at a flow rate of 2 000 m³/hr. The annual operating costs such as O&M, G&A etc. is estimated to be $2 million and does not include power costs. Assume power cost of $0.10 per kWh for the electric motor driven pumps. The project will be financed at a debt equity ratio 70/30. The interest rate on debt is 7% and the rate of return allowed by regulators is 14%. Assume a project life of 25 years and overall tax rate of 35%.

12.6.2  Compare pipeline sizes of 12 -inch, 14 -inch and 16-inch for an application that requires shipments of crude oil from a tank farm to a refinery 30 miles away at 5 000 bbl/hr. The tank farm is at an elevation of 350 ft while the refinery is at 675 ft elevation. The pipe MAOP is limited to 1400 psi. Consider 5LX-65 pipe and 72% design factor. The suction pressure at the tank farm may be assumed at 50 psi and the delivery pressure at the refinery is 30 psi. The pipeline will be operated 355 days a year, 24 hrs per day. Electricity cost is 6 cents/kWh.

12.6.3  In the previous problem, assuming the steady state flow rate of 5 000 bbl/hr remains constant for the first ten years, what is the estimated tariff? Determine the reduction in tariff if the flow rate increased by 20% for the next 10 years.

# APPENDIX A

## Tables and Charts

# A.1. Units and Conversions

| Item | English Units | SI Units | Conversion - English to SI |
|---|---|---|---|
| Mass | slugs (slugs) | kilograms (kg) | 1 lb = 0.45359 kg |
| | Pound mass (lbm) | | 1 slug = 14.594 kg |
| | US tons | metric tonnes (t) | 1 US ton = 0.9072 t |
| | long tons | | 1 long ton = 1.016 t |
| Length | inches (in) | millimeters (mm) | 1 in = 25.4 mm |
| | feet (ft) | meters (m) | 1 ft = 0.3048 |
| | miles (mi) | kilometers (km) | 1 mi = 1.609 km |
| Area | square feet ($ft^2$) | Square meters ($m^2$) | 1 $ft^2$ = 0.0929 $m^2$ |
| Volume | cubic inch ($in^3$) | cubic millimeters ($mm^3$) | 1$in^3$ = 16387.0 $mm^3$ |
| | cubic feet ($ft^3$) | cubic meters ($m^3$) | 1$ft^3$ = 0.02832 $mm^3$ |
| | US gallons (gal) | liters (L) | 1 gal = 3.785L |
| | barrel (bbl) | | 1 bbl = 42 US gal |
| Density | slugs per cubic foot ($slug/ft^3$) | Kilograms per cubic meter ($kg/m^3$) | 1 $slug/ft^3$ = 515.38 $kg/m^3$ |
| Specific weight | Pound per cubic foot ($lb/ft^3$) | Newton per cubic meter ($N/m^3$) | 1 $lb/ft^3$ = 157.09 $N/m^3$ |
| Viscosity (Kinematic) | $ft^2/s$ | $m^2/s$ | 1 $ft^2/s$ = 0.092903 $m^2/s$ |
| Flow rate | gallons/minute (gal/min) | liters/minute (L/min) | 1 gal/min = 3.7854 L/min |
| | barrels/hour (bbl/hr) | cubic meters/hour ($m^3/hr$) 1 bbl/hr = 0.159 $m^3/hr$ | |
| | barrels/day (bbl/day) | | |

| Force | pounds (lb) | Newton (N) | 1 lb = 4.4482 N |
|---|---|---|---|
| Pressure | pounds/square inch (psi) | kiloPascal (kPa) | 1 psi = 6.895 kPa |
| | lb/in² | kilograms/ square centimeter (kg/cm²) | 1 psi = 0.0703 kg/cm² |
| Velocity | feet/second (ft/s) | meters/second (m/s) | 1 ft/s = 0.3048 m/s |
| Work and Energy | British Thermal Units (Btu) | Joule (J) | 1 Btu = 1055.0 J |
| Power | Btu/hour | Watt (W) | 1 Btu/hr = 0.2931W |
| | | Joules/second (J/s) | |
| | Horsepower (HP) | kiloWatt (kW) | 1 HP = 0.746 kW |
| Temperature | degree Fahrenheit (°F) | degree Celsius (°C) | 1°F = 9/5 °C +32 |
| | degree Rankin (°R) | degree Kelvin (°K) | 1°R = °F +460 |
| | | | 1°K = °C +273 |
| Thermal Conductivity | Btu/hr/ft/°F | W/m/°C | 1 Btu/hr/ft/ °F = 1.7307 W/m/°C |
| Specific Heat | Btu/lb/°F | kJ/kg/°C | 1 Btu/lb/°F = 4.1869 kJ/kg/°C |

# A.2. Common Properties of Petroleum Fluids

| Product | Viscosity | °API | Specific Gravity | Reid Vapor |
|---|---|---|---|---|
| | cSt @ 60°F | Gravity | @ 60°F | Pressure |
| | | | | |
| **Regular Gasoline** | | | | |
| Summer Grade | 0.70 | 62.0 | 0.7313 | 9.5 |
| Inter-seasonal Grade | 0.70 | 63.0 | 0.7275 | 11.5 |
| Winter Grade | 0.70 | 65.0 | 0.7201 | 13.5 |
| | | | | |
| **Premium Gasoline** | | | | |
| Summer Grade | 0.70 | 57.0 | 0.7467 | 9.5 |
| Inter-seasonal Grade | 0.70 | 58.0 | 0.7165 | 11.5 |
| Winter Grade | 0.70 | 66.0 | 0.7711 | 13.5 |
| | | | | |
| No.1 Fuel Oil | 2.57 | 42.0 | 0.8155 | |
| No.2 Fuel Oil | 3.90 | 37.0 | 0.8392 | |
| Kerosene | 2.17 | 50.0 | 0.7796 | |
| Jet Fuel JP-4 | 1.40 | 52.0 | 0.7711 | 2.7 |
| Jet Fuel JP-5 | 2.17 | 44.5 | 0.8040 | |

# A.3. Specific Gravity and API Gravity

| Liquid | Specific Gravity | API |
|:---:|:---:|:---:|
| | @ 60 °F | Gravity @ 60 °F |
| | | |
| Propane | 0.5118 | N/A |
| Butane | 0.5908 | N/A |
| Gasoline | 0.7272 | 63.0 |
| Kerosene | 0.7796 | 50.0 |
| Diesel | 0.8398 | 37.0 |
| Light Crude | 0.8348 | 38.0 |
| Heavy Crude | 0.8927 | 27.0 |
| Very Heavy Crude | 0.9218 | 22.0 |
| Water | 1.0000 | 10.0 |

# A.4. Viscosity Conversions

| Viscosity(SSU) | Viscosity(cSt) | Viscosity(SSF) |
|:---:|:---:|:---:|
| 31.0 | 1.00 | |
| 35.0 | 2.56 | |
| 40.0 | 4.30 | |
| 50.0 | 7.40 | |
| 60.0 | 10.30 | |
| 70.0 | 13.10 | 12.95 |
| 80.0 | 15.70 | 13.70 |
| 90.0 | 18.20 | 14.44 |
| 100.0 | 20.60 | 15.24 |
| 150.0 | 32.10 | 19.30 |
| 200.0 | 43.20 | 23.50 |
| 250.0 | 54.00 | 28.0 |
| 300.0 | 65.00 | 32.5 |
| 400.0 | 87.60 | 41.9 |
| 500.0 | 110.00 | 51.6 |
| 600.0 | 132.00 | 61.4 |
| 700.0 | 154.00 | 71.1 |
| 800.0 | 176.00 | 81.0 |
| 900.0 | 198.00 | 91.0 |
| 1000.0 | 220.00 | 100.7 |
| 1500.0 | 330.00 | 150 |
| 2000.0 | 440.00 | 200 |
| 2500.0 | 550.00 | 250 |
| 3000.0 | 660.00 | 300 |
| 4000.0 | 880.00 | 400 |
| 5000.0 | 1100.00 | 500 |
| 6000.0 | 1320.00 | 600 |
| 7000.0 | 1540.00 | 700 |
| 8000.0 | 1760.00 | 800 |
| 9000.0 | 1980.00 | 900 |
| 10000.0 | 2200.00 | 1000 |
| 15000.0 | 3300.00 | 1500 |
| 20000.0 | 4400.00 | 2000 |

# A.5. Thermal Conductivities

| Substance | Thermal Conductivity |
|---|---|
| | (Btu/hr/ft/ °F) |
| | |
| Fire Clay Brick (burnt at 2426 °F) | 0.60 to 0.63 |
| Fire Clay Brick (burnt at 2642 °F) | 0.74 to 0.81 |
| Fire Clay Brick (Missouri) | 0.58 to 1.02 |
| | |
| Portland Cement | 0.17 |
| Mortar Cement | 0.67 |
| Concrete | 0.47 to 0.81 |
| Cinder concrete | 0.44 |
| Glass | 0.44 |
| | |
| Granite | 1.0 to 2.3 |
| Limestone | 0.73 to 0.77 |
| Marble | 1.6 |
| Sandstone | 0.94 to 1.2 |
| | |
| Corkboard | 0.025 |
| Fiber insulating board | 0.028 |
| | |
| Aerogel, Silica | 0.013 |
| Coal, anthracite | 0.15 |
| Coal, powdered | 0.067 |
| Ice | 1.28 |
| Sandy Soil, Dry | 0.25 to 0.40 |
| Sandy Soil, Moist | 0.50 to 0.60 |
| Sandy Soil, Soaked | 1.10 to 1.30 |
| Clay Soil, Dry | 0.20 to 0.30 |
| Clay Soil, Moist | 0.40 to 0.50 |
| Clay Soil, Moist to Wet | 0.60 to 0.90 |
| River Water | 2.00 to 2.50 |
| Air | 2.00 |

# A.6. Absolute Roughness of Pipe

| Pipe Material | Roughness | Roughness |
|---|---|---|
| | mm | in |
| Riveted steel | 0.9 to 9.0 | 0.0354 to 0.354 |
| Concrete | 0.3 to 3.0 | 0.0118 to 0.118 |
| Wood stave | 0.18 to 0.9 | 0.0071 to 0.0354 |
| Cast iron | 0.26 | 0.0102 |
| Galvanized iron | 0.15 | 0.0059 |
| Asphalted cast iron | 0.12 | 0.0047 |
| Commercial steel | 0.045 | 0.0018 |
| Wrought iron | 0.045 | 0.0018 |
| Drawn tubing | 0.0015 | 0.000059 |

# A.7. Moody Diagram

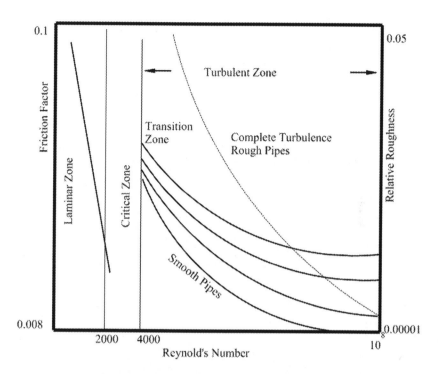

# A.8. Typical Hazen-Williams C-factors

| Pipe Material | C-factor |
|---|---|
| Smooth Pipes (All metals) | 130-140 |
| Smooth Wood | 120 |
| Smooth Masonry | 120 |
| Vitrified Clay | 110 |
| Cast Iron (Old) | 100 |
| Iron (worn/pitted) | 60-80 |
| Polyvinyl Chloride (PVC) | 150 |
| Brick | 100 |

# A.9. Friction Loss in Valves

## Resistance Coefficient K

| Description | L/D | Nominal Pipe Size - inches | | | | | | | | | | | |
|---|---|---|---|---|---|---|---|---|---|---|---|---|---|
| | | 1/2 | 3/4 | 1.0 | 1-1/4 | 1-1/2 | 2 | 21/2 to 3 | 4 | 6 | 8 to 10 | 12 to 16 | 18 to 24 |
| Gate Valve | 8 | 0.22 | 0.20 | 0.18 | 0.18 | 0.15 | 0.15 | 0.14 | 0.14 | 0.12 | 0.11 | 0.10 | 0.10 |
| Globe Valve | 340 | 9.2 | 8.5 | 7.8 | 7.5 | 7.1 | 6.5 | 6.1 | 5.8 | 5.1 | 4.8 | 4.4 | 4.1 |
| Ball Valve | 3 | 0.08 | 0.08 | 0.07 | 0.07 | 0.06 | 0.06 | 0.05 | 0.05 | 0.05 | 0.04 | 0.04 | 0.04 |
| Butterfly Valve | | | | | | | 0.86 | 0.81 | 0.77 | 0.68 | 0.63 | 0.35 | 0.30 |
| Plug Valve Straightway | 18 | 0.49 | 0.45 | 0.41 | 0.40 | 0.38 | 0.34 | 0.32 | 0.31 | 0.27 | 0.25 | 0.23 | 0.22 |
| Plug Valve 3 way thru-flo | 30 | 0.81 | 0.75 | 0.69 | 0.66 | 0.63 | 0.57 | 0.54 | 0.51 | 0.45 | 0.42 | 0.39 | 0.36 |
| Plug Valve branch - flo | 90 | 2.43 | 2.25 | 2.07 | 1.98 | 1.89 | 1.71 | 1.62 | 1.53 | 1.35 | 1.26 | 1.17 | 1.08 |

# A.10. Equivalent Lengths of Valves and Fittings

| Description | L/D |
|---|---|
| Gate Valve | 8 |
| Globe valve | 340 |
| Ball valve | 3 |
| Swing check valve | 50 |
| Standard Elbow - 90° | 30 |
| Standard Elbow - 45° | 16 |
| Long Radius Elbow - 90° | 16 |

Example: 14 inch Gate valve has L/D ratio = 8
Equivalent length = 8x14 inch = 112 inches = 9.25 feet

# A.11. Seam Joint Factors for Pipes

| Specification | Pipe Class | Seam Joint Factor (E) |
|---|---|---|
| ASTM A53 | Seamless | 1.00 |
| | Electric Resistance Welded | 1.00 |
| | Furnace Lap Welded | 0.80 |
| | Furnace Butt Welded | 0.60 |
| ASTM A106 | Seamless | 1.00 |
| ASTM A134 | Electric Fusion Arc Welded | 0.80 |
| ASTM A135 | Electric Resistance Welded | 1.00 |
| ASTM A139 | Electric Fusion Welded | 0.80 |
| ASTM A211 | Spiral Welded Pipe | 0.80 |
| ASTM A333 | Seamless | 1.00 |
| ASTM A333 | Welded | 1.00 |
| ASTM A381 | Double Submerged Arc Welded | 1.00 |
| ASTM A671 | Electric - Fusion - Welded | 1.00 |
| ASTM A672 | Electric - Fusion - Welded | 1.00 |
| ASTM A691 | Electric - Fusion - Welded | 1.00 |
| API 5L | Seamless | 1.00 |
| | Electric Resistance Welded | 1.00 |
| | Electric Flash Welded | 1.00 |
| | Submerged Arc Welded | 1.00 |
| | Furnace Lap Welded | 0.80 |
| | Furnace Butt Welded | 0.60 |
| API 5LX | Seamless | 1.00 |
| | Electric Resistance Welded | 1.00 |
| | Electric Flash Welded | 1.00 |
| | Submerged Arc Welded | 1.00 |
| API 5LS | Electric Resistance Welded | 1.00 |
| | Submerged Arc Welded | 1.00 |

# A.12. ANSI Pressure Ratings

| Class | Allowable Pressure (psi) |
|-------|--------------------------|
|       |                          |
| 150   | 275                      |
| 300   | 720                      |
| 400   | 960                      |
| 600   | 1 440                    |
| 900   | 2 160                    |
| 1500  | 3 600                    |

# A.13 Centrifugal Pump Performance
# Viscosity Correction

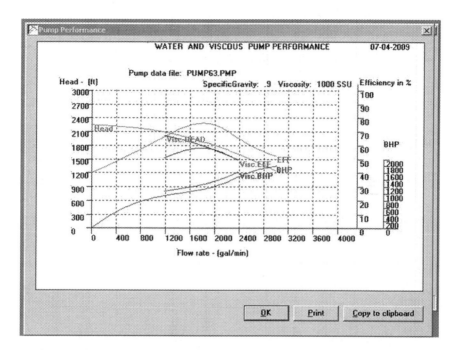

# A.14. Approximate Pipeline Construction Cost

| Pipe Diameter (inches) | Average Cost $/in-dia/mi |
|:---:|:---:|
| | |
| 8 | 18 000 |
| 10 | 20 000 |
| 12 | 22 000 |
| 16 | 14 900 |
| 20 | 20 100 |
| 24 | 33 950 |
| 30 | 34 600 |
| 36 | 40 750 |

# A.15 Thermal Hydraulics Summary

********** TEMPERATURE AND PRESSURE PROFILE **********

| Distance (mi) | FlowRate (bbl/day) | Temperature (degF) | SpGrav -- | Viscosity CST | Pressure (psi) | MAOP (psi) | Location |
|---|---|---|---|---|---|---|---|
| 0.00 | 100,000.00 | 190.00 | 0.9355 | 68.72 | 50.00 | 1400.00 | Compton |
| | | | | | | | |
| 0.00 | 100,000.00 | 192.74 | 0.9345 | 63.93 | 1181.18 | 1400.00 | Compton |
| 18.00 | 100,000.00 | 186.76 | 0.9366 | 74.96 | 792.70 | 1400.00 | |
| 22.00 | 100,000.00 | 185.50 | 0.9371 | 77.59 | 298.41 | 1400.00 | |
| 23.00 | 100,000.00 | 185.19 | 0.9372 | 78.26 | 299.60 | 1400.00 | |
| 24.00 | 100,000.00 | 184.88 | 0.9373 | 78.93 | 300.74 | 1400.00 | |
| 29.00 | 100,000.00 | 183.35 | 0.9378 | 82.36 | 306.20 | 1400.00 | |
| 35.00 | 100,000.00 | 181.56 | 0.9385 | 86.60 | 248.57 | 1400.00 | |
| 40.00 | 100,000.00 | 180.12 | 0.9390 | 90.24 | 50.00 | 1400.00 | Dimpton |
| | | | | | | | |
| 40.00 | 100,000.00 | 183.34 | 0.9378 | 82.39 | 1371.78 | 1400.00 | Dimpton |
| 40.58 | 100,000.00 | 183.16 | 0.9379 | 82.79 | 1348.95 | 1400.00 | |
| 50.00 | 100,000.00 | 180.38 | 0.9389 | 89.56 | 977.68 | 1400.00 | |
| 54.00 | 100,000.00 | 179.25 | 0.9393 | 92.52 | 868.04 | 1400.00 | |
| 55.00 | 100,000.00 | 178.97 | 0.9394 | 93.28 | 865.82 | 1400.00 | |
| 60.00 | 100,000.00 | 177.58 | 0.9398 | 97.10 | 813.82 | 1400.00 | |
| 70.00 | 100,000.00 | 174.91 | 0.9408 | 105.06 | 666.49 | 1400.00 | |
| 80.00 | 100,000.00 | 172.37 | 0.9417 | 113.38 | 391.64 | 1400.00 | |
| 90.00 | 100,000.00 | 169.97 | 0.9425 | 122.06 | 213.27 | 1400.00 | |
| 100.00 | 100,000.00 | 167.69 | 0.9433 | 131.08 | 50.00 | 1400.00 | Terminus |

# A.16 Thermal Hydraulics Pressure Gradient

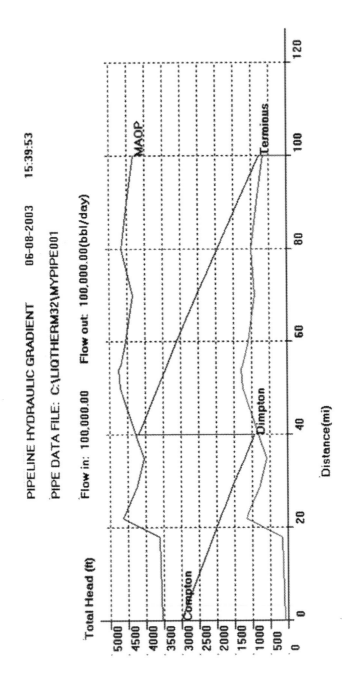

# A.17 Thermal Hydraulics Temperature Gradient

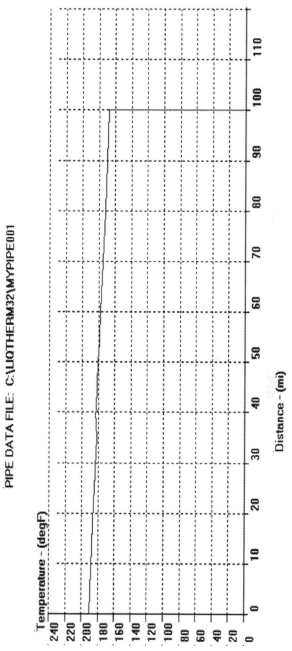

Temperature Gradient    06-08-2003    15:40:21

PIPE DATA FILE: C:\LIQTHERM32\MYPIPE001

Temperature – (degF)

Distance – (mi)

# APPENDIX B

# Answers to Selected Problems

## Chapter 2

Problem 2.9.1   Specific weight = 52.88 lb/ft$^3$ Specific gravity = 0.8474

Problem 2.9.2   Specific gravity at 85$^0$F = 0.845

Problem 2.9.3   Specific gravity = 0.7428

Problem 2.9.7   Blended viscosity = 145 SSU

Problem 2.9.8   75% of Liquid A and 25% of Liquid B

## Chapter 3

Problem 3.15.1   Average velocity = 8.03 ft/s.
                 Reynold's number = 203 038

Problem 3.15.2   Darcy friction factor f = 0.0164.
                 Modified Colebrook-White friction factor = 0.0167.
                 Pressure drop = 101.66 psi

Problem 3.15.4   Total Pressure = 36.72 MPa. Four pump stations

## Chapter 4

Problem 4.5.2   MAOP = 1 380 psi.
                Hydrotest range = 1 725 psi to 1 820 psi.

Problem 4.5.4   Volume per mile = 926.19 bbl

# Chapter 5

Problem 5.13.1   (a) 914 psi     (b) BHP = 2 201    Motor HP = 2 293 HP
                 (c) 2 500 HP  (d) 6 510 bbl/hr
Problem 5.13.2   640 mm diameter, 10 mm wall thickness.
Problem 5.13.3   8 177 bbl/hr gasoline flow rate, Gasoline head = 4 277 ft
                 Diesel head = 3 724 ft

# Chapter 6

Problem 6.7.1    (a) 0.375 inch wall thickness     (b) Two pump station
                 (c) Two additional pump station
                 (d) 6 888 HP and 18 530 HP at 75% pump efficiency.

# Chapter 7

Problem 7.14.1   (a) Trim impeller in Pump A or Pump B
                 (b) 9.2 inch impeller for Pump A    (c) 3 282 RPM

Problem 7.14.2

| Q (gal/min) | 0 | 500 | 1 000 | 1 500 | 2 000 |
|---|---|---|---|---|---|
| P (psi) | 130 | 208 | 419 | 758 | 1 224 |

# Chapter 8

Problem 8.5.1    (a) 122 psi  (b) $111 542 based on 350 days operation/yr
Problem 8.5.2    Speed of VSD pump = 3 126 RPM

# Chapter 9

Problem 9.4.1    Minimum flow rate = 1 050 bbl/hr
                 HP = 298
Problem 9.4.2    Minimum flow rate = 1 500 bbl/hr
                 HP = 895

# Chapter 10

Problem 10.9.1  (a) R = 212 750     (b) 3.84 inch throat diameter
(c) R = 221 288
Problem 10.9.3  Flow rate = 959 gal/min

# Chapter 11

Problem 11.6.1  Potential surge = 903 ft Wave speed = 3 913 ft/s
Problem 11.6.3  Thin wall:    3 967 ft/s      3 926 ft/s      3 868 ft/s
Thick wall:  3 942 ft/s      3 903 ft/s      3 847 ft/s

# APPENDIX C

## Summary of Formulas

## Chapter 2

### API gravity/Specific gravity

Specific gravity Sg = 141.5/(131.5 +API)            (2.1)

API = 141.5/Sg - 131.5            (2.2)

### Specific gravity versus Temperature

$$S_T = S_{60} - a\ (T-60)$$            (2.3)

where

$S_T$       - Specific gravity at temperature T
$S_{60}$    - Specific gravity at 60°F
T       - Temperature, °F
a       - A constant that depends on the liquid

### Specific gravity blending

$$S_b = \frac{(Q_1 \times S_1) + (Q_2 \times S_2) + (Q_3 \times S_3) + \ldots}{Q_1 + Q_2 + Q_3 + \ldots}$$            (2.4)

where

S$_b$               - Specific gravity of the blended liquid
Q$_1$, Q$_2$, Q$_3$ etc   - Volume of each component
S$_1$, S$_2$, S$_3$   etc   - Specific gravity of each component

## Viscosity Conversion

$$\text{Centistokes} = 0.226(\text{SSU}) - 195/(\text{SSU}) \tag{2.8}$$
$$\text{for } 32 \leq \text{SSU} \leq 100$$

$$\text{Centistokes} = 0.220(\text{SSU}) - 135/(\text{SSU}) \tag{2.9}$$
$$\text{for SSU} > 100$$

$$\text{Centistokes} = 2.24(\text{SSF}) - 184/(\text{SSF}) \tag{2.10}$$
$$\text{for } 25 < \text{SSF} \leq 40$$

$$\text{Centistokes} = 2.16(\text{SSF}) - 60/(\text{SSF}) \tag{2.11}$$
$$\text{for SSU} > 40$$

## Viscosity versus Temperature

$$\text{Log Log } (Z) = A - B\text{Log}(T) \tag{2.15}$$

where

Log is the logarithm to base 10

Z     - depends on viscosity of the liquid $v$
$v$     - Viscosity of liquid, cSt
T     - Absolute temperature, °R or °K

A and B are constants that depend on the specific liquid.

The variable Z is defined as follows:

$$Z = (v + 0.7 + C - D) \tag{2.16}$$
$$C = \exp[-1.14883 - 2.65868(v)] \tag{2.17}$$
$$D = \exp[-0.0038138 - 12.5645(v)] \tag{2.18}$$

## Viscosity Blending

$$\sqrt{V_b} = \frac{Q_1 + Q_2 + Q_3 + \ldots}{(Q_1 / \sqrt{V_1}) + (Q_2 / \sqrt{V_2}) + (Q_3 / \sqrt{V_3})} \tag{2.21}$$

where

$V_b$            - Viscosity of blend, SSU
$Q_1, Q_2, Q_3$   etc    - Volumes of each component
$V_1, V_2, V_3$   etc    - Viscosity of each component, SSU

$$H = 40.073 - 46.414 \, \text{Log}_{10} \, \text{Log}_{10} \, (V+B) \tag{2.22}$$
$$B = 0.931 \, (1.72)^V \text{ for } 0.2 < V < 1.5 \tag{2.23}$$
$$B = 0.6 \text{ for } V >= 1.5 \tag{2.24}$$
$$Hm = [H1(pct1) + H2(pct2) + H3(pct3) + \ldots]/100 \tag{2.25}$$

where

H, H1, H2 ....    - Blending index of liquids.
Hm             Blending index of mixture.
B               Constant in Blending Index Equation.
V               Viscosity, cSt.
pct1, pct2, ....    - Percentage of liquids 1, 2, .... in blended
                    mixture.

## Bulk Modulus

Adiabatic Bulk Modulus

$$Ka = A + B(P) - C(T)^{\frac{1}{2}} - D(API) - E(API)^2 + F(T)(API) \tag{2.27}$$

where

A = 1.286x10$^6$        B = 13.55        C = 4.122x10$^4$
D = 4.53x10$^3$         E = 10.59        F = 3.228

Isothermal Bulk Modulus

$$Ki = A + B(P) - C(T)^{1/2} + D(T)^{3/2} - E(API)^{3/2} \qquad (2.28)$$

where

A = 2.619x10$^6$        B = 9.203
C = 1.417x10$^5$        D = 73.05
E = 341.0

P       - Pressure in psig
T       - Temperature in °R
API     - API gravity of liquid

## Bernoulli's Equation

$$Z_A + P_A/\gamma + V_A^2/2g + H_P = Z_B + P_B/\gamma + V_B^2/2g + \Sigma h_L \qquad (2.37)$$

where

$Z_A$, $P_A$ and $V_A$ represent the elevation, pressure and liquid velocity at point A.
$Z_B$, $P_B$ and $V_B$ represent the elevation, pressure and liquid velocity at point B
$\gamma$ is the specific weight of liquid.
$H_P$ is the pump head input at point A.

and

$h_L$ is the head lost in friction between point A and point B

# Chapter 3

## Pressure and Head

$$Head = 2.31(psig)/Spgr \text{ ft - English units} \qquad (3.7)$$
$$Head = 0.102(kPa)/Spgr \text{ m - SI units.} \qquad (3.8)$$

where

Spgr  - Liquid specific gravity

## Velocity of Flow

$$V = 0.0119(bbl/day)/ D^2 \qquad (3.10)$$
$$V = 0.4085(gal/min)/ D^2 \qquad (3.11)$$
$$V = 0.2859(bbl/hr)/D^2 \qquad (3.12)$$

where

V      - Velocity, ft/s
D      - Inside diameter, in

$$V = 353.6777 \ (m^3/hr)/ D^2 \qquad (3.13)$$

where

V      - Velocity, m/s
D      - Inside diameter, mm.

## Reynold's Number

$$R = VD\rho/\mu \qquad (3.14)$$
$$R = VD/\nu \qquad (3.15)$$

where

V     - Average velocity, ft/s
D     - Pipe internal diameter, ft
$\rho$     - Liquid density, slugs/ft$^3$
$\mu$     - Absolute viscosity, lb-s/ft$^2$
R     - Reynold's number is a dimensionless value.
$\nu$     - Kinematic viscosity, ft$^2$/s

$$R = 92.24 \, Q/(\nu D) \tag{3.16}$$

where

Q     - Flow rate, bbl/day
D     - Internal diameter, in
$\nu$     - Kinematic viscosity, cSt.

$$R = 3\,160 \, Q/(\nu D) \tag{3.17}$$

where

Q     - Flow rate, gal/min

$$R = 353\,678 \, Q/(\nu D) \tag{3.18}$$

where

Q     - Flow rate, m$^3$/h
D     - Internal diameter, mm
$\nu$     - Kinematic viscosity, cSt.

## Darcy-Weisbach Equation for Head Loss

$$h = f(L/D)(V^2/2g) \tag{3.19}$$

where

h      - head loss, ft of liquid
f      - Darcy friction factor, dimensionless,
L      - Pipe length, ft
D      - Pipe internal diameter, ft
V      - Average liquid velocity, ft/s
g      - acceleration due to gravity, 32.2 ft/s$^2$ - English units.

## Darcy Friction Factor

For laminar flow, with Reynold's number R < 2 000

$$f = 64/R \tag{3.20}$$

For turbulent flow, with Reynold's number R > 4 000

(Colebrook-White Equation)
$$1/\sqrt{f} = -2 \, Log_{10}[(e/3.7D) + 2.51/(R \sqrt{f})] \tag{3.21}$$

where

f      - Darcy friction factor, dimensionless
D      - Pipe internal diameter, in
e      - Absolute pipe roughness, in
R      - Reynold's number of flow, dimensionless

$$f_f = f_d / 4 \tag{3.25}$$

where

$f_f$      - Fanning friction factor
$f_d$      - Darcy friction factor

## Pressure drop due to friction

$$Pm = 0.0605 \, f \, Q^2 \, (Sg/D^5) \tag{3.27}$$
$$Pm = 0.2421(Q/F)^2 \, (Sg/D^5) \tag{3.28}$$

where

Pm   - Pressure drop due to friction, psi per mile
Q     - Liquid flow rate, bbl/day
f      - Darcy friction factor, dimensionless
F     - Transmission factor, dimensionless
Sg   - Liquid specific gravity
D     - Pipe internal diameter, in

$$F = 2/\sqrt{f} \tag{3.29}$$

where

F     - Transmission factor
f      - Darcy friction factor

$$F = -4 \, Log \, [(e/3.7D) + 1.255(F/R)] \tag{3.30}$$
for Turbulent flow R > 4 000

In SI units

$$P_{km} = 6.2475 \times 10^{10} \, f \, Q^2 \, (Sg/D^5) \tag{3.31}$$
$$P_{km} = 24.99 \times 10^{10} \, (Q/F)^2 \, (Sg/D^5) \tag{3.32}$$

where

$P_{km}$   - Pressure drop due to friction, kPa/km
Q     - Liquid flow rate, m³/hr
f      - Darcy friction factor, dimensionless
F     - Transmission factor, dimensionless
Sg   - Liquid specific gravity
D     - Pipe internal diameter, mm

## Colebrook-White Equation

Modified Colebrook-White Equation

$$F = -4 \, Log \, [(e/3.7D) + 1.4125(F/R)] \tag{3.34}$$

## Hazen-Williams Equation

$$h = 4.73 \, L \, (Q/C)^{1.852} / D^{4.87} \tag{3.35}$$

where

h  - Head loss due to friction, ft
L  - Length of pipe, ft
D  - Internal diameter of pipe, ft
Q  - Flow rate, ft³/s
C  - Hazen-Williams coefficient or C-Factor, dimensionless

$$Q = 0.1482(C) \, (D)^{2.63}(Pm/Sg)^{0.54} \tag{3.36}$$

where

Q   - Flow rate, bbl/day
D   - Pipe internal diameter, in
Pm  - Frictional pressure drop, psi/mi
Sg  - Liquid specific gravity
C   - Hazen-Williams C-factor

$$GPM = 6.7547 \times 10^{-3} \, (C) \, (D)^{2.63} \, (H_L)^{0.54} \tag{3.37}$$

where

GPM  - Flow rate, gal/min
$H_L$    - Friction loss, ft per 1000 ft

In SI units

$$Q = 9.0379 \times 10^{-8} \, (C)(D)^{2.63}(P_{km}/Sg)^{0.54} \tag{3.38}$$

where

Q    - Flow rate, m³/hr
D    - Pipe internal diameter, mm
$P_{km}$  - Frictional pressure drop, kPa/km

Sg     - Liquid specific gravity

C     - Hazen-Williams C-factor

## Shell-MIT Equation

$$R = 92.24(Q)/(Dv) \qquad (3.39)$$
$$Rm = R/(7\ 742) \qquad (3.40)$$

where

R     - Reynold's number, dimensionless

Rm     - Modified Reynold's number, dimensionless

Q     - Flow rate, bbl/day

D     - Internal diameter, in

v     - Kinematic viscosity, cSt.

$$f = 0.00207/Rm \qquad \text{for Laminar flow} \qquad (3.41)$$
$$f = 0.0018 + 0.00662(1/Rm)^{0.355} \qquad (3.42)$$
for Turbulent flow

$$Pm = 0.241\ (f\ SgQ^2)/D^5 \qquad (3.43)$$

where

Pm     - Frictional pressure drop, psi/mi

f     - Friction factor, dimensionless

Sg     - Liquid specific gravity

Q     - Flow rate, bbl/day

D     - Pipe internal diameter, in

In SI units

$$Pm = 6.2191 \times 10^{10}\ (f\ SgQ^2)/D^5 \qquad (3.44)$$

where

Pm     - Frictional pressure drop, kPa/km

f     - Friction factor, dimensionless

Sg    - Liquid specific gravity
Q    - Flow rate, m³/hr
D    - Pipe internal diameter, mm

## Miller Equation

$$Q = 4.06 \,(M) \,(D^5 P_m/Sg)^{0.5} \tag{3.45}$$
$$M = Log_{10}(D^3 SgP_m/cp^2) + 4.35 \tag{3.46}$$

where

Q    - Flow rate, bbl/day
D    - Pipe internal diameter, in
$P_m$    - Pressure drop, psi/mi
Sg    - Liquid specific gravity
cp    - Liquid viscosity, centipoise

In SI Units

$$Q = 3.996 \times 10^{-6} \,(M) \,(D^5 P_m/Sg)^{0.5} \tag{3.47}$$
$$M = Log_{10}(D^3 SgP_m/cp^2) - 0.4965 \tag{3.48}$$

where

Q    - Flow rate, m³/hr
D    - Pipe internal diameter, mm
Pm    - Frictional pressure drop, kPa/km
Sg    - Liquid specific gravity
cp    - Liquid viscosity, centipoise

$$P_m = (Q / 4.06M)^2 \,(Sg / D^5) \tag{3.49}$$

## T.R. Aude Equation

$$P_m = [Q(z^{0.104}) \,(Sg^{0.448}) / (0.871(K) \,(D^{2.656}))]^{1.812} \tag{3.50}$$

where

P$_m$  - Pressure drop due to friction, psi/mi
Q  - Flow rate, bbl/hr
D  - Pipe internal diameter, in
Sg  - Liquid specific gravity
z  - Liquid viscosity, centipoise
K  - T.R. Aude K-factor, usually 0.90 to 0.95

In SI Units

$$P_m = 8.888 \times 10^8 \left[ Q(z^{0.104}) (Sg^{0.448}) / (K (D^{2.656})) \right]^{1.812} \qquad (3.51)$$

where

Pm  - Frictional pressure drop, kPa/km
Sg  - Liquid specific gravity
Q  - Flow rate, m$^3$/hr
D  - Pipe internal diameter, mm
z  - Liquid viscosity, centipoise
K  - T.R. Aude K-factor, usually 0.90 to 0.95

**Head Loss in Valves and Fittings**

$$h = K V^2/2g \qquad (3.52)$$

where

h  - Head loss due to valve or fitting, ft
K  - Head loss coefficient for the valve or fitting, dimensionless
V  - Velocity of liquid through valve or fitting, ft/s
g  - Acceleration due to gravity, 32.2 ft/s$^2$ - English units

**Gradual Enlargement**

$$h = K (V_1 - V_2)^2 / 2g \qquad (3.53)$$

Head loss coefficient K depends upon the diameter ratio $D_1/D_2$ and the different cone angle due to the enlargement.

**Sudden Enlargement**

$$h = (V_1 - V_2)^2 / 2g \qquad (3.54)$$

**Drag Reduction**

$$\text{Percentage Drag Reduction} = 100(DP_0 - DP_1)/DP_0 \qquad (3.55)$$

where

$DP_0$ - Friction drop in pipe segment without DRA, psig
$DP_1$ - Friction drop in pipe segment with DRA, psig

**Explicit Friction Factor Equations**
**Churchill Equation**

This equation proposed by Stuart Churchill for friction factor was reported in Chemical Engineering magazine in November 1977. Unlike the Coelbrook-White Equation that requires trial and error solution, this equation is explicit in f as indicated below.

$$f = [(8 / R)^{12} + 1/ (A+B)^{3/2}]^{1/12} \qquad (3.56)$$

where

$$A = [2.457 Log_e (1 / ((7/R)^{0.9} + (0.27e / D)))]^{16} \qquad (3.57)$$

$$B = (37\,530 / R)^{16} \qquad (3.58)$$

The above equation for friction factor appears to correlate well with the Colebrook-White equation

**Swamee-Jain Equation**

P.K. Swamee and A.K. Jain presented this equation in 1976 in the Journal of the Hydraulics Division of ASCE. It is found to be the best and easiest of all explicit equations for calculating the friction factor.

$$f = 0.25 / [Log_{10} (e / 3.7D + 5.74 / R^{0.9})]^2 \qquad (3.59)$$

It correlates well with the Colebrook-White equation.

# Chapter 4

**Barlow's Equation for Internal Pressure**

$$S_h = PD/2t \qquad (4.1)$$

where

$S_h$     - Hoop stress, psi
P     - Internal pressure, psi
D     - Pipe outside diameter, in
t     - Pipe wall thickness, in

$$S_a = PD/4t \qquad (4.2)$$

where

$S_a$     - Axial (or longitudinal) stress, psi

Internal design pressure in a pipe in English units

$$P = \frac{2 \, T \times S \times E \times F}{D} \qquad (4.3)$$

where

P     - Internal pipe design pressure, psig
D     - Nominal pipe outside diameter, in
T     - Nominal Pipe wall thickness, in
S     - Specified Minimum Yield Strength (SMYS) of pipe material, psig
E     - Seam Joint Factor, 1.0 for seamless and Submerged Arc Welded (SAW) pipes. See Table in Appendix A.11
F     - Design Factor,

In SI units

$$P = \frac{2\,T\,x\,S\,x\,E\,x\,F}{D} \tag{4.4}$$

where

P     - Pipe internal design pressure, kPa
D     - Nominal pipe outside diameter, mm
T     - Nominal pipe wall thickness, mm
S     - Specified Minimum Yield Strength (SMYS) of pipe material, kPa
E     - Seam Joint Factor, 1.0 for seamless and Submerged Arc Welded (SAW) pipes. See Table in Appendix A.11
F     - Design Factor

## Line Fill Volume

In English Units

$$V_L = 5.129(D)^2 \tag{4.7}$$

where

$V_L$     - Line fill volume of pipe, bbl/mile
D     - Pipe inside diameter, in

In SI Units

$$V_L = 7.855 \times 10^{-4} \, D^2 \tag{4.8}$$

where

$V_L$    - Line fill volume, $m^3/km$
$D$     - Pipe inside diameter, mm

# Chapter 5

## Total pressure required

$$P_t = P_{friction} + P_{elevation} + P_{del} \tag{5.1}$$

where

Pt          - Total pressure required at A
$P_{friction}$    - Total friction pressure drop between A and B
$P_{elevation}$   - Elevation head between A and B
$P_{del}$      - Required delivery pressure at B

## Pump station discharge pressure

$$P_d = (P_t + P_s) / 2 \tag{5.3}$$

where

$P_d$    - Pump station discharge pressure
$P_s$    - Pump station suction pressure

## Equivalent length of series piping

$$L_E / (D_E)^5 = L_A / (D_A)^5 + L_B / (D_B)^5 \tag{5.7}$$

where

$L_E$     - Equivalent length of diameter $D_E$
$L_A, D_A$ - Length and diameter of pipe A
$L_B, D_B$ - Length and diameter of pipe B

## Equivalent diameter of parallel piping

$$Q^2 / D^5_E = Q^2_{BC} / D^5_{BC} \qquad\qquad (5.14)$$
$$Q^2_{BC} / D^5_{BC} = (Q - Q_{BC})^2 / D^5_{BD} \qquad (5.15)$$

where

$Q$     - Total flow through both parallel pipes.
$Q_{BC}$ - Flow through pipe branch BC
$(Q$    $- Q_{BC})$- Flow through pipe branch BD
$D_{BC}$ - Inside diameter of pipe branch BC
$D_{BD}$ - Inside diameter of pipe branch BD
$D_E$     - Equivalent pipe diameter to replace both parallel pipes BC and BD

## Horsepower required for pumping

$$BHP = Q\,P / (2\,449E) \qquad\qquad (5.16)$$

where

$Q$     - Flow rate, bbl/hr
$P$     - Differential Pressure, psi
$E$     - Efficiency, expressed as a decimal value less than 1.0

$$BHP = (GPM)(H)(Spgr) / (3\,960E) \qquad (5.17)$$

$$BHP = (GPM)P / (1\,714E) \qquad\qquad (5.18)$$

where

GPM - Flow rate, gal/min
H    - Differential head, ft
P    - Differential Pressure, psi
E    - Efficiency, expressed as a decimal value less than 1.0
Spgr - Liquid specific gravity, dimensionless

In SI units:

$$\text{Power (kW)} = \frac{Q\ H\ Spgr}{367.46\ (E)} \qquad (5.19)$$

where

Q    - Flow rate, m³/hr
H    - Differential head, m
Spgr - Liquid specific gravity
E    - Efficiency, expressed as a decimal value, less than 1.0

$$\text{Power (kW)} = \frac{Q\ P}{3\ 600\ (E)} \qquad (5.20)$$

where

P    - Pressure in kPa
Q    - Flow rate in m³/hr
E    - Efficiency, expressed as a decimal value less than 1.0

# Chapter 6

## Pump station discharge pressure

$$P_D = (P_T - P_S) / N + P_S \qquad (6.4)$$

where

$P_D$     - Pump station discharge pressure
$P_T$     - Total pressure required
$P_S$     - Pump station suction pressure
N      - Number of pump stations

## Line fill volume

$$\text{Line fill volume} = 5.129 L(D)^2 \tag{6.5}$$

where

D     - Pipe inside diameter, in
L     - Pipe length, mi

# Chapter 7

## Brake Horsepower

$$\text{Pump BHP} = \frac{Q\,H\,Sg}{3\,960\,(E)} \tag{7.1}$$

where

Q     - Pump flow rate, gal/min
H     - Pump head, ft
E     - Pump efficiency as a decimal value, less than 1.0
Sg    - Liquid specific gravity (for water Sg = 1.0)

In SI units

$$\text{Power kW} = \frac{Q\,H\,Sg}{367.46\,(E)} \tag{7.2}$$

where

Q    - Pump flow rate in $m^3/hr$
H    - Pump head in meters
E    - Pump efficiency as a decimal value, less than 1.0
Sg    - Liquid specific gravity (for water Sg = 1.0)

## Specific Speed

$$N_S = N \, Q^{1/2} / H^{3/4} \qquad\qquad (7.3)$$

where

$N_S$    - Pump specific speed
N    - Pump impeller speed, RPM
Q    - Flow rate or capacity, gal/min
H    - Head, ft

## Suction Specific Speed

$$N_{SS} = N \, Q^{1/2} / (NPSH_R)^{3/4} \qquad\qquad (7.4)$$

where

$N_{SS}$    - Suction specific speed
N    - Pump impeller speed, RPM
Q    - Flow rate or capacity, gal/min
$NPSH_R$    - NPSH required at BEP

## Impeller Diameter Change

$$Q_2 / Q_1 = D_2 / D_1 \qquad\qquad (7.5)$$
$$H_2 / H_1 = (D_2 / D_1)^2 \qquad\qquad (7.6)$$

where

| | |
|---|---|
| $Q_1, Q_2$ | - Initial and final flow rates |
| $H_1, H_2$ | - Initial and final heads |
| $D_1, D_2$ | - Initial and final impeller diameters |

## Impeller Speed Change

$$Q_2 / Q_1 = N_2 / N_1 \qquad\qquad (7.7)$$
$$H_2 / H_1 = (N_2 / N_1)^2 \qquad\qquad (7.8)$$

where

| | |
|---|---|
| $Q_1, Q_2$ | - Initial and final flow rates |
| $H_1, H_2$ | - Initial and final heads |
| $N_1, N_2$ | - Initial and final impeller speeds |

## Suction Piping

| | | |
|---|---|---|
| Suction Head | $= H_S - H_{fs}$ | (7.10) |
| Discharge Head | $= H_D + H_{fd}$ | (7.11) |

where

| | |
|---|---|
| $H_S$ | - Static suction head |
| $H_D$ | - Static discharge head |
| $H_{fs}$ | - Friction loss in suction piping |
| $H_{fd}$ | - Friction loss in discharge piping |

## NPSH

$$(P_a - P_v)(2.31/Sg) + H + E1 - E2 - h \qquad\qquad (7.12)$$

where

| | |
|---|---|
| $P_a$ | - Atmospheric pressure, psi |
| $P_v$ | - Liquid vapor pressure at flowing temperature, psi |

Sg    - Liquid specific gravity
H     - Tank head, ft
E1    - Elevation of tank bottom, ft
E2    - Elevation of pump suction, ft
H     - Friction loss in suction piping, ft

# Chapter 8

## Control Pressure and Throttle Pressure

$$P_{case} = P_s + \Delta P_1 + \Delta P_2 \qquad (8.1)$$

where

$P_{case}$  - Case pressure in Pump2 or upstream pressure at control valve

$$P_{thr} = P_{case} - P_d \qquad (8.2)$$

where

$P_{thr}$  - Control valve throttle pressure
$P_d$    - Pump station discharge pressure

# Chapter 9

## Thermal Conductivity

$$H = K (A)(dT/dx) \qquad (9.1)$$

where

H    - Heat flux perpendicular to the surface area, Btu/hr
K    - Thermal conductivity, Btu/hr/ft/ °F.
A    - Area of heat flux, ft$^2$

dx    - Thickness of solid, ft

dT    - Temperature difference across the solid, °F

In SI units

$$H = K (A)(dT/dx) \tag{9.2}$$

where

H    - Heat flux, W

K    - Thermal conductivity, W/m / °C.

A    - Area of heat flux, $m^2$

dx    - Thickness of solid, m

dT    - Temperature difference across the solid, °C

## Heat Transfer Coefficient

$$H = U (A)(dT) \tag{9.3}$$

where

U    - Overall heat transfer coefficient, Btu/hr/ft$^2$/ °F

Other symbols in Equation (9.3) are the same as in Equation (9.1)

In SI units

$$H = U (A)(dT) \tag{9.4}$$

where

U    - Overall heat transfer coefficient, W/m$^2$/ °C

Other symbols in Equation (9.4) are the same as in Equation (9.2)

## Heat Content Balance

$$Hin - DeltaH + Hw = Hout \tag{9.6}$$

where

Hin        - Heat content entering line segment, Btu/hr
DeltaH     - Heat transferred from line segment to surrounding medium(soil or air), Btu/hr
Hw        - Heat content from frictional work, Btu/hr
Hout      - Heat content leaving line segment, Btu/hr

In SI units, Equation (9.6) will be the same, with each term expressed in Watts instead of Btu/hr.

## Logarithmic Mean Temperature

$$Tm - Ts = \frac{(T1 - Ts) - (T2 - Ts)}{Log_e\,[(T1 - Ts)/(T2 - Ts)]} \tag{9.7}$$

where

Tm   - Logarithmic mean temperature of pipe segment, °F
T1    - Temperature of liquid entering pipe segment, °F
T2    - Temperature of liquid leaving pipe segment, °F
Ts    - Sink temperature(soil or surrounding medium), °F

In SI units, Equation (9.7) will be the same, with all temperatures expressed in °C instead of °F.

## Heat Content Entering and Leaving a Pipe Segment

$$H_{in} = w(C_{pi})(T_1) \tag{9.8}$$
$$H_{out} = w(C_{po})(T_2) \tag{9.9}$$

where

$H_{in}$    - Heat content of liquid entering pipe segment, Btu/hr
$H_{out}$  - Heat content of liquid leaving pipe segment, Btu/hr
$C_{pi}$    - Specific heat of liquid at inlet, Btu/lb/ °F
$C_{po}$   - Specific heat of liquid at outlet, Btu/lb/ °F
w     - Liquid flow rate, lb/hr

$$T_1 \quad \text{- Temperature of liquid entering pipe segment, } °F$$
$$T_2 \quad \text{- Temperature of liquid leaving pipe segment, } °F$$

In SI units,

$$H_{in} = w(C_{pi})(T_1) \tag{9.10}$$
$$H_{out} = w(C_{po})(T_2) \tag{9.11}$$

where

$H_{in}$    - Heat content of liquid entering pipe segment, J/s (W)
$H_{out}$   - Heat content of liquid leaving pipe segment, J/s (W)
$C_{pi}$    - Specific heat of liquid at inlet, kJ/kg/ °C
$C_{po}$   - Specific heat of liquid at outlet, kJ/kg/ °C
w     - Liquid flow rate, kg/s
$T_1$     - Temperature of liquid entering pipe segment, °C
$T_2$     - Temperature of liquid leaving pipe segment, °C

## Heat Transfer - Buried Pipeline

$$H_b = 6.28 \, (L) \, (T_m - T_s) \, / \, (\text{Parm1} + \text{Parm2}) \tag{9.12}$$
$$\text{Parm1} = (1/K_{ins}) \, \text{Log}_e (R_i /R_p) \tag{9.13}$$
$$\text{Parm2} = (1/K_s) \, \text{Log}_e [2S/D + ((2S/D)^2 -1)^{1/2}] \tag{9.14}$$

where

$H_b$     - Heat transfer, Btu/hr
$T_m$    - Log mean temperature of pipe segment, °F
$T_s$     - Ambient soil temperature, °F
L      - Pipe segment length, ft
$R_i$     - Pipe insulation outer radius, ft
$R_p$    - Pipe wall outer radius, ft
$K_{ins}$   - Thermal conductivity of insulation, Btu/hr/ft/ °F
$K_s$     - Thermal conductivity of soil, Btu/hr/ft/ °F
S      - Depth of cover to pipe centerline, ft
D     - Pipe outside diameter, ft

In SI units

$$H_b = 6.28 \ (L) \ (T_m - T_s) \ / \ (Parm1 + Parm2) \qquad (9.15)$$
$$Parm1 = (1/K_{ins}) \ Log_e(R_i /R_p) \qquad (9.16)$$
$$Parm2 = (1/K_s) \ Log_e[2S/D + ((2S/D)^2 -1)^{1/2}] \qquad (9.17)$$

where

| | |
|---|---|
| $H_b$ | - Heat transfer, W |
| $T_m$ | - Log mean temperature of pipe segment, °C |
| $T_s$ | - Ambient soil temperature, °C |
| L | - Pipe segment length, m |
| $R_i$ | - Pipe insulation outer radius, mm |
| $R_p$ | - Pipe wall outer radius, mm |
| $K_{ins}$ | - Thermal conductivity of insulation, W/m/ °C |
| $K_s$ | - Thermal conductivity of soil, W/m/ °C |
| S | - Depth of cover to pipe centerline, mm |
| D | - Pipe outside diameter, mm |

## Heat Transfer - Above Ground Pipeline

$$H_a = 6.28 \ (L) \ (T_m - T_s) \ / \ (Parm1 + Parm3) \qquad (9.18)$$
$$Parm3 = 1.25/[R_i \ (4.8 + 0.008(T_m - T_s))] \qquad (9.19)$$
$$Parm1 = (1/K_{ins}) \ Log_e(R_i /R_p) \qquad (9.20)$$

where

| | |
|---|---|
| $H_a$ | - Heat transfer, Btu/hr |
| $T_m$ | - Log mean temperature of pipe segment, °F |
| $T_s$ | - Ambient soil temperature, °F |
| L | - Pipe segment length, ft |
| $R_i$ | - Pipe insulation outer radius, ft |
| $R_p$ | - Pipe wall outer radius, ft |
| $K_{ins}$ | - Thermal conductivity of insulation, Btu/hr/ft/ °F |
| $K_s$ | - Thermal conductivity of soil, Btu/hr/ft/ °F |
| S | - Depth of cover to pipe centerline, ft |
| D | - Pipe outside diameter, ft |

In SI units

$$H_a = 6.28 \, (L) \, (T_m - T_s) / (Parm3 + Parm1) \tag{9.21}$$
$$Parm3 = 1.25/[R_i \, (4.8 + 0.008(T_m - T_s))] \tag{9.22}$$
$$Parm1 = (1/K_{ins}) \, Log_e(R_i /R_p) \tag{9.23}$$

where

$H_a$     - Heat transfer, W
$T_m$     - Log mean temperature of pipe segment, °C
$T_s$     - Ambient soil temperature, °C
L     - Pipe segment length, m
$R_i$     - Pipe insulation outer radius, mm
$R_p$     - Pipe wall outer radius, mm
$K_{ins}$     - Thermal conductivity of insulation, W/m/°C
$K_s$     - Thermal conductivity of soil, W/m/°C
S     - Depth of cover to pipe centreline, mm
D     - Pipe outside diameter, mm

## Frictional Heating

$$H_w = 2\,545 \, (HHP) \tag{9.24}$$
$$HHP = (1.7664 \times 10^{-4}) \, (Q)(Sg) \, (h_f)(L_m) \tag{9.25}$$

where

$H_w$     - Frictional Heat gained, Btu/hr
HHP     - Hydraulic horsepower required for pipe friction
Q     - Liquid flow rate, bbl/hr
Sg     - Liquid specific gravity
$h_f$     - Frictional head loss, ft/mi
$L_m$     - Pipe segment length, mi

In SI units

$$H_w = 1\,000 \, (Power) \tag{9.26}$$
$$Power = (0.00272) \, (Q)(Sg)((h_f)(L_m)) \tag{9.27}$$

where

$H_w$     - Frictional Heat gained, W
Power - Power required for pipe friction, kW
Q      - Liquid flow rate, m³/hr
Sg     - Liquid specific gravity
$h_f$     - Friction loss, m/km
$L_m$    - Pipe segment length, km

**Pipe Segment Outlet Temperature**

For buried pipe:

$$T_2 = (1/wC_p)[2\,545\,(HHP) - H_b + (wC_p)T_1] \tag{9.28}$$

For above ground pipe:

$$T_2 = (1/wC_p)[2\,545\,(HHP) - H_a + (wC_p)T_1] \tag{9.29}$$

where

$H_b$     - Heat transfer for buried pipe, Btu/hr from Equation (9.12)
$H_a$     - Heat transfer for above ground pipe, Btu/hr from Equation (9.18)
$C_p$     - Average specific heat of liquid in pipe segment

In SI units

For buried pipe:

$$T_2 = (1/wC_p)[1\,000\,(Power) - H_b + (wC_p)T_1] \tag{9.30}$$

For above ground pipe:

$$T_2 = (1/wC_p)[1\,000\,(Power) - H_a + (wC_p)T_1] \tag{9.31}$$

where

$H_b$     - Heat transfer for buried pipe, W
$H_a$     - Heat transfer for above ground pipe, W
Power - Frictional Power defined in Equation (9.27), kW

# Chapter 10

## Mass flow

$$Q = AV \tag{10.2}$$

where

A     - cross sectional area of flow
V     - velocity of flow
$\rho$     - liquid density

## Venturi Meter

Velocity in Main Pipe Section:

$$V_1 = C \sqrt{[2g(P_1 - P_2)/\gamma] / [(A_1/A_2)^2 - 1]} \tag{10.6}$$

Velocity in throat,

$$V_2 = C \sqrt{[2g(P_1 - P_2)/\gamma] / [1 - (A_2/A_1)^2]} \tag{10.7}$$

Volume flow rate:

$$Q = CA_1 \sqrt{[2g(P_1 - P_2)/\gamma] / [(A_1/A_2)^2 - 1]} \tag{10.8}$$
$$Q = CA_1 \sqrt{[2g(P_1 - P_2)/\gamma] / [(1/\beta)^4 - 1]} \tag{10.9}$$

where

P1     - Pressure in main section of diameter D and area $A_1$
P2     - Pressure in throat section of diameter d and area $/A_2$

γ    - Specific weight of liquid
C    - Coefficient of discharge that depends on Reynolds number and diameters

Beta ratio $\beta - d/D$ and $A_1/A_2 = (D/d)^2$

## Flow nozzle discharge coefficient:

$$C = 0.9975 - 6.53 \sqrt{(\beta/R)} \qquad (10.10)$$

where

$$\beta - d/D$$

and    R is the Reynold's number based on the main pipe diameter D.

## Turbine Meter

$$Q_b = Q_f \times M_f \times F_t \times F_p \qquad (10.11)$$

where

$Q_b$    - Flow rate at base conditions, such as 60 °F and 14.7 psi
$Q_f$    - Measured Flow rate at operating conditions, such as 80 °F and 350 psi
$M_f$    - Meter factor for correcting meter reading, based on meter calibration data.
$F_t$    - Temperature correction factor for correcting from flowing temperature to the base temperature.
$F_p$    - Pressure correction factor for correcting from flowing Pressure to the base pressure.

# Chapter 11

Pressure rise due to sudden closure of valve

$$\Delta H = aV / g \qquad (11.1)$$

where

a      - Velocity of propagation of pressure wave, ft/s
V      - Velocity of liquid flow, ft/s
g      - Acceleration due to gravity, ft/s$^2$

$$\Delta H = a\,(V_1 - V)\,/g \tag{11.2}$$

Wave speed

$$a = \frac{(K/\rho)^{1/2}}{[1+ C\,(K/E)\,(D/t)]^{1/2}} \tag{11.3}$$

where

a      - Wave speed, ft/s
K      - Bulk modulus of liquid, psi
$\rho$      - Density of liquid, slugs/ft$^3$
C      - Restraint factor, dimensionless
D      - Pipe outside diameter, in
t      - Pipe wall thickness, in
E      - Young's modulus of pipe material, psi

In SI units

$$a = \frac{(K/\rho)^{1/2}}{[1+ C\,(K/E)\,(D/t)]^{1/2}} \tag{11.4}$$

where

a      - Wave speed, m/s
K      - Bulk modulus of liquid, kPa
$\rho$      - Density of liquid, kg/m$^3$
C      - Restraint factor, dimensionless
D      - Pipe outside diameter, mm
t      - Pipe wall thickness, mm
E      - Young's modulus of pipe material, kPa

The restraint factor C depends on the type of pipe condition as follows:

Case1: Pipe is anchored at the upstream end only.
Case2: Pipe is anchored against any axial movements
Case3: Each pipe section anchored with expansion joints

Restraint factor C for thin-walled elastic pipes,

$$C = 1 - 0.5\mu \qquad \text{for Case 1} \qquad (11.5)$$
$$C = 1 - \mu^2 \qquad \text{for Case 2} \qquad (11.6)$$
$$C = 1.0 \qquad \text{for Case 3} \qquad (11.7)$$

Where $\mu$ = Poisson's ratio for pipe material, usually in the range of 0.20 to 0.45 and for steel pipe $\mu = 0.30$

For thick-walled pipes with D/t ratio less than 25, C values are as follows:

Case1

$$C = \frac{2t (1 + \mu)}{D} + \frac{D}{D+t} \frac{(1 - \mu)}{2} \qquad (11.8)$$

Case2

$$C = \frac{2t (1 + \mu)}{D} + \frac{D(1 - \mu^2)}{D+t} \qquad (11.9)$$

Case3

$$C = \frac{2t (1 + \mu)}{D} + \frac{D}{D+t} \qquad (11.10)$$

# Chapter 12

## Pipe material cost

$$PMC = 28.1952 \, L \, (D\text{-}t) \, t \, (Cpt) \qquad\qquad (12.1)$$

where

PMC  - Pipe material cost, $
L       - Pipe length, mi
D       - Pipe outside diameter, in
t        - Pipe wall thickness, in
Cpt    - Pipe cost, $/ton

In SI units

$$PMC = 0.02463 \, L \, (D\text{-}t) \, t \, (Cpt) \qquad\qquad (12.2)$$

where

PMC  - Pipe material cost, $
L       - Pipe length, km
D       - Pipe outside diameter, mm
t        - Pipe wall thickness, mm
Cpt    - Pipe cost, $/metric ton

# REFERENCES

1. Vennard & Street. Elementary Fluid Mechanics. Sixth Edition. John Wiley and Sons, 1982.

2. N. Cheremisinoff. Fluid Flow. Ann Arbor Science, 1982

3. Brater & King. Handbook of Hydraulics. McGraw-Hill, 1982

4. R. Benedict. Fundamentals of Pipe Flow. John Wiley and Sons, 1980

5. Pipeline Design for Hydrocarbons, Gases and Liquids. American Society of Civil Engineers, 1975.

6. Cameron Hydraulic Data. Ingersoll-Rand, 1981

7. Flow of Fluids through Valves, Fittings and Pipes. Crane Company, 1976

8. V.S. Lobanoff and R.R. Ross. Centrifugal Pumps Design & Application. Gulf Publishing, 1985

9. Hydraulic Institute Engineering Data Book - Hydraulic Institute, 1979

10. R. W. Miller. Flow Measurement Engineering Handbook. McGraw-Hill, 1983.

JOHN WINSTON
LENNON
LOVES
CYNTHIA POWELL

TR UE $\times \times \times$
TRUE $\times \times \times$
JOHN TRUE $\times \times \times$

CYNTHIA
I love you Cyn xx

# John

## CYNTHIA LENNON

HODDER &
STOUGHTON

Every reasonable effort has been made to contact the copyright holders, but if there are any errors or omissions, Hodder & Stoughton will be pleased to insert the appropriate acknowledgement in any subsequent printing of this publication.

First published in Great Britain in 2005 by Hodder & Stoughton
A division of Hodder Headline

A Hodder & Stoughton Book

3

A CIP catalogue record for this title is available from the British Library

Hardback ISBN 034089511X
Trade paperback ISBN 0340896558
Limited Edition ISBN 0340920912

Typeset in Monotype Sabon by
Palimpsest Book Production Limited, Polmont, Stirlingshire

Printed and bound by
Mackays of Chatham Ltd, Chatham, Kent

Hodder Headline's policy is to use papers that are natural, renewable
and recyclable products and made from wood grown in sustainable forests.
The logging and manufacturing processes are expected to conform to the
environmental regulations of the country of origin.

Hodder & Stoughton Ltd
A division of Hodder Headline
338 Euston Road
London NW1 3BH

For my son Julian and for John's sisters Julia and
Jackie, all three of whom have had to cope with the
pain that being part of the Lennon legend imposed.
And for my husband Noel, with love and thanks.

## Acknowledgements

Thank you to my family, my parents Charles and Lilian Powell and brothers Tony and Charles for giving me the love and grounding that helped me survive the roller-coaster ride my life has turned out to be.

Also to my son Julian, my best friend and the love of my life, who has never failed to support me in every possible way.

Heartfelt thanks to all the friends who have stood by me through thick and thin, giving me sound advice, laughter and unwavering support, especially Phyl, my soul sister.

Thanks also to Julia Baird for your memories and your friendship and to my sisters-in-law, Marjorie and Penny.

Warm thanks to my editor Caro Handley for holding my hand through every step of the joy and the pain of writing this book – I couldn't have done it without you.

Also to everyone at Hodder & Stoughton, especially Rowena Webb, Briar Silich and Kerry Hood, for guiding me so smoothly, with humour and generosity, through what could have been a rocky passage.

Grateful thanks to John Cousins, my business manager and friend, for planting the seed of the book and whose advice and encouragement has been invaluable. Also to Celia Quantrill, for so much superb research.

Finally, love and special thanks to Noel, my husband, whose constant love and support has been with me through my darkest moments and greatest joys.

Thank you all for encircling my life.

I've included these lyrics here, because of all John and Paul's songs this is the one that speaks to me most. The sentiments in the song feel especially appropriate as I've spent so much time, while writing this book, reflecting on the places and people who've been in my life. But it's not just the feelings expressed in the song that make it special to me, it also reminds me of the time when John, Julian and I were at our very happiest. After the ups and downs of the early years and of Beatlemania, all our dreams had come true, we were happy, healthy, successful and secure in our new home with our beautiful son.

In My Life

There are places I remember all my life,
Though some have changed,
Some forever, not for better,
Some have gone and some remain.

All these places had their moments
With lovers and friends I still can recall.
Some are dead and some are living.
In my life I've loved them all.

But of all these friends and lovers,
There is no one compares with you,
And these mem'ries lost their meaning
When I think of love as something new.

Though I know I'll never lose affection
For people and things that went before,
I know I'll often stop and think about them,
In my life I'll love you more.

Though I know I'll never lose affection
For people and things that went before,
I know I'll often stop and think about them,
In my life I'll love you more.
In my life I'll love you more.

Cynthia Lennon,
July 4 2005

## Foreword
by Julian Lennon

Growing up as John Lennon's son has been a rocky path. All my life I've had people coming up to me saying 'I loved your Dad'. I always have very mixed feelings when I hear this. I know that Dad was an idol to millions who grew up loving his music and his ideals. But to me he wasn't a musician or a peace icon, he was the father I loved and who let me down in so many ways. After the age of five, when my parents separated, I saw him only a handful of times, and when I did he was often remote and intimidating. I grew up longing for more contact with him but felt rejected and unimportant in his life.

Dad was a great talent, a remarkable man who stood for peace and love in the world. But at the same time he found it very hard to show any peace and love to his first family – my mother and me. In many accounts of Dad's life Mum and I are either dismissed, or at best treated as insignificant bit players in his life, which sadly is something that continues to this day. Yet Mum was his first real love and she was with him for half his adult life, from art college, to the genesis of the Beatles, to their overwhelming worldwide success. That's why I'm so happy that she's decided to write her side of the story. For far too long now, Mum has put up with being relegated to a puff of smoke in Dad's life and that simply is not the

xi

truth. Now it's time to set the record straight. There's so much that has never been said, so many tales that have never been told. If there is to be a balanced picture of Dad's life, then Mum's side of the story is long overdue.

I'm immensely proud of her. She's always been there for me; she was the one who kept it all together, taught me what matters in life and stayed strong when our world was crumbling. While Dad was fast becoming one of the wealthiest men in his field, Mum and I had very little and she was going out to work to support us. Mum has always acted with dignity and I have her to thank for who I am. I love her honesty and her courage, and I know it's taken a great deal of both for her to write her story. That's why I offer her my full support and recommend this book to anyone who wants to know the truth, the real truth about Dad's life.

Spain, 2005

# Introduction

For ten years I shared my life with a man who was a huge figure in his lifetime, and who has become a legend since his death. Through the years in which the Beatles came together and went on to delight and astound the world, I was with him, sharing the highs and lows of his public and private lives.

Since John's death I've watched shelves full of books come and go, most by people who never knew him and who painted a one-sided, flawed picture of him and of our relationship. Many consigned me to a brief walk-on part in John's life, notable only because we had a son. I was usually dismissed as the impressionable young girl who fell for him, then trapped him into marriage.

That was a long way from the truth. I was at John's side throughout the most exciting, extraordinary and eventful ten years of his life. It was a time when he was at his creative best. A time when he was witty, passionate, honest and open, when he loved his family and loved the Beatles. A time before drugs and fame led him towards the destruction of so much that he had valued.

After my marriage to John fell apart I tried to escape the world of celebrity and the Lennon label by going off to find my own life. I wanted security for our son, and a life that was real and purposeful, out of the limelight.

Both my privacy and my dignity were important to me, so I preferred to let others do the talking.

But somehow I was never able to escape completely. The public interest always caught up with me and I was frequently sought out for various Beatles-related projects, interviews or books. Far from fading, fascination with the Beatles, and John in particular, increased over the years.

In the early days I said no to most of the offers and requests I received. But in the end I realised there was no escaping the Lennon legend, or that I had been a part of it. So occasionally, when the project was worthwhile or I needed to earn a living, I said yes to the requests and opportunities that came my way. I even talked about my relationship with John a few times – which I had refused to do for several years after we split up. I wrote a book back in the seventies, and after John's death I helped out with a biography about him and gave a couple of magazine interviews.

What I never did was tell the full and truthful story of my life with John. After our divorce I was so desperately hurt, angry and lost that the only way I could cope was to push my feelings to one side and try to detach myself from them. I succeeded so well that whenever I talked about John and our split I sounded calm, rational, accepting and even cheerful. 'Oh, well, these things happen' was the approach I adopted. But, of course, the pain of the break-up stayed with me, even though I buried it as deeply as I could.

Now the time has come when I feel ready to tell the truth about John and me, our years together and the years since his death. There is so much that I have never said, so many incidents I have never spoken of and so many feelings I have never expressed: great love on one hand; pain, torment and humiliation on the other. Only I know what really happened between us, why we stayed together, why we parted, and the price I paid for having been John's wife.

Why now? Because, having tried to live an ordinary life for so many years since John and I parted, I have come to realise that I will always be known as John's first wife. And because I have a powerful story to tell, which is part of John's history.

John was an extraordinary man. Our relationship has shaped much of my life. I have always loved him and never stopped grieving for him. That's why I want to tell the real story of the real John – the infuriating, lovable, sometimes cruel, funny, talented and needy man who made such an impact on the world. John believed in the truth and he would want nothing less.

# 1

ONE EARLY DECEMBER AFTERNOON IN 1980 my friend Angie and I were in the little bistro we ran in north Wales, putting up the Christmas decorations. It was a cold, dark afternoon, but the atmosphere inside was bright and warm. We'd opened a bottle of wine and were hanging baubles on the tree and festive pictures on the walls. Laughing, we pulled a cracker and the toy inside fell on to the floor. I bent to pick it up and shivered when I saw it was a small plastic gun. It seemed horribly out of place among the tinsel and paper chains.

## John

The next day I went to stay with my friend Mo Starkey in London. I couldn't really spare the time during the busy pre-Christmas season, but my lawyer had insisted I go to sign some legal papers, so I took the train, planning to return the following day. I left my husband and Angie to look after things in my absence. Angie was the ex-wife of Paul McCartney's brother, Mike, and after her marriage broke up she'd come to work for us, living in the small flat above the bistro.

It was always good to see Mo. We'd been friends since 1962, when I was John's girlfriend and she was the teenage fan who fell in love with Ringo at the Cavern. Ringo and Mo had married eighteen months after us, and in the days when the Beatles were travelling all over the world, she and I had spent a lot of time together. Her oldest son, Zak, was fifteen, a year and a half younger than my son Julian, and the boys had always been playmates.

When Mo and Ringo parted in 1974 she had been so heartbroken that she got on a motorbike and drove it straight into a brick wall, badly injuring herself. She had been in love with him since she was fifteen and his public appearances with his new girlfriend, American actress Nancy Andrews, had devastated her.

After the split Mo, still only twenty-seven, had moved into a house in Maida Vale with her three children, Zak, eight, Jason, six, and Lee, three. Because of the injuries she'd received in the motorbike accident she had plastic surgery on her face and was delighted with the result, which she felt made her look better than she had before. Gradually she'd begun to get over Ringo, and she had a brief fling with George Harrison before she began to see Isaac Tigrett, millionaire owner of the Hard Rock Café chain.

The evening I arrived Mo had her usual houseful of people. Her mother, Flo, lived with her, as well as the children and their nanny. Mo always had an open house and that evening some old friends of ours, Jill and Dale Newton, had joined us for dinner. The nanny had cooked a huge meal, and later, Jill and Dale, Maureen and I sat over a couple of bottles of wine and talked about old times. After a while the conversation turned to the death of Mal Evans, the Beatles' former road manager. Mal had been a giant of a man, generous and soft-hearted. We'd known him since the early days when he'd worked for the post office and moonlighted as a bouncer at the Cavern Club. When the Beatles began to be successful they took him on to work for them.

Mal had been a faithful friend to the boys and was especially close to John: they got on incredibly well and, with the Beatles' other loyal roadie, Neil Aspinall, he had been on every tour, organising, trouble-shooting, protecting and looking after them.

When the Beatles broke up Mal had been lost. He'd gone to live in Los Angeles where he began drinking and taking drugs. It was there, on 4 January 1976, that the police had been called by his girlfriend during a row. She claimed that Mal had pulled a gun on her, and when they burst into the apartment the officers found Mal holding a gun. Apparently he pointed it at them before they shot him. It was only after he died that they found the gun wasn't loaded. It was a tragic story, and we could only imagine that Mal had been under the influence of drugs. The Mal we knew could no more have shot someone than flown to the moon. Whatever the true story, his death had shocked us all and that night, our talk around Mo's fireplace was of what a good man he had been and how awful his premature death was. To us, the idea of being shot was almost unimaginable – how could it have happened to such a good friend?

After a while I went to bed. I knew the others would carry on talking and drinking until the early hours, but I wanted a good night's sleep as I had to get up early in the morning to catch the train home.

I was asleep in the spare room when screams woke me. It took me a few seconds to realise that they were Mo's. At that moment she burst into my room: 'Cyn, John's been shot. Ringo's on the phone – he wants to talk to you.'

I don't remember getting out of bed and going down the stairs to the phone. But Ringo's words, the sound of his tearful voice crackling over the transatlantic line, is crystal clear: 'Cyn, I'm so sorry, John's dead.'

The shock engulfed me like a wave. I heard a raw, tearing sob and, with that strange detachment that sudden shock can trigger, realised I was making the noise. Mo took the phone, said goodbye to Ringo, then put her arms round me. 'I'm so sorry, Cyn,' she sobbed.

In my stunned state I had only one clear thought. My son – our son – was at home in bed: I had to get back so that I could tell him about his father's death. He was seventeen and history was repeating itself in a hideous way: both John and I had lost a parent at that age.

I rang my husband and told him I was on the way and not to tell Julian what had happened. My marriage – the third – had been strained for some time and, in my heart of hearts, I knew it was going to end, but he was supportive. 'Of course,' he said. 'I'll do my best to keep it from him.' By the time I was dressed and had gathered my things, Mo had organised a car and a driver to take me to Wales. She insisted on coming too, with Zak. 'I'll bring Julian back to stay with us if he needs to get away from the press,' she promised.

John had been shot in New York at 10.50 p.m. on 8 December. The time difference meant it was 3.50 a.m. on 9 December in Britain. Ringo had rung us barely two hours after it had happened, and we were on the road by seven. It was a four-hour drive to north Wales, and during the journey I stared out of the window in the grey dawn and thought of John.

In the jumble of thoughts whirring round my mind two kept recurring. The first was that nine had always been a significant number for John. He was born on 9 October and so was his second son, Sean. His mother had lived at number nine; when we met my house number had been eighteen (which adds up to nine) and the hospital address Julian was born in was number 126 (nine). Brian Epstein had first heard the Beatles play on the ninth of the month, they had got their first record contract on the ninth and John had met Yoko on the ninth. The number had cropped up in John's life in numerous other ways, so much so that he wrote three songs around it – 'One After 909', 'Revolution 9' and '#9 Dream'. Now he had died on the ninth – an astonishing coincidence by any reckoning.

My second thought was that for the past fourteen years John had lived with the fear that he would be shot. In 1966 he'd received a letter from a psychic, warning that he would be shot while he was in the States. We were both upset by that: the Beatles were about to do their last tour of the States and, of course, we thought the warning referred to that trip. He had just made his infamous remark about the Beatles being more popular than Christ and the world was in uproar about it – cranky letters and warnings arrived by every post. But that one had stuck in his mind.

Afraid as he was, he went on the tour, and apologised reluctantly for the remark. When he got home in one piece we were both relieved. But the psychic's warning remained in his mind and from then on it seemed that he was looking over his shoulder, waiting for the gunman to appear. He often used to say, 'I'll be shot one day.' Now, unbelievably, tragically, he had been.

We reached Ruthin mid-morning, and as we rounded the corner into what was normally a sleepy little town, my heart sank. There was no way that my husband could have kept the news from Julian: the town was packed with press. Dozens of photographers and reporters filled the square, the streets to our house and the bistro.

Amazingly we managed to park a few streets away and slip in through the back door, without being spotted by the crowd at the front. Inside my husband was pacing up and down restlessly. My mother, who lived above the bistro with Angie, was peering anxiously at the crowd from behind a drawn curtain. She was seventy-seven and suffering from the early stages of Alzheimer's. Confused by the crowds outside, she had no idea what was going on.

I looked at my husband, the question unspoken. Did Julian know? He nodded towards the stairs. A minute later Julian came running down. I held out my arms to him. He came over to me and his lanky teenage frame crumpled into my lap. He wrapped his arms round my neck and sobbed on to my shoulder. I hugged him and we cried together, both heartbroken at the awful, pointless waste that his father's death represented.

Mo had busied herself making tea, while Zak sat quietly nearby, not knowing what to say or do. While we drank the tea we talked about what to do. Maureen offered to take Julian back to London, but he said, 'I want to go to New York, Mum. I want to be where Dad was.' Although the idea alarmed me, I understood.

Maureen and Zak hugged us and left, then Julian and I went up to the bedroom to ring Yoko. We were put straight through to her, and she agreed that she would like Julian to join her. She said she would organise a flight for him that afternoon. I told her I was worried about the state he was in, but Yoko made it clear that I was not welcome. 'It's not as though you're an old schoolfriend of mine, Cynthia.' It was blunt, but I accepted it: there is no place for an ex-wife in public grieving.

A couple of hours later my husband and I drove Julian

to Manchester airport. The press spotted us as we left home, but when they saw our faces they drew back and let us pass. I was grateful. We sat through the two-hour drive in virtual silence. I was exhausted by the depth of my emotions and by the need to hold back my pain and attend to the necessary practicalities, for Julian's sake.

At the airport I watched him being led off by a flight attendant, his shoulders bowed, his face chalk white. I knew he would sit on the plane surrounded by people reading newspapers with headlines about his father's death splashed across their front pages and I longed to run after him. Before he disappeared through the gate he turned back and waved. He looked painfully young and I ached at having to let him go.

Back in Wales the press were still camped outside our door in huge numbers – there wasn't a spare room left in town. Years later, when she was presenting *This Morning*, with her husband Richard Madeley, Judy Finnegan told me that she had been a young reporter among that throng. 'I felt for you,' she told me. 'You looked absolutely shattered.'

I was furious when my husband let one of the more persuasive journalists, a man who said he was writing a book about John, into our home. Later he claimed that I gave him a lengthy interview, but in fact I said just a few words, then asked him to leave. I was in no state and no mood to give an interview. I fell into bed and lay, numb and exhausted, too wrung out for any more tears, trying to take in the enormity of what had happened.

That night, after I drifted into a shallow sleep, there

was a terrible crash. I leapt up, screaming – it was as though a bomb had gone off. I ran outside in my night-dress and saw that the chimney-pot on our roof had crashed through the ceiling into Julian's attic bedroom. A high wind had blown up, as if from nowhere. It seemed ominous and I thanked God that Julian hadn't been there.

The next day Julian rang to tell me he had arrived safely and was in the Dakota apartment with Yoko, Sean and various members of staff. Hundreds of people were camped outside the building, but Sean didn't yet know of John's death so those inside were trying to keep up the pretence of normality until Yoko felt ready to tell him. Julian sounded tired, but he said that John's assistant, Fred Seaman, had met him at the airport and had been very kind to him. It was a relief to know that someone was looking out for my son.

In Wales, life had to go on. We couldn't afford to close the bistro and John and Angie couldn't manage in the busy season without me, so we opened for business. I cleaned, cooked, served customers and looked after my mother, all the while feeling numb and disconnected. While I got on with the business of life I had to contain my grief, but as headlines about John continued to dominate the news and his music soared up the charts, memories of him, our life together and all we had shared played constantly through my mind. The many hundreds of sympathy cards and messages I received from those who had known John, and those who had simply loved the man and his music, helped. But as I struggled through a disjointed, empty couple of weeks in the lead-up to

## John

Christmas, with my son away and my marriage on the rocks, I felt overwhelmed with sadness, frustration and loss. How could the man I had loved for so long and with such fierce, passionate intensity be gone? How could his vibrant life energy and his unique creativity have been snuffed out by a madman's bullet? And how could he have left his two sons without a father when they both needed him so much?

# 2

THE LATE FIFTIES WAS A wonderful time to be young and setting out in the world. The grim days of the war and post-war deprivation were over; national service had been lifted and the young were allowed to be youthful and unafraid. It was as though the grey austerity of the forties had been replaced by a brilliant spectrum of opportunities and possibilities. Britain was celebrating survival and freedom, and the time was ripe for dreams, hopes and creativity.

I started at Liverpool College of Art in September 1957.

I had just turned eighteen and could hardly believe my luck. A year earlier my father had died, after a painful battle with lung cancer. My two older brothers had left home, and my mother and I had little money. Before he died Dad, who was desperately worried about providing for us, told me that I wouldn't be able to go to college: I'd have to get a job and help Mum. I promised I would, but it was hard to accept that my college hopes were at an end.

Mum said nothing at the time, but she knew how much college meant to me, and after Dad's death she said, 'You go to college, love. We'll manage somehow.' She took in lodgers to make ends meet: she crammed four beds into the master bedroom for four working lads, young apprentice electricians who were happy to share. From then on home was more like a boarding-house – there were always queues for the bathroom and I had to get up at dawn if I wanted to be first in, but I was hugely grateful to Mum and determined not to let her down.

When I got into art college, I set out to be a model student. I turned up promptly every day, neat in my best twinsets and tweed skirts with my pencils sharpened, ready to be the hardest-working girl in the place. My dream was to be an art teacher. Art was the only subject I'd ever liked at school and I was thrilled when, at the age of twelve, I got into the junior art school, which was down the street from the art college. It was there that I became best friends with a girl called Phyllis McKenzie. We planned to go on to college together, but Phyl's father refused to let her go and insisted she get a job. She had to settle for evening classes in life drawing, after spending the day working as a commercial artist for a local corn merchant.

A couple of other girls from the junior art school, Ann Mason and Helen Anderson, started college with me. We were thrilled to be there, and in awe of the older students, many of whom wore the kind of Bohemian, beatnik clothes we considered incredibly daring and could only stare at with a mixture of envy and admiration.

Most of us starting college then had been born just before or during the war – in my case a week after war was declared. My mother, with a group of other pregnant women, had been sent to the relative safety of Blackpool, where she gave birth in a tiny cell of a room in a bed-and-breakfast on the seafront on 10 September 1939. It was a nightmare birth: she was left alone, in labour, for a day and a night, and when the midwife finally got to her it was clear that, without immediate help, neither my mother nor I was going to make it. The midwife locked the door, swore my mother to secrecy and dragged me into the world by my hair, ears and any other part of me she could get hold of. My father, who had arrived hours earlier and burst into tears at the sight of my exhausted, terrified mother, had been sent for a walk. He returned to find that his wife had survived and he had a daughter.

My parents both came from Liverpool, but at the outbreak of war they decided to leave the city for the relative safety of the Wirral, across the Mersey in Cheshire. They moved with me and my brothers – Charles, then eleven, and Tony, eight – to a two-bedroom semi-detached house in a small seaside village called Hoylake. My father worked for GEC, selling electrical appliances to shops, and had to travel into the city each day to make his rounds, but at home we were away from the worst of the relentless bombing that ravaged so much of Liverpool. When the bombers flew overhead my mother would scoop us into the cupboard under the stairs, where the force of the explosions jolted us off our seats.

I grew up with rationing as a way of life. Like all the other families around us, we dug for Britain, with an allotment where we grew our vegetables and a little hen coop in the back garden. As in so many households in those days, the boys generally took precedence over the girls. When my brothers got bacon, I got the rind, and when they got scraps of meat from a bone, I got the bone to chew. It was my job to clean their shoes and help my mother look after them and Dad. I was a quiet, timid child and I accepted my role in the house, as the youngest and the only girl, without question.

Rationing went on for some years after the war, so for most of my childhood scarcity was normal. I used to shop for two old ladies in our street and in return one gave me her sweet coupons and the other gave me old clothes that had belonged to her children. Both the clothes and the sweets were rare treats. Charles left when he was sixteen and I was five, so I have few memories of him living at home. He went to work for GEC, first in Birmingham, then London. He was a wonderful pianist – the whole street used to listen to him.

I was closer to Tony, and when he was called up for national service in 1950, at the age of eighteen, I missed him dreadfully. After the army he joined the police to please his girlfriend, who wanted the accommodation that went with the job. He hated being a policeman and was relieved when she left him and he could resign.

By the time I was ten it was just my parents and me at home. They were opposites in many ways, but they loved each other and I never heard them argue. My father, also Charles, was easy-going, kind, robust and jolly. I remember him losing his temper with me only once, when I came home from school and used a swear word. I adored him and after I got into the junior art school I travelled into Liverpool on the train with him in the mornings and evenings. He used to carry a bag of sweets for his customers, and he'd slip me a couple on the way home.

My mother, Lilian, was unusual for her day: she had no interest in housework and remembered to clean our home about once a month – the rest of the time it gathered dust. But Mum had a strong artistic streak: she always had a vase of flowers in the window, which she took pleasure in arranging, and she knitted fantastic Fair Isle sweaters. Her real passion, though, was the auction rooms to which she would head every Monday to spot the latest bargains.

On Monday evenings Dad and I would arrive home to find the front room changed. There might be a new sofa, carpet, curtains, table or even all of them, the old ones already dispatched to the same sale rooms. We didn't mind:

it was always fun to see what she'd done and, most importantly, it made Mum happy.

When Dad became ill, at the age of fifty-six, everything changed. Like so many others in those days, he smoked untipped cigarettes, unaware of the damage it was doing to him. When he developed lung cancer he went downhill rapidly: his solid frame wasted away and his breathing was laboured. Before long all he could do was sit in his chair in the bedroom, where I would sit with him after school each day. After his death only Mum and I were left, grieving for him and wondering how we would manage. Art college gave me a new focus, something to be excited about, to work for, and to take me out of our quiet little house of mourning into the world.

Watching the older, more confident kids at college, I longed to be like them. I envied their casual, arty style and their long hair. I had arrived with my short mousy hair in a neat perm, courtesy of my mother's friend who was a hairdresser. The trouble was, most of her clientele were over fifty and she made me look middle-aged and dowdy. Every few weeks she would experiment, giving me a different style, but they were all ghastly. And, to make things worse, I wore glasses. I'd arrived at college thrilled to be rid of school uniform and pleased with my smart new clothes. But I soon felt frumpy and dull, with my matronly hair and conventional outfits. I longed to be more daring, but in those early days I didn't have the courage.

To add to my problems I was saddled with the 'over

the water' posh image that Scousers had of anyone who lived across the Mersey. I spoke differently, and to them this meant I was stuck-up, even though many of them were better off than I was. My shyness didn't help: it made me seem aloof, when most of the time I was going through agonies, trying to think of the right thing to say. I was hopeless at sparkling conversation and witty repartee and watched enviously as others bantered while I remained tongue-tied. But despite the drawbacks I loved college. It gave me a sense of independence and freedom I had never experienced before.

During my first year I was seeing a boyfriend I'd met while I was still at school. Barry was a bit of a catch: he was the son of a window-cleaner but he looked Spanish and exotic, and he was the Romeo of Hoylake. I was the envy of the local girls when he asked me out. He'd seen me in my white duffel coat, walking my dog Chummy on the beach, and one day he followed me and asked me to the pictures. I was just seventeen and he was five years older. Flattered, I said yes.

By the time we'd been together for a year I was starting college and we were thinking of getting engaged. Barry was working for his dad and saving in the building society for our future. One day he persuaded me to make love with him on the sofa in my parents' front room when Mum was out. It took him hours to talk me into it, promising we'd get married and telling me how much he loved me, but when I finally agreed I didn't think much of it: over in a flash and no fun. I went on seeing Barry, but I made sure we never got the chance to be alone in the house again. One day he announced that he'd fallen for a red-haired girl who lived up the road and I was heart-broken. It was the first betrayal I had experienced and I vowed I'd never forgive him. But, a few months later, when he begged me to go back to him, swearing he'd made a mistake and I was his true love, I relented.

Two-thirds of the way through my foundation year Phyl arrived at college. She had won a grant, and had finally persuaded her father to let her attend full-time. We were both delighted and in between classes we hung around together most of the time.

At the end of that year we had to choose which areas we wanted to specialise in. I went for graphics, but I also signed up for a twice-weekly class in lettering. Phyl decided on painting and lettering, and we were glad of the chance to do a class together.

I arrived for my second year in college just as keen as I had been in the first, but I'd softened my appearance a little. I'd plucked up the courage to say no to Mum's hair-dresser friend and was growing my hair. I'd acquired some rather hip black velvet pants to replace the tweed skirts, and I'd begun to ditch my glasses as often as I could. I could hardly see without them – I'm very short-sighted – so this caused me all kinds of problems: I'd frequently get off the bus at the wrong stop or misread notices in college – but I didn't care. I hated my glasses so much that it was worth the odd hiccup. I only put them on when I was working in class, because without them I couldn't see the board or even what I was drawing on the paper in front of me.

We had all taken our seats for the first lettering class when a teddy-boy slouched into the room, hands stuffed deep into his coat pockets, looking bored and a shade defiant. He sat at an empty desk behind me, tapped me on the back, twisted his face into a ludicrous grimace and said, 'Hi, I'm John.' I couldn't help smiling. 'Cynthia,' I whispered, as the teacher, who had begun to talk, frowned at me.

I'd seen John around the college but had never spoken to him: we moved in completely different circles. I was surprised to see him in the lettering class – he didn't seem the type for the painstaking, detailed work involved. He hadn't even brought any equipment. As soon as we started work he tapped my back again and asked to borrow a pencil and a brush, which I reluctantly handed over. After that he always sat behind me, borrowing whatever he needed from me. Not that he used it much: most of the time he did no work at all. He spent his time fooling around, making everyone in the class laugh,

It turned out that John hadn't chosen to do lettering: he'd been ordered into the class when most of the other teachers had refused to have him. He made it clear he didn't want to be there and did his best to disrupt the class. When he wasn't teasing someone he'd give us a wicked commentary on the teacher, or provoke hoots of laughter with his cruelly funny and uncannily accurate cartoons of teachers, fellow students or of twisted, grimacing, malformed figures.

When I'd first looked at John I'd thought, Yuck, not my type. With his teddy-boy look – DA (duck's arse)

haircut, narrow drainpipe trousers and a battered old coat that was too big for him – he was very different from the clean-cut boys I was used to. His outspoken comments and caustic wit were alarming, I was terrified he might turn on me, and he soon did, calling me 'Miss Prim' or 'Miss Powell' and taking the mickey out of my smart clothes and posh accent.

The first time he did it I rushed out of the room, red-faced, at the end of the class, wishing he'd disappear. But as the weeks went by I began to look forward to seeing him. We never met anywhere but the lettering class, but I found myself hurrying to it, looking out for him. He made me laugh and his manner fascinated me. I had always been in awe of authority, anxious to please and do well, but John was the opposite: he was aggressive, sarcastic and rebellious. He didn't seem to be afraid of anyone, and I envied the way he could laugh about everything and everyone.

A mutual friend told me that his mother had been killed in a car accident at the end of the previous term. I missed my father desperately, so I felt for him. He never mentioned it and neither did anyone else, but the knowledge that he was hiding grief behind the acerbic front made me look at him more closely.

One morning the students in the lettering class were testing each other's eyesight for fun. It turned out that John and I were equally short-sighted: just like me he couldn't see a thing and hated wearing glasses, most of all, ironically, the little round lenses you got on the National Health. Instead he had horn-rimmed black ones, which had cost quite a bit. Laughing about our rotten luck and the blunders we'd made when we couldn't see gave us our first real connection, and after that we often chatted during class.

John usually had a guitar slung across his back when he arrived and he told me he was in a group, the Quarrymen, named after his old school, Quarry Bank High. Sometimes when we were sitting around after class he would get it out and strum the pop tunes of the day, by Bo Diddley, Chuck Berry or Lonnie Donegan. As soon as he began to play I saw a different side of him. It was plain that he loved his music: his face softened and he lost his usually cynical expression.

Half-way through the term I realised I was falling for him and scolded myself. I was being ridiculous: he wasn't at all the type of boy I'd imagined myself with and, in any case, I couldn't see him being interested in me. But that changed one day when everyone else had left the class and I was packing up my things. John was sitting a few feet away with his guitar. He began to play 'Ain't She Sweet', a song that was popular at the time and which the Beatles were later to record:

> Oh ain't she sweet?
> Well, see her walking down that street.
> Yes, I ask you very confidentially:
> Ain't she sweet?
> Oh, ain't she nice?
> Well, look her over once or twice.
> Yes, I ask you very confidentially:

## John

*Ain't she nice?*
*Just cast an eye in her direction.*
*Oh me, oh my, ain't that perfection?*
*Oh I repeat, well, don't you think that's kind*
  *of neat?*
*Yes, I ask you very confidentially: ain't she sweet?'*

I blushed scarlet, made an excuse, and fled before the end of the song. But I'd seen the look in his eyes, which he'd kept fixed on me as he sang – could it be that John fancied me too?

I confided in Phyl, who told me he wasn't my type and not to be so daft. She knew John: they lived near each other and travelled together to college on the seventy-two bus. Although she often had to lend him the fare, she liked him – but she didn't think he was for me. She reminded me that I was thinking of getting engaged to Barry . . . but my plans with Barry were taking a back seat. I saw less and less of him as I continued to moon over John, and the lettering class was the highlight of my week.

One lunchtime I saw John staring at a girl as she walked up the staircase. She was dressed in a tight black skirt and had long blonde hair. John whistled. 'She looks just like Brigitte Bardot,' I heard him say to a friend.

I wasn't about to be outdone. The following Saturday I went out, got the latest Hiltone blonde dye and got to work on my hair. On Monday I arrived in college by several shades blonder. I was delighted when John noticed: 'Get

25

you, Miss Hoylake!' He laughed, but I could see he liked it.

One afternoon all the intermediate students were asked to be in the lecture theatre for a discussion. John was a few seats away from me, and my friend Helen Anderson, who was also friendly with John, suddenly leant forward and stroked his hair. Helen didn't fancy John – it was a friendly gesture in response to something he'd said. But when I realised how jealous I was it brought me up with a jolt.

Although John and I chatted in lettering classes we spent our free time in college with our different groups of friends and virtually ignored each other. I thought of him as unattainable and, despite my fantasies, still didn't think for a minute that we might actually get together.

We were all getting excited about the holidays, when someone suggested we hold a party one lunchtime before we broke up. One of the staff, an ex-boxer named Arthur Ballard, a tough but excellent teacher, gave us permission to use his room, provided he could come too. We happily agreed, found a record-player and chipped in for the beers.

I was looking forward to the party, not because I thought John would be there – I felt sure a tame little students' do wouldn't be his style – but because I thought it would take my mind off him. After that we'd be on holiday. I was looking forward to the break and was determined to get over my crush on John.

The day of the party was warm and the sun streamed through the grubby windows of Arthur Ballard's first-floor room, where we gathered once a week to produce paintings on a chosen theme. We pushed the tables and chairs to one side, set out the food and drink and put on a pile of records. The usual gang were there, a group of ten or fifteen of us who'd been friends since our foundation year. I arrived feeling good: I was wearing a new baggy black cotton top over a short black and white skirt, with black tights and my best black winklepicker shoes.

By now several romances were budding so the atmosphere was heady. Ann Mason was getting together with Geoff Mohammed, a close friend of John's. They smooched away – Phyl and I glanced knowingly at each other. Then John walked in. My face was hot and my stomach contorted as I pretended not to notice him. Like me, he was in black – his usual drainpipe trousers with a sweater and suede shoes. He made a beeline for me and said. 'D'you want to get up?' I blushed, but leapt to my feet to dance with him.

While we were dancing to Chuck Berry John shouted, 'Do you fancy going out with me?'

I was so flustered that I came out with, 'I'm sorry but I'm engaged to this fellow in Hoylake.' The moment I said it I wanted the ground to swallow me – I knew I sounded stuck-up and prim.

'I didn't ask you to fucking marry me, did I?' John shot back. He walked off and, convinced I'd blown it, I was plunged into gloom. But a couple of hours later, as the party was breaking up, John and his friends asked me and Phyl to the pub. This was good news – perhaps all was not lost.

I persuaded Phyl we should go and we followed them to Ye Cracke, a pub where the students often hung out. The place was packed and we had to yell to each other above the hubbub. We'd never been there before, we'd always headed straight home like the good girls we were, and this was our first taste of student social life. We loved the noise, the laughter and the buzzy atmosphere – and realised what we'd been missing.

John was with a couple of his cronies, Geoff Mohammed and Tony Carricker, on the other side of the pub, and made no move to come over to us. Phyl and I had found some friends and were chatting with them, but after a couple of black velvets – the mix of Guinness and cider that all the students drank – I felt a little wobbly and decided I'd better head for my train home. I was disappointed that John hadn't talked to me, and wondered if, after all, he had been laughing at me when he invited me to the pub.

As I made for the door he called me over, teased me about being a nun and asked me to stay. Phyl said she had to get her bus home and asked if I was coming. I knew she didn't approve of John, but I was hooked: if he wanted me to stay I was staying. I smiled apologetically at her. She gave a helpless shrug and headed for the door. John and I had another couple of drinks and then he whispered, 'Let's go.' The two of us slipped away from the crowd.

By this time it was evening and the street outside was quiet. Almost as soon as we'd left the pub John kissed me, a long, passionate, irresistible kiss. He whispered that his friend, Stuart, had a room we could go to, grabbed my hand and pulled me down the road. I was happy, hugely happy, to be with him and that he felt the same. At that moment I would have gone anywhere with him.

Stuart's place was a large room at the back of a shared house, with no curtains, a mattress on the floor and clothes, art materials, empty cigarette packets and books scattered around it. We couldn't have cared less about the mess and headed for the mattress where we made love for the next hour. For me it was special and very different from my previous brief experience. And I think it was equally special for John, whose cockiness and tough-guy demeanour melted away as we lay wrapped in each other's arms.

Afterwards John said, 'Christ, Miss Powell, that was something else. What's all this about being engaged, then?' I told him my romance in Hoylake was over. John grinned and said he thought I was incredibly sexy and he'd been lusting after me all term. 'By the way,' he added, 'no more Miss Powell. From now on, you're Cyn.'

We snapped back to reality when I realised I was about to miss my last train home. We pulled on our clothes and raced to the station, where we managed a hasty goodbye kiss before I leapt into a carriage. 'What are you doing tomorrow, and the next day, and the next?' John called, as I waved out of the window.

'Seeing you,' I shouted back.

Others might have seen us as an unlikely couple, but I knew from the outset that we had made a deep connection. My feelings for John were very different from those I'd had for any other boy: more powerful, more exciting and totally unshakeable. And I sensed in John the same

29

strong feelings. Perhaps each of us recognised and was drawn to a deep need in the other. But at the time I didn't analyse it. I simply felt certain that this was no passing fling. It was real love.

# 3

WE'D HAD OUR FIRST PHONE installed at home just before I started at art college. It was a bulky black contraption fitted to the wall at the bottom of the stairs and you had to put two pennies into the box beneath it to make a call. It was still a bit of a novelty and its shrill peal always made me jump. When it rang the following morning I couldn't grab the receiver fast enough.

John asked me to meet him the next day. But I couldn't: it was the start of the holidays and Mum and I were off to stay with my brother Charles in Buckinghamshire for a couple of weeks. I'd been looking forward to this for ages, but suddenly it was an obstacle in the way of my being with John. There was nothing I could do, though, it was all arranged, and although I was nineteen Mum wouldn't have considered letting me stay behind. I promised John I'd write.

As soon as I got back John and I met in a café in the centre of Liverpool and gazed into each other's eyes over a cup of coffee. It lasted us two hours, because neither of us had any money for a second, but we didn't care.

From then on we spent all our spare time together. We were always broke: our small daily allowances went on fares, lunch and, in John's case, the ciggies he smoked – Park Drive, Woodbines or Embassy because they were the cheapest. Not to mention the pints of black velvet in Ye Cracke at lunchtime or after college. If we had enough money we went to the pictures, where we sat in the double seats at the back and kissed and cuddled, mostly ignoring whatever film was on and often sitting through a couple of showings. More often we didn't have any money at all so we just walked and talked, or stretched out one drink in a pub or café.

The friends who'd thought us an unlikely match soon got used to seeing us together – we were joined at the hip most of the time so they had no choice. Only Phyl worried about me and, with a best friend's concern, said, 'Cyn, you're too good for him, he's not right for you. I don't trust him.' She was afraid that John wasn't serious and would drop me when he got bored. I wouldn't listen: I was far too besotted with John to give him up.

## John

Before I started going out with John I had been a conscientious student, completing all my work on time and putting in hours of effort. But John was a demanding lover, who insisted that I put him before everything else, including college work, my friends and my mother. Inevitably my work took a nose-dive, although I did my best to keep up.

John had been out with plenty of girls before me, but none had lasted long: this was his first serious relationship. If we went to the pub for a drink at lunchtime he would often insist that we bunk off college in the afternoon. When the weather was warm, we'd take the ferry across the Mersey to New Brighton, where there was a funfair beside the sea. Up on the deserted sand dunes behind the beach we'd make love, braving the chill winds and the sand. We'd catch the ferry back, with sand under our clothes, horribly uncomfortable but giggling as we imagined what everyone else would think if they knew what we'd been up to.

Most of the time I went along with what John wanted. We laughed a lot, the attraction between us was powerful and exciting and he constantly came up with new escapades for us. But there was friction too. His insistence that I stayed with him until the last train from Liverpool Central to Hoylake, which got me home at midnight, upset me. I knew my mother would fret about me and I worried about leaving her alone so much of the time when she was still grieving for my father. Besides, I often had college work to catch up on. But John didn't give a damn about any of that, and if I tried to go home before the last train he threw a fit. He wanted me with him for as much of the time as possible, which meant that very early in our relationship I had to choose between him and my other needs and responsibilities.

John's temper could be frightening and at times I felt torn to pieces by him. All sense of reason disappeared and his tantrums were awesome: he would batter away at me verbally until I gave in, overwhelmed by the force of his determination. Then he was back to his usual self, apologetic and loving.

He was full of contradictions and confusion. He wanted proof, daily, that he mattered most to me. He was jealous of my close friendship with Phyl and even of my work, if I chose to spend an evening catching up with it instead of with him. Yet despite John's aggression and jealousy I felt protective towards him. To me he was a lost soul and I wanted to give him understanding, acceptance and the security of being loved to ease his pain and bitterness.

In college I had a few admirers, although I wasn't aware of this until John pointed it out. He was incredibly jealous of any boy who came near me and wouldn't hesitate to warn them off. Not long after we got together we went to a party at another student's flat. There was plenty of loud music, beer and cider, and we were having a good time until a very tall student I recognised from the sculpture department came over and asked me to dance. Before I could answer all hell broke loose as John, in a blind fury, launched himself at the guy. The sculpture student was big enough to hold him off with one hand, but in the end everyone piled in to pull John away, and eventually we calmed him down. The other guy, baffled by the uproar he had unwittingly caused, apologised for upsetting John and backed away.

I felt frustrated by incidents like this. John had no need to worry – I would never have been unfaithful to him – and his overreaction embarrassed me. I tried repeatedly to reassure him, but it made no difference: John was provoked to fury if another boy paid any attention to me, however innocent.

Much as we wanted to, we never spent a night together – there was nowhere to go. As often as he could John persuaded Stuart to lend us his room for a few hours and we'd grab the chance to be alone together to make love. When Stuart's room wasn't available John would try to talk me into 'quickies' in dark alleys or shop doorways. Much as I loved him I didn't enjoy these snatched encounters, so mostly we stuck to kissing and cuddling anywhere and everywhere we could.

Frightening and demanding as John could be, he was also romantic, a side of him I saw more often as our relationship deepened. He wrote love poems on scraps of paper and passed them to me at college. For our first Christmas he drew a card with a picture of me in my new shaggy coat, standing opposite him, our heads together, his hand on my arm. It was covered with kisses and hearts and he wrote, 'Our first Christmas, I love you, yes, yes yes.' A few years later he used the same idea in one of the Beatles' first hits, 'She loves you, yeah, yeah, yeah'. On the back it said, 'I hope it won't be our last.' I loved that card and kept it in pride of place in my bedroom.

I was totally absorbed by John and wanted to be with him whenever I could. Despite the conflict I felt as I neglected my studies and my mother, I was blissfully happy that we were together and that he loved me too. When John was at his warmest and most loving I felt sure we would last for ever. At these times he would let his guard down and tell me over and over again that he loved me.

Yet it was neither an easy nor a comfortable relationship. There was an air of danger about John and he could terrify me. I lived on a knife edge. Not only was he passionately jealous but he could turn on me in an instant, belittling or berating me, shooting accusations, cutting remarks or acid wisecracks at me that left me hurt, frustrated and in tears. He would push me away with some taunt, almost daring me to leave him. It was as if he wanted to prove that a girl like me would never stay with a boy like him.

Hurt as I was, many times, my response to John's provocative and cruel behaviour was to stick by him more solidly than ever. Although I thought about leaving him, I felt that if he could trust me and believe that I loved him he might soften.

We had been together for a few months when John took me home to meet his Aunt Mimi. He lived with her in a smart house in Menlove Avenue in the well-off district of Woolton. The house was called Mendips and had a big garden; at one time the Mayor of Liverpool had lived next door. The joke was that although John called me posh, he came from a far better-off family than I did. Our little semi over the water in Hoylake was half the size of Mendips.

John

I was nervous about meeting Mimi. I knew she had brought John up and that it was important she liked and approved of me, so I wore my smartest skirt and jumper and prepared to be on my best behaviour.

Mimi was a striking woman, not tall but with presence. She was slim with the fine bone structure characteristic of John's family. When we arrived she smiled at me and invited us into the breakfast room, next to the kitchen. I saw instantly that Mimi was a woman who didn't miss a trick. Sharply observant, she sized me up throughout the visit.

We sat at the dining-table watching Mimi make us the standard Liverpool tea of egg and chips, with a mountainous plate of bread and butter and a huge pot of tea. While we ate Mimi asked questions. She was friendly but cool. More than once I caught her looking at me so penetratingly that I was unnerved. I was glad when the meal was over and John walked me to the bus stop.

'Do you think she liked me?' I asked John.

'Yeah, sure,' he said. 'Don't worry about Mimi. If I like you she'll like you.'

I thought he was wrong about that. I was sure that Mimi hadn't liked me, although she had taken care not to show it. I wondered what it was about me that she didn't like, but later I grasped that it wasn't personal: Mimi didn't think any girl was good enough for her boy.

Mimi's manner was almost regal. She spoke without a hint of Scouse and I thought John must have adopted his working-class Liverpool accent as a rebellion against her. Early on it became apparent to me that Mimi was something of a snob; she was middle class with upper-class aspirations and one of her favourite words was 'common'. She used it to condemn most of John's interests and friends – including, I suspect, me. In fact, my family were middle class too, but with no upward aspirations.

John's first meeting with Mum was more successful. He wasn't the respectable, hard-working young man she had dreamt of for me, but she knew I was in love and wisely kept quiet. To my delight John was polite and respectful to her and they seemed to get on well. If they had reservations about each other they didn't mention them, for which I was grateful.

An incident early in our relationship showed me a side of John that I would see again at many crucial moments in our life together. Mum suggested that we invite Mimi to come for tea with John, so that she and Mimi could meet. John agreed and on the day of the visit Mum, determined to impress, got out her best china and made sandwiches and cakes.

It started well. Mimi and Mum were both polite, and as the four of us sat down together John and I exchanged 'It's going OK' glances and began to relax. Too soon. Mimi made a remark about me distracting John from his studies and Mum leapt to my defence. Before we knew it they were arguing, Mimi telling Mum why I was wrong for John and Mum telling Mimi that John was lucky to find a girl like me.

John and I were aghast. After a few moments, he got up and fled from the house and I ran after him. I raced down the street to catch him up, and found him in tears.

Eventually I persuaded him to come back, and when we reappeared Mum and Mimi had called a frosty truce.

John couldn't stand conflict or confrontation and his reaction was invariably to escape. It was in stark contradiction to his often aggressive manner, but in fact he was only confrontational when he had been drinking. He was often cutting and critical, but mostly he went out of his way to avoid direct conflict.

The incident also told me a great deal about John and Mimi's relationship. When she was openly critical of me it hurt and humiliated John, but she either didn't notice or didn't care because she carried on. Time after time I saw her upset him with negative remarks about him or someone he cared about. John would become angry and embarrassed, then run.

That disastrous tea was one of only two occasions when Mum and Mimi met. After the first we stuck to seeing them separately and, despite undercurrents of disapproval from them both, I got on well superficially with Mimi, and John was friendly with Mum.

John liked my brothers. When Charles was home one day visiting us, he realised John had little money and offered him a pile of his jumpers. John was delighted and ever after had a soft spot for him.

John met Tony and his fiancée, Marjorie, at their wedding in April 1960. I was a bridesmaid so I had to be there early and John agreed to come along later. As my entire family would be present, I prayed he would make an effort and turn up on time, looking reasonable. When he appeared my jaw dropped: he was the epitome of respectability, in a dark suit, white shirt and smart tie, with his hair sleeked neatly back. He had his glasses on, which made him look like an office clerk, but at least he could see the other guests. John seldom hid his dislike of social convention and I knew he had made an enormous effort for my sake.

The whole thing went off beautifully. John was charming and polite to everyone, chatting to Charles and his girl-friend Katie, to Mum, Tony and Marjorie. In fact, it was the beginning of a friendship between John and Tony, who shared John's dry sense of humour.

It struck me early on that John had developed his hard outer shell – the cynicism, cruel wit, aggression and posses-siveness – to deal with his painful childhood and the deep insecurity that had resulted from it. But in those days he told me little, only that Mimi had brought him up after his father disappeared when he was five, and that his mother had died a few months before we got together. He met any questions with a shrug.

The loss of a parent was one of the things we had in common and proved a powerful bond. Sometimes we talked to each other about how we felt; we both missed our parents terribly. But while I had come to terms with Dad's death, John was still angry about his mother Julia's. I'm certain that his bitter rage at his mother's death, and espe-cially at the way she died, was behind so much of his aggression during that period.

Music had been an important part of John's relation-ship with his mother and when she died he used it to blot out the pain and anger he felt. Julia had bought him his first guitar and loved music. She played the piano and banjo, and sat with him patiently for hours, showing him over and over again how to play the chords. She had also introduced John to rock and roll. She would play Elvis Presley records at top volume, grabbing John's hand to jive round the kitchen to them. She always encouraged John's musical dreams.

By the time John and I got together, he talked, ate and breathed music. When he wasn't playing his guitar, he was writing lyrics or talking about the latest Lonnie Donegan, Elvis Presley, Buddy Holly or Chuck Berry record. Almost every lunchtime he met the two other Quarrymen to rehearse. They were both younger and went to the Liverpool Institute, next door to the art college, the best known of Liverpool's boys' grammar schools: distinguished judges and politicians had been educated there, its pupils were expected to do well. But John's friends, Paul McCartney and George Harrison, were more interested in playing music than passing exams.

John had met Paul at a fête at St Peter's Church in Woolton when the Quarrymen had played there almost three years earlier, on 6 July 1957. At fourteen, Paul was a year and eight months younger than John, a fairly big age gap for teenagers. But when John realised that Paul was a talented musician who knew the words to dozens of hit songs and who could even tune his guitar – which John couldn't – his age was irrelevant. Soon afterwards John sent Paul a message via a friend: you're in the group.

George was a friend of Paul's. They'd started playing guitar together some time previously and practised at each other's houses. Eventually Paul introduced him to John. He was eight months younger than Paul, but a talented guitarist, and in early 1958 John let him join the group.

When I first met them, I was nineteen, John was eighteen, Paul was seventeen and George sixteen. It was about as inconsequential as a meeting can be. Paul and George had come over to the art college one lunchtime and John said, 'This is Paul and George and this is Cyn.' They both said, 'Hi.' I said, 'Hi,' and that was it. As John's new girlfriend they gave me a few curious glances, but then everyone got on with the serious business of making music.

When we knew the two boys were coming over at lunchtime, John and I would go across the road for fish and chips. Back in college we'd slip behind the curtain separating the tiny stage from the canteen, which was always packed. A few minutes later Paul and George would arrive. They'd have stripped off their caps and ties and put the collars of their blazers up to look cool as they made their way, as casually as they could, through the crowds of students and teachers. Paul always appeared nonchalant, George furtive, as they did their best not to look like the schoolboys they were. When they joined us behind the curtain, we'd lay out the mound of chips and scallops in their paper on the floor and the four of us would dive in. Then the boys started to play.

From the start I loved their music: I never minded spending lunchtimes on that little stage because it was fascinating to watch them teach each other new chords, work out the tunes to popular songs and begin to put together their own. Paul had been first to write a song, but John leapt on the idea and soon they were writing more and more new stuff.

George was the kid who tagged along. He was always serious and his shy, toothy grin only ever flicked for a moment before it disappeared. He was quiet and seemed troubled as he trailed behind John and Paul, deferring to them even though he was a fantastic guitar player. They tolerated him because he was good, but they patronised and often ignored him when they were absorbed in something together.

George's strength was his tenacity: he would spend hours working out a chord sequence or practising a song until it was perfect. John and Paul were fired with ambition. They wanted to make the group stand out and get as many gigs as they could.

They had one of their first real breaks at the Casbah Club, a venue in the cellar of a house in a Liverpool suburb called West Derby. It belonged to Mona Best, who started the club as a meeting-place for her elder son Pete and his friends. They charged a shilling membership to keep out the rough element and served coffee and sweets. Before they opened, Pete had suggested to his mum that they ask one of the beat groups that were springing up all over town to come and play there. She agreed, and they invited the Quarrymen – a girl who knew the group had told the Bests how good they were. John, Paul and George went round to see Mona, who told them they were welcome to play but she was still painting the cellar for the club's opening the following week. The three boys grabbed paintbrushes and helped her finish it off. John mistook gloss for emulsion – because of his short sight – which took days to dry.

The boys played at the club's opening on 29 August 1959, and I was there to watch them. They played with another lad, Ken Brown, on guitar, but without a drummer, as they couldn't find one. About three hundred people came along that night, and the boys played rock and roll hits for a couple of hours. The place heaved, with kids jiving and swinging, and the temperature soared until it was hard to breathe.

That was the evening when we first met the Beatles' future roadies, Neil Aspinall and Mal Evans, both friends of Pete, but Neil was also his mother's boyfriend and the father of his younger brother, Roag.

After that the Quarrymen played in the Casbah regularly, to audiences of up to four hundred. It was hot, sweaty and noisy in that cellar, but we loved it. They earned fifteen shillings each – seventy-five pence today – every time they played.

Eventually Pete Best decided he wanted to be in a group, got himself a drum kit and formed the Blackjacks with Ken Brown, who left the Quarrymen. They became the Casbah's resident group, and after that the Quarrymen only played there occasionally.

In November 1959 the boys got an audition at the Manchester Hippodrome with a man called Carroll Levis. Known as the Star-maker, Levis had a lot of successful acts on his books and John was wild with excitement. By this time John, Paul and George had renamed themselves Johnny and the Moondogs and their hopes were high, but they came back despondent: they had failed the audition, mainly because they lacked a drummer. It was a setback, but it didn't put them off. When something didn't work out John would be down for a day or so, then he'd carry on, determined to be the best and to show anyone who didn't believe in the group how wrong they'd been.

But while John was keen for the group to do well, he was also easily distracted and would flip from one project or plan to another. He needed Paul's drive and determination to keep him focused on the pursuit of success.

By the time I got to know Paul, he and John had formed a close partnership. They had agreed that any songs they wrote, together or separately, would be by Lennon and McCartney. It was as though, even then, they had a strong sense that their success depended on the connection between them. Paul's organised, conscientious way of going about things – he wrote down all the lyrics in a notebook he carried with him – was in sharp contrast to John's 'anything goes' style. Paul turned up for appointments on time, looking well turned out; he was a perfectionist and you always knew he'd washed behind his ears. John arrived late, looking as though he'd just fallen out of bed. But they complemented each other. John needed Paul's attention to detail and persistence; Paul needed John's anarchic, lateral thinking. When they wrote songs together, Paul's gentler melodies blended beautifully with John's more rousing, challenging tunes and lyrics.

In those days Paul tried hard to impress John, posing and strutting with his hair slicked back to prove that he was cool, because John was very much the leader: it was his band, he had the final say about who got in and who didn't, and what they played. Then, he was everything Paul wanted to be – laid-back, self-assured and in charge. As the schoolboy he still was, Paul could only aspire to these things.

As the two became closer this changed. John recognised Paul's musical talent and that he could learn from him. Paul responded by becoming more confident and they came to share decisions and eventually ran the group together. But John, I suspect, always had the edge because he had formed the group, and neither of them forgot it.

Loss had played a part in Paul's make-up too. His mother, Mary, had died of breast cancer when he was fourteen and his brother Mike twelve. Mary, a midwife, had never complained about the pain she suffered, and died just a couple of months after diagnosis. Her boys were left in the care of their father, Jim, who was a devoted dad.

Jim McCartney had been a keen musician in his youth, with his own successful amateur jazz band. He had played the piano to Paul and Mike when they were small boys and even though Paul's music was very different from the sounds Jim loved, he encouraged Paul's musical dreams. However, Jim was cautious about John; he worried that he might be a bad influence on his son. He told Paul that he wasn't to wear the narrow teddy-boy drainies – drain-pipe trousers – that John loved (in fact, John wasn't allowed to wear them either: he wore baggy trousers over them when he went home to Mimi). Eventually, though, he saw that John and Paul had a strong musical connection, which won him round.

We often went over to Paul's house in Forthlin Road, close to Woolton where John lived, for the boys to practise. Jim was usually at work, but if he was there he always welcomed us. He'd greet us with his sleeves rolled up, a tea-towel in his hand and an apron tied round his waist. Then, while John, Paul and George twanged away at their guitars in the front room, he busied himself in the kitchen until he called us all for tea.

Paul was one of the three people John was closest to. Although he had plenty of cronies, he only really let down his guard with Paul, me and Stuart Sutcliffe. Slight, dark and intense, Stuart was serious and hard-working, unlike John, who played the lazy loon. Everyone knew that Stuart was a gifted artist who would go far – he'd even had a private commission for a painting, unheard-of for a student. When John saw his large, colourful canvases, he loved them and longed to try something similar. Before that he had limited himself to the cartoons and caricatures he drew so brilliantly and which had got him into college, despite his lack of O levels.

What Stuart and John had in common was a restless, speedy nature. But while John bounced from one idea to the next, never able to settle or concentrate for long, Stuart put all his energy into being a star pupil. He was only three months older than John, but he was a year ahead in college and encouraged John to experiment in his art. Stuart was so wrapped up in his work that he didn't have a girlfriend and often forgot to eat. Most days he would stay after classes to paint. John was good for him because he reminded Stuart to take time out to have fun, and taught him to play the guitar. Unlike many of John's cronies, Stuart didn't look up to John or try to ape him. He respected John and treated him as an equal, which was something John valued a great deal.

Like Paul and me, Stuart saw something special in John. Most people thought John was destined to be a drop-out and a bum, who would never knuckle down to a decent job or make anything of himself. All they saw was the fool who clowned around in class and gave all the serious students wicked – and very irritating – nicknames. Our teachers said that he would be either a genius or a tramp – there was no in-between with him. In the same way, people either liked or loathed him and it was impossible to ignore him. Those of us who loved him knew that he could go off the rails, but we also saw in him raw talent and the potential for real creativity.

Stuart and I got on well. I was in awe of his talent, but he was amusing and good company too. I was glad John had a friend who took art seriously, but I never saw him as anything more than a mate who hung around with John and me. One night we were at a party and John went mad when someone told him Stuart and I were dancing together. As soon as I saw the look on John's face we stopped and, as so often before, I reassured him that it was him I loved. He seemed to accept it. But the next day at college he followed me to the girls' loos in the basement. When I came out he was waiting, with a dark look on his face. Before I could speak he raised his arm and hit me across the face, knocking my head into the pipes that ran down the wall behind me. Without a word he walked away, leaving me dazed, shaky and with a very sore head. I was shocked, really shocked, that John had been physically violent. I could put up with his outbursts, the jealousy and possessiveness, but violence was a step too far. Phyl had been right: I knew I had to end our relationship.

# 4

FOR WEEKS I REMAINED DETERMINED not to take John back. I even went on a few dates with a boy who lived near me and did my best to catch up with my college work. I put John firmly out of my mind.

It wasn't easy, though, because I saw him around college every day. He would look over at me in the canteen or in the lettering class, and when our eyes met, in the moment before I turned away, I knew we still cared for each other as much as ever. Friends told me John missed me, and I certainly missed him – his humour, his music and passion.

After three months, he phoned me and asked me to go back to him. It had taken him that long to pluck up the courage. He apologised for hitting me and said it would never happen again. I hesitated for a whole second before I said yes.

John was true to his word. He was deeply ashamed of what he had done: I think he had been shocked to discover he had it in him to hit me. So, although he was still verbally cutting and unkind, he was never again physically violent to me. As we grew closer, in the second phase of our romance, even his verbal put-downs and attacks diminished. It was as though the more certain he was that I was there for him and loved him, the easier he found it to remove his protective armour.

As John began to trust me, he talked more about his upbringing, letting me see the hurt, lost little boy inside him.

His childhood had been enormously difficult. His father had abandoned him. He opted for and he was taken away from his mother by his aunt and brought up in a cold, austere home with little affection or comfort. Then, in his teens, the two loving figures in his life, his uncle and his mother, had died.

John's mother had been the fourth of five sisters born into a middle-class, well-to-do family. The Stanleys lived in an elegant four-storey Georgian house in a smart residential area of Liverpool, close to the cathedral. Julia's father, George Stanley, went to sea, and later became an insurance investigator. He and his wife Annie had had a son and daughter, both of whom died as babies, before their five healthy girls arrived. Mary Elizabeth, known as Mimi, was the eldest; then came Elizabeth, later nick-named Mater, Ann Georgina, called Nanny, Julia, or Judy, and finally Harriet – Harrie.

All five were strong characters, bright, determined and unconventional – traits that Julia passed on to John.

In those days girls were expected to marry and have families, but the Stanley girls had their own ideas. Mimi married, but refused to have children. She'd helped to look after her younger sisters, which had put her off having any of her own. Nanny became a career woman, unusual in the 1930s: she was a civil servant and didn't marry until her mid-thirties, when she had one son, Michael.

Harrie scandalised everyone by marrying an Egyptian student and going off to Cairo, where she had a daughter, Leila. After her husband, Ali, died suddenly, she headed back to Liverpool to escape his parents, who wanted custody of the baby. During the war Harrie, as the widow of a foreign national, was deemed an alien and had to report to the police station every day. Later she married a kind, gentle man called Norman, with whom she had a son, David.

Elizabeth married Charles Parkes and had a son, Stanley. She refused to be called 'Mummy' – far too conventional – and settled on 'Mater', which soon became the name by which the whole family knew her. Her son, though, seldom had a chance to use it: Stanley was a delicate baby and she found it difficult to cope. When he was a few weeks old she handed him over to her mother, who brought him up until he was old enough to go to boarding-

Mum, in 1918 when she was 15

Me, outside the back door of our house in Hoylake, aged three

With Dad and my brother Charles, not long before he left home – he was 16 and I was five

Dressed for a party, a shy 16 (right), with my brother Tony's girlfriend Marjorie

Madly in love: with John and art college friends Tony Carricker and John Hague, sitting on John Hague's mother's car in 1959. I was 19, John was 18 and we'd been together a few months

A rare picture of John wearing heavy black-framed glasses, at my brother Tony's wedding to Marjorie in April 1960. I was bridesmaid and John is standing behind me

With my friends at art college, aged 21 in June 1961. From left, me, Phyl, Ann Mason, Annette and Jenny

John, aged 22, in playful mood with Mimi

John with George and Stuart on their first visit to Hamburg, in 1960

The Beatles in Hamburg 1960. From left: Paul, Pete, Stuart, George and John

Inside the Cavern, where the Beatles played regularly from March 1961 to August 1963

Mum, far left, with friends on deck, setting sail for Canada in July 1961

John, photographed by Astrid in Hamburg

Astrid and Stuart in Hamburg, 1960

There were no photos of my wedding to John on August 23 1962, so I
drew the scene in the register office – complete with pneumatic drill
outside the window. From left: George, Marjorie and Tony, Paul, me,
John, Brian and the registrar

On our way to
New York for the
first US tour,
1964

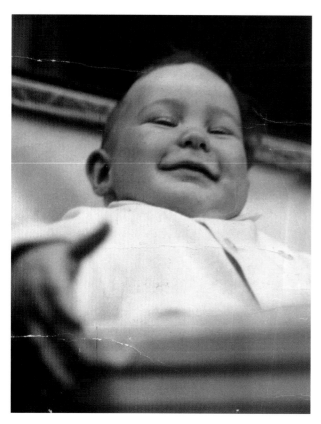

Our son Julian, born on April 8 1963, the spitting image of his dad

John, with Julian, aged around 18 months

school. Later she divorced Charles and married a dentist, Robert Sutherland, known as Bert. After the war they moved to Edinburgh, where Bert set up a practice. John got on well with his cousin Stanley, who was a little older than he was, and they spent summer holidays together in Scotland.

Julia, John's mother, was the prettiest and most unusual of the sisters. Five foot two with shoulder-length auburn hair, she was only fourteen when she met Alf Lennon, a fifteen-year-old office boy. Alf's mother had died giving birth to her youngest son, and his father died soon afterwards, so at five Alf, with his two younger brothers, was in an orphanage. He met Julia a week after he had left the orphanage. He was a good-looking boy, but as far as Julia's family were concerned, he was simply not good enough for her. Despite this she adored him and continued to see him. Soon after they met he went to sea, as a ship's waiter, but the two met whenever he was home.

They married, eleven years later, in December 1938 – for 'a laugh', according to Alf, who claimed that it had been Julia's idea. It was a register-office wedding, which none of Julia's family attended. They spent their honeymoon at the cinema – Julia was so mad about movies that she put 'cinema usherette' on her marriage certificate, although she never actually was one. Afterwards she went back to her parents' house and Alf to his rented room because they had nowhere to live together.

The next day Alf went back to sea, on a trip to the West Indies, for three months. Julia had little choice but to stay at home with her parents. Every now and then he would come home and spend a few weeks with her, then take off again, much to her parents' dismay.

It was after Alf's Christmas visit in 1939 that Julia found she was pregnant, and John was born on 9 October 1940, at six thirty in the evening; patriotically she gave him Winston as a middle name. Mimi, who was at the hospital with her for the birth, suggested the first.

Julia took her son home to the house where she and her father had moved after her mother's death. It was in the Penny Lane area of the city and it was there that John spent his first few years. Alf continued to appear from time to time until, when John was about eighteen months old, Julia was told that he had jumped ship and disappeared. Until then he had sent her money each month from his wages, but now it stopped. When he finally reappeared he said he had been jailed for three months for theft – later he claimed this was untrue – but the marriage was faltering. Julia had had enough of waiting for him: she was young and exuberant, she loved life and wanted more than waiting around for a feckless husband who might or might not appear and who was no longer helping to support his son.

In late 1944 she had an affair with a young soldier who was home on leave. When he went back to the front Julia found she was pregnant. Her father insisted that the baby was adopted and arrangements were made through the Salvation Army. Julia's daughter, Victoria Elizabeth, was born in 1945, just before the end of the war, and it was believed that she was taken to Norway by her adoptive family. A few months later, in early 1946, Julia was working as a waitress in a café in Penny Lane. There, she met John Dykins, known as Bobby, who was a hotel manager. Once again, the Stanleys felt he wasn't good enough for her and, in any case, she was still legally married to Alf. They were horrified when she moved into a tiny flat with Bobby and John.

After their mother's death Mimi had become the family matriarch. Soon after Julia moved in with Bobby, Mimi went to visit them and told Julia that she was unfit to be five-year-old John's mother because she was living with a man to whom she wasn't married and had given birth to an illegitimate child. She told Julia to hand John over to her and she would bring him up.

Julia and Bobby refused: John was Julia's child and should stay with her. They sent Mimi away, but she was a determined woman. She told social services that Julia and Bobby weren't married and demanded that John be given to her. But the social worker came down on Julia's side: there were no grounds for John to be taken away from a loving home.

When Mimi discovered that John didn't have his own room and shared a bed with Julia and Bobby, she insisted that social services inspect Julia's home again. This time Julia was told that John must live with Mimi until she and Bobby had found a bigger flat. Mimi had won.

Although Mimi undoubtedly cared for John she was not a woman for cuddles and praise. She was strict and insisted on rigid rules and absolute order. If he tried to put his arms round her, she shrugged him off, saying, 'Get away, go on with you.'

Life with his aunt must have been hard for little John, who was used to the warm, easy-going style of his mother. His saving grace was Mimi's husband, Uncle George, who loved children and was delighted when John arrived – the son he had always longed for. George, who worked as a dairyman on a nearby farm, was a tall, kindly man who was never angry and gave John the 'squeakers' – kisses – that Mimi could not.

It's hard to see why Mimi wanted John, since she had always said she didn't want children. Perhaps she regretted her decision. Perhaps she was jealous of Julia. Or perhaps she genuinely felt that she was doing the best thing for John and was giving him a stable home.

Shortly after John went to live with Mimi and George, Alf reappeared. He had landed at Southampton, phoned home and heard that John was with Mimi. Armed with this information, he made his move: he told Mimi he wanted to take John on holiday to Blackpool and, somewhat surprisingly, Mimi agreed. When Julia heard what had happened she was afraid that Alf would never bring John back and went after them. She found them in a Blackpool boarding-house, where Alf admitted to her that he had planned to take John to live in New Zealand. He asked Julia to go too, to try to repair their marriage, but for her it was too late: she only wanted John.

Alf called John into the room and asked him to choose between his parents. John, faced with a heart-breaking decision no five-year-old should ever have to make, chose his father. In tears, Julia agreed to let him go. But as she left John jumped up, sobbing, and ran after her. She took him back to Liverpool and that was the last John saw or heard of his father for many years.

John returned to Mimi's, where he remained. In time Julia and Bobby found a bigger home, but it was decided that John had settled and should not be disrupted again. Had anyone asked John, he would undoubtedly have chosen to live with his mother. He told me that as a little boy he often dreamed of running away from Mimi to his mother, but in those days no one considered asking a child what he wanted. Mimi had decided to keep John and that was that.

Julia and Bobby went on to have two daughters, Julia and Jacqui, and stayed together, unmarried but stable, until Julia's death twelve years later. When John was small he saw his mother only when she visited him at Mimi's, which she did regularly, but later, at eleven or twelve, he started going to Julia and Bobby's house on his own, which opened up a whole new world. He loved being there and often stayed the night. The two girls slept in the double bed in his sister Julia's room, and he had Jacqui's.

Bobby was always kind to him, slipping him a bit of extra pocket money, and he got to know his sisters – who were six and eight years younger than him and soon worshipped him. Most important of all, he began to get to know his mother better. But despite this he never lost the feeling that he was an outsider in his mother, step-father and sisters' happy family unit.

John adored his mother. Her high spirits and love of life were so different from Mimi's strictness and rigidity. Julia believed in enjoying herself, while Mimi thought her frivolous and self-indulgent. To Mimi life was about hard work and self-restraint, and she constantly berated John for not working hard enough at school.

John grew up bouncing between these two very different women: Mimi, the firm 'mother', and Julia, the more playful 'aunt'. With Mimi he was expected to be neatly groomed, dutiful and obedient. With Julia he could laugh, play and fool around. Julia encouraged him in his music. Mimi wouldn't allow him to play the guitar in the house and insisted he practise in the glass-walled front porch. John's sister Julia told me, much later, that as a little girl she would lie in bed and listen to John and their mother chatting, playing music and dancing while Bobby was working the evening shift at the hotel.

When John started the Quarrymen, Julia was delighted. Skiffle had been made popular by Lonnie Donegan and the great thing about it was that anyone could have a go on instruments like the washboard and tea chest – you didn't need musical knowledge or even talent to play them. Even the other instruments, the bass and the guitar, weren't hard if you got the hang of a few chords. Julia happily allowed the group to practise in her house. In fact, she often joined them, playing washboard, when they shut themselves into the tiny bathroom – where the acoustics were best – to rehearse for hours on end. She was there at their first gig, with John's little sisters, when they performed as Johnny and the Rainbows from the back of a lorry at a street party to celebrate Empire Day in May 1956 when John was fifteen. John's sister remembered sitting, open-mouthed, to watch her big brother and his friends play as people jived in the street or hung out of windows to listen, clap and cheer. After this they played regularly at parties and weddings, usually for no more than a few free beers.

Quarry Bank High School was an old-fashioned boys' grammar. John had passed his eleven-plus to get in but, once there, he did no academic work and was put in the bottom stream. Despite Mimi's admonishments, he had no respect for his teachers and no interest in passing exams. He was disruptive and constantly in trouble at school, but even the countless canings he received made no difference.

John's teenage years were punctuated by tragic losses. The first was his beloved uncle George, Mimi's husband. He collapsed at home one Sunday afternoon and died instantly from a massive liver haemorrhage. His death came without any warning and to John, who was four-teen and had looked on George as a father, it was a terrible shock. He missed his uncle badly – the big old overcoat he wore at college had been George's and even when it became tatty and threadbare John insisted on wearing it.

His mother's death, three years later, also came out of the blue. One July evening she had been round to Mimi's for a cup of tea and a chat. Despite their past disagree-ments, the two still got on well and, like the other Stanley sisters, loved gossiping over a pot of tea. John wasn't there: Julia had left him at her own house, with Bobby and the girls.

When Julia left to catch her bus home, Mimi waved her off. Sometimes she walked with her to the bus stop, but that night she didn't. Seconds later Julia was hit by a car as she crossed the road outside Mimi's house. She died instantly. That night a policeman broke the news to John and Bobby. John's little sisters were in bed, and as Bobby's mother was there, John and Bobby went to Sefton General Hospital, where Julia had been taken. John couldn't bear to see her, but Bobby went in to say goodbye. When he came out he collapsed in tears in John's arms.

John went to her funeral in a daze, unable to believe that his vibrant, laughing mother was inside the wooden coffin everyone was filing past. Her death was made worse for everyone by the fact that the man who killed her was a drunk off-duty policeman and a learner driver who had apparently hit the accelerator instead of the brake when he saw Julia crossing. When he went to court he was acquitted of the charges and the only punishment he received was a reprimand and a period of suspension from duty. After Julia's funeral he bottled up his grief, anger, shock and pain, and didn't talk about it to anyone. Mimi wasn't the kind of person who discussed things, she simply carried on with her life, and John tried to do the same.

He couldn't even share the loss with his little sisters. Their aunts took over and would not let anyone tell the girls, who were then eleven and eight, that their mother was dead. They were told only that she was ill and bundled off to stay with Mater, who was living in Edinburgh with Bert. Bobby was distraught and the aunts had decided that the girls couldn't be left with him. In those days it was rare to find a single father, and Julia's family felt that, working shift hours and grief-stricken, Bobby would find it impossible to cope.

After the girls came home they went to live with their mother's younger sister, Harrie, her husband Norman, and their cousins Leila and David, who was ten. Even then they weren't told. It was at least two months after Julia's death that Norman, to Harrie's fury, sat them down and told them that their mother was dead. After that John could talk to them about Julia, but even then they seldom had the chance: Julia had become a taboo subject and no one mentioned her. The family dealt with pain by keeping it under wraps. If a subject was difficult, it was not aired. It was an attitude that hurt John, but which he often adopted as an adult. During and after our marriage I discovered that he had an astonishing ability to ignore anything that distressed him. For example, he didn't contact his sisters when he heard that their father had died, as Julia had, in a car accident, and he ignored our son for several years after our divorce.

John was close to his sisters and used to visit them at Harrie and Norman's house, The Cottage, which wasn't far from Mimi's. Soon after he and I got together he took me to Harrie's to meet them. Mimi's and Harrie's homes were completely different: While Mimi's spick-and-span house was cold and unwelcoming, Harrie's was warm, loving and buzzing with activity. Harrie made us feel instantly welcome: she sat us down on a comfy old sofa and plied us with tea and biscuits. When Julia and Jacqui arrived home from school, rosy-cheeked and chattering, their faces lit up when they saw John, and so did David's, when he came in moments later, his shock of sandy hair sticking out in all directions.

After that it was non-stop shrieks of 'I'm starving, Harrie, can I have a piece of toast, just one more biscuit, pleeeease.' John and I sat, hand in hand, and watched. Full of curiosity, the girls scrutinised me from top to toe. To them I was clearly a bit of a glamour figure, with my blonde hair and the tight black clothes I had taken to wearing. Julia was a chatterbox who asked a thousand questions, while Jacqui was much quieter but clearly just as interested in the new girl in John's life.

Harrie, smiling and slightly harassed, rushed in and out with plates of food while Norman, thin as a whippet, moustached and in a tweed jacket with grey flannels, pottered around filling the coal bucket. Like all the Stanley women, Harrie was in charge.

After that first visit John and I went to The Cottage quite often to see them all. Harrie seemed to understand young love: it didn't matter what time of day we turned up, she always welcomed us. While Mimi conveyed the impression that she grudged every mouthful you ate, Harrie was generous and you wanted to be around her. We loved the warmth and life of The Cottage, and I was glad that John had somewhere to go where he could be part of a family and truly at home.

John also took me to meet his aunt Mater in Scotland. We had a day off college to hitchhike to Edinburgh, travelling most of the way in an enormous truck with a driver who had clearly exceeded his hours. John and I spent the entire journey trying to keep him awake.

When we got there it was pouring with rain and there was no one in at Mater's house – John hadn't thought to tell them we were coming. We waited in the doorway of the church across the road, and after a couple of hours Stanley, Mater's son, arrived. He was delighted to see John, took us in and told us that Mater and Bert were at their croft in the north but would be back the next day.

When I was introduced to Mater, we clicked immediately. She was as warm and welcoming as Harrie, and we all talked for hours that night, over a marvellous meal. By the end of our two-day stay I felt I'd joined another family and John was delighted that I'd got on so well with them. It brought us even closer.

Early in 1960, John moved in with Stuart Sutcliffe. A second mattress was laid on the floor – there was plenty of space because the room was enormous – and his chaos, combined with Stuart's, created a pad that was at best unusual, at worst a shambles. The boys had a Belisha beacon and an empty coffin as part of their furniture and revelled in their outrageousness.

Mimi was deeply upset at John's defection and asked me to persuade him to move back home. But I knew John was happy where he was, and, besides, it meant we had somewhere to be alone together. I told Mimi – truthfully – that I could do nothing: I knew that John wouldn't be persuaded back. So that she could see John, Mimi agreed to do his washing. He'd take a pile round to her every few days and stay long enough to eat a huge meal, but he was in no danger of moving back in: he and Stuart loved living together, and when Stuart won sixty pounds in a prestigious art competition, John persuaded him to spend the lot on a bass guitar and join the Quarrymen. Stuart wasn't a natural musician and struggled to master the guitar, his fingers calloused and bleeding as he practised for hour after hour. When the boys did a gig he often turned away from the audience so that they couldn't see how little he was playing. John knew Stuart was the group's weak link, but he didn't care: he wanted him along. It was the perfectionist Paul who found such an inexperienced guitarist hard to accept and this led to rows and even fights between him and Stuart. I think Paul was also a bit jealous of Stu: until then he had had most of John's attention.

The boys had started playing at a club called the Jacaranda, on Slater Street, not far from college. The Jac, as we called it, was in a cellar, like the Casbah. Condensation dripped off the walls and it was so dark inside that we could hardly see each other, but this added to the slightly dangerous, exciting atmosphere. The Quarrymen got their first gig there after John begged the owner, a bluff Welshman called Allan Williams, to give them a chance. Some of Liverpool's biggest groups played at the Jac, including Rory Storm and the Hurricanes, whose drummer was a lad called Ringo Starr.

By now the boys had smartened themselves up a bit: they all wore black jeans and polo necks, which gave them an edge over some of the other fledgling beat groups then doing the rounds in Liverpool. Allan Williams was closely involved with Liverpool's budding beat scene and he regularly organised gigs for groups at venues all over the city.

After much nagging on John's part, he got the boys a session at the Grosvenor Ballroom in Seacombe. It was a rough spot, where local gangs often fought, and that night I kept a low profile, head down to avoid trouble. The place was so noisy you could barely hear them play, and as soon as they'd finished we shot out.

The image of the group was changing rapidly. They'd dropped the old skiffle sound and the ballads by Cliff Richard and the Everly Brothers in favour of the far more exciting rock and roll numbers that were racing up the charts.

To fit with their changing image the boys decided it was time for a new name. We had a hilarious brainstorming session over a beer-soaked table in the Renshaw Hall bar, where we often drank. John loved Buddy Holly and the Crickets, so they toyed with insect names. It was John who came up with Beetles. He changed it to Beatles because he said if you turned it round it was 'les beat', which sounded French and cool. They settled on the Silver Beatles.

Soon afterwards Allan Williams got the boys an audition with a man called Larry Parnes, who had a reputation for creating stars: he had discovered Tommy Steele, who was a huge name then, Marty Wilde and Billy Fury among others. He was in Liverpool looking for a backing group to go on tour with Billy Fury, and John and Paul were wild with excitement. As they still didn't have a drummer, a guy from one of the other bands stood in at the audition to help them out.

The boys failed the audition, and Paul blamed Stuart's poor playing. But Parnes offered them an alternative, backing one of his lesser-known singers, Johnny Gentle, on a tour of northern Scotland for two weeks, for a fee of eighteen pounds each. This was their biggest break by far and they were delighted. It meant John and Stuart bunking off college for the week, but John was barely turning up anyway, and Stuart would easily make up anything he missed. George was in his first job so he took his annual two-week holiday. It was a bit tougher for Paul, who was due to sit his A levels in a few weeks' time and had already applied to teacher-training college. Understandably his dad was reluctant to let him go. But Paul told Jim that a break would refresh him for the exams and, in the end, he agreed.

The final problem they had to face was their lack of a drummer. Luckily a local jazz drummer named Tommy Moore agreed to go with them. He was quite a bit older than the other four, but that didn't matter. They piled into a van driven by Johnny Gentle and set off.

During those two weeks I missed John like mad, but he sent me postcards and I had a chance to catch up on my collegework. We had exams coming up and I really wanted to pass.

When the boys got back from the tour, they were exhausted, filthy, sick of the sight of the van – and high on their success. They'd played ballrooms and nightspots in out-of-the-way places and had received an enthusiastic welcome. It was the first time any of them had been asked for their autograph, which had made them feel like stars – and they loved it. John and I couldn't wait to be alone together, so Stuart went out for a few hours. We began to make love – until a sharp abdominal pain made me cry out and brought things to an abrupt halt.

John became more and more concerned as I writhed in agony and begged him to get me home. We got dressed and he took me to the station where I crawled on to the train. As soon as I walked through the door my mother took one look at my white, drawn face and called an ambulance.

It was appendicitis and I was stuck in hospital for two weeks. After a couple of days John came to visit me, dragging George with him. I had been so desperate to see him, and was so frustrated when I saw George, that I burst into tears. Shocked, John told George to hop it and held my hand for an hour to mollify me. After a while my mum arrived, and later took John and George back to our house for tea.

I recovered in time to sit my exams and was delighted when I passed. But John, despite my best efforts to help him at the last minute, failed dismally. I had sat over his lettering for hours while he and Stuart cracked jokes and peered over my shoulder, but it wasn't enough to save him. He had done virtually no work that year and was thrown out of the college before his final year.

Not that he cared. After the Scottish tour the boys had regular gigs at local clubs and nightspots at least twice a week, and John was certain that music was his future. He and the others were longing for another big break, and while they waited, they dropped 'Silver' and the group became known as the Beatles.

Their next break wasn't long in coming. In August 1960

Allan Williams offered them a six-week stint playing in a nightclub in Hamburg, Germany. He had been exporting groups over there since a German seaman had heard a steel band playing in the Jacaranda and told people back home how good they were. A Hamburg nightclub had engaged them and Allan had followed them there and got to know a man called Bruno Koschmider, owner of the Kaiserkeller nightclub. Bruno agreed to try out a Liverpool beat group called Derry and the Seniors. When they proved a success, he asked Allan for another group. He suggested the Beatles.

# 5

THE BOYS WERE DUE TO leave for Hamburg in two weeks' time, on 16 August. There was just one problem: they still didn't have a drummer. They racked their brains until someone mentioned Pete Best.

The Blackjacks had fallen apart when a couple of the members decided to move on, so Pete had a great set of drums and no group. Like Paul, he'd left school that summer with good enough qualifications to get into college, but by this time he had a taste for the entertainment business. Paul gave him a ring and invited him to join the Beatles for the Hamburg trip. He leapt at the chance, and he, Stu, John, Paul and George were on their way.

Pete was good-natured but very quiet, with the kind of sultry, moody looks that had girls flocking round him. While the other boys mucked around, teased each other mercilessly and wisecracked their way through rehearsals, Pete said little and often seemed to be in his own little world. He seldom showed any emotion; he was as happy as the others to be going to Hamburg, but while they leapt around shouting, 'Look out, you Krauts, here we come,' Pete shrugged, smiled and got on with setting up his drums.

Most of the boys' parents were fine about the trip. George's mum and dad were a lovely couple who, although he was still only seventeen, wanted him to follow his dream. His mum warned him to be careful, made Allan promise to look after him, then sent him off with clean clothes and a tin of home-made scones.

Paul's dad was reluctant at first – he still hoped his son would go to teacher-training college. But Paul was no longer interested in teaching. Jim saw that and, despite his worries for his elder son, let him go to Hamburg with his blessing.

Pete's mum had always encouraged his dreams of going into entertainment, so she was happy too, and Stuart, the oldest Beatle, had no problems with his mum, Millie.

The only person who disapproved was Mimi. She was horrified and did everything she could think of to stop John going. He had largely hidden his passion for music from Mimi. She had told him so often that playing the guitar was all very well but he'd never earn a living at it that he was sick of hearing it. When he was younger she'd forbidden him to play in a group, so most of the time John lied to her about what he was doing. He practised at friends' homes or in college, and when he and the group played in clubs, he told Mimi he was having a night out with friends. So, she was shocked to discover that John was not only in a group but one that had got far enough to be invited to play abroad for several weeks. Nevertheless, she didn't change her hardline position: she clucked, admonished and pleaded with John to get back to his studies. But nothing and no one could have swayed him and in the end Mimi had no choice but to accept it.

Not long before the boys left John phoned to tell me that Mimi had gone out for the day to visit her sister Nanny in Birkenhead. 'Come over as fast as you can,' he urged. He had borrowed a camera and wanted to take some photos of me to take with him to Germany. He insisted I try out various seductive poses while he snapped away, so I put my hair up, let it down, hitched up my skirt and thrust out my chest in an attempt to do my best Brigitte Bardot.

After the photo-session we made love, then lit a fire and lay on the sofa in front of the television, eating anything we could find in the fridge. It was all the more exciting for being so illicit. Mimi was due back that evening and eventually we agreed reluctantly that I'd better go. Then the phone rang. It was Mimi: the fog was so bad that the buses weren't running so she wouldn't be back until morning.

We couldn't believe our luck. I called Phyl, then my mother to say I was staying at Phyl's. For the first time John and I spent a night together. Squeezed into his single bed we slept curled up in each other's arms.

By morning the fog had cleared and I left early – we weren't about to take any chances. Our night together had been blissful and we didn't want it ruined by Mimi's sudden appearance.

John and I were incredibly close and I was terrified at the thought of parting from him. We had been together almost every day for nearly two years, so the thought of a six-week separation was unimaginable. And we both knew it might be for longer: if they did well their contracts might be extended. John didn't want to leave me either. We promised each other that we'd be faithful and write to each other every day, and he made me promise not to smoke too many ciggies. Despite his own twenty-a-day habit, compared with my two a week, he hated women smoking and wanted me to stop altogether.

Allan drove the boys to Hamburg. After a few last kisses I waved John off, tears running down my cheeks as the van disappeared round the corner. I felt sure of him in Liverpool, but I didn't know anything about Hamburg or the girls there. Would John still love me when he got back?

I felt really despondent and threw myself into my neglected art. My mother didn't say anything, but I could see that she was delighted to have me back at home for so much of the time. She wasn't over the moon about my relationship with John: although she had said little I knew she thought he was too rough, not the sort of steady boy she had in mind for me. No doubt she was hoping that, with John so far away, things would cool off and I'd forget about him.

But there was no chance of that. I was madly in love and when John's letters began to arrive I was reassured that he still felt the same about me. He wrote to me almost every day, letters that were ten pages long, covered with kisses, cartoons and declarations of love. Even the envelopes bore poems, kisses and messages like 'Postman, Postman, don't be slow, I'm in love with Cyn so go, man, go.' His letters could be as lewd as they were passionate: he made me blush with references to the 'massive throbber' he had when he thought of me. I had to find a good hiding-place for them – my mother would have been scandalised if she'd read them.

At one point John wrote that he'd bought me 'a fantastic pair of leather pants'. I thought he meant trousers and I was thrilled, imagining how cool I'd look in them at college. I was soon disillusioned: he must have realised I'd interpret what he'd said wrongly and in his next letter he explained that he'd bought me a minute pair of black leather knickers – he even drew them, and added that he couldn't wait to see me in them.

In turn, I wrote to John, enclosing what I hoped were sexy photographs, taken in the Woolworths photo booth – the latest in new technology. I was determined to do what I could to keep him faithful, so I would put on my most provocative outfit and pout as seductively as I knew how. I'd contort myself into sexy poses in the cramped little cabin – one hand behind my head, chest thrust forward and what I hoped were come-to-bed eyes. The results, when the photos popped out, were sadly disappointing. Far from the enigmatic temptress I'd hoped to become, I looked more like one of John's ludicrous cartoons. Nonetheless I sent them off to John. I had no photo of him and begged him to send me one that I could carry around with me. Plenty arrived, but nothing I could display publicly, or even look at without laughing. There were insanely grinning hunchbacks, grotesque poses and ghastly leers. John could never play anything straight.

A few weeks after John left, I spent my twenty-first birthday quietly with Mum, celebrating in front of the television with a cup of tea. I didn't mind: I'd promised not to go out and have fun without John and I didn't even want to. Phyl was twenty-one a few days later, so I took a bottle of champagne and a silver key into college. We drank the champagne in our life class, broke the key in half, and each kept a piece as a symbol of our friendship.

The story of the boys' first trip to Hamburg has now passed into legend, but I had my own unique perspective on it through John's many long, detailed letters. He wrote about every aspect of their stay, including those he would never have wanted his family to know about. I kept his letters for several years, but destroyed many in the aftermath of our divorce. I sold others when I was desperate for money. Of course, I regret it now because John's letters were precious to me.

He was graphic about their living conditions. They had

expected to play in the Kaiserkeller, but when they arrived they were told they'd be playing in another of Bruno Koschmider's clubs, the much smaller Indra. The previous act had been a risqué girlie show and most of the punters made it clear that they didn't think much of the Beatles and wanted the girls back. The boys had to work hard to win them over. The audience expected a real show and the boys weren't about to let them down. Until then they'd only ever played one-hour sets but now they were expected to play for seven or eight hours, so they spun out each song for twenty minutes, cavorting on the tiny stage as they played, and stomping their feet to emphasise the beat. John, a natural showman, was brilliant at dancing, rolling around or leaping in the air, and the audience were soon enjoying the cocky young group from Liverpool.

The boys were put up in a cinema called the Bambi, which had once been a theatre. The dirty, disused dressing rooms were their living quarters. They slept on camp beds with ancient blankets and used the grubby public toilets, with only washbasins, no showers or baths. After playing late into the night they'd be woken each day by the sound of the first film screening. They'd head for the ladies' toilets, which were the cleanest, but would usually be thrown out by an outraged woman.

John often complained in his letters about how awful it was but, reading between the lines, it was obvious that he was enjoying the whole experience, especially after they had begun to make an impact. At first there were few customers in the club, but when word got round about the show it was crowded every night and stayed open past its usual closing time of twelve thirty until two a.m.

After two months the Indra was closed because of complaints about the noise, and the Beatles were moved to the Kaiserkeller where another Liverpool group, Rory Storm and the Hurricanes, were playing. The Beatles and Rory Storm's group played alternately through the evening, an hour at a time for twelve hours, and the crowds loved them. The boys knew Rory Storm's group – they were bigger in Liverpool than the Beatles, who had only got the Hamburg job because Rory Storm was busy and turned it down. Later, his band decided to give Hamburg a go, and it was in Hamburg that the boys met Rory's drummer Ringo Starr. He seemed pleasant enough, but they didn't get to know him well.

The Kaiserkeller, like the Indra, was situated on the Reeperbahn, Hamburg's Soho, where the streets were lined with clubs offering live sex shows, strippers and all-hours alcohol. Drugs and violence were rife. John wrote that they often saw people draw flick-knives and the waiters carried a cosh to keep order. It was a world away from Liverpool, but luckily their parents had no idea of it and I certainly never mentioned it to Mum or anyone else.

It didn't occur to me that they'd take drugs. It was only after they got back that John told me they'd learnt to stay awake through the night by swallowing slimming pills. Pete refused to try them but the others thought they were great and soon graduated to stronger pills, amphetamines called Black Bombers and Purple Hearts. They barely slept and didn't eat properly, but they were having the time of their lives and their contracts were extended several times.

I was glad of their success, but all my energies were concentrated on willing John to avoid the temptation of other girls. I began to get worried when he started to talk in his letters about a girl called Astrid Kirchherr. She was a photographer's assistant who had seen the Beatles with her boyfriend, Klaus Voormann. She had been reluctant to go, but Klaus had wandered into the club alone one night when they were playing and told her they were sensational.

Once Astrid had seen them in action, she agreed with Klaus and they went back almost every night. Astrid was a middle-class girl, who had come across the Beatles in an area where most respectable Hamburg citizens would never go. But once she and Klaus had told their friends about the group, more and more of them went to the club, and gradually students took over from the previous, somewhat rougher clientele. Between sets the boys would sit with Astrid, Klaus and their friends. Only Pete, who'd taken German O level, could speak any of the language, but some of the students had a little English.

Astrid was fascinated by the boys' teddy-boy style and they were flattered by her interest: she wanted to photograph them and invited them to her home for tea. They were astonished by her black and white candle-lit room – and delighted by the ham sarnies she served, the most 'normal' food they'd had in Germany. They began to go most days to Astrid's for a meal.

To John and the other boys, Astrid was glamorous and

sophisticated. They were bowled over by her hip black clothes, her *avant garde* way of life, her photography and her sense of style.

As John's letters became filled with 'Astrid this' and 'Astrid that' I was consumed with jealousy, certain it would be only a matter of time before he wrote to tell me he was in love with her. Eventually I was put out of my misery: John wrote that Astrid had fallen in love with Stuart. She'd been drawn to his James Dean looks, which were enhanced by the dark glasses he wore almost all the time. Like John and me, he was short-sighted, and got round the problem by wearing hip frames with dark prescription lenses.

When Astrid broke off her relationship with Klaus – who must have regretted ever taking her to see the Beatles – and got together with Stuart, she was twenty-two and he was twenty, but two months later they were engaged and bought each other a gold ring, according to the German custom. Not long after this the boys were offered a contract with a bigger, better club, called the Top Ten. They'd been sneaking over there to watch a singer called Tony Sheridan, whom they thought was great, and had had a couple of jamming sessions with him. But this was strictly against the terms of their contract with Bruno Koschmider, who was furious.

Just as they were about to move to the Top Ten the police found out – or were tipped off – that George was only seventeen and had no work or resident's permit. The clubs they were playing in didn't admit under-eighteens. George had got away with it for four months, but his luck had run out and he was ordered to leave the country. Astrid and Stu drove him to the station and put him on a train home with a bag of sweets and some apples. Days later the other four had hit trouble. When Paul and Pete were clearing their things out of the Bambi, to move into nicer accommodation at the Top Ten, they accidentally started a small fire when a candle fell on to some old sacking. There was little damage, but Koschmider called the police and they were ordered to leave the country.

A couple of days later John was told that his work permit was being revoked, so he, too, set off for home. He told me later that he'd been terrified travelling alone, wondering if he'd ever find England. He had never been abroad before and had been driven to Hamburg by Allan Williams, so it was his first journey alone. No wonder, as he rattled along in trains for a couple of days, he got nervous. Stu was ordered to leave as well but as he had tonsillitis Astrid gave him the air-fare.

As soon as John got back he called me and I rushed over to his place. He was delighted to see me but the way things had ended in Germany had left him dispirited. After all their success they'd drifted back to England one or two at a time, with no prospects and no idea of what to do next. As far as he knew, it might be the end of the road for the Beatles.

Despite their reasonable pay in Germany John came back with little money. He'd sent back some of his earnings to Mimi, proud that he could help her at last. He had also saved enough to buy me a handsome present: he'd seen all kinds of fantastic leather clothes in Germany and he wanted me to have a leather coat. We headed for C & A Modes, a large department store in the centre of town, where we scoured the long racks of coats. Black was John's first choice of colour for me, but as there wasn't one in my size, we chose a gorgeous three-quarter length chocolate brown one for seventeen pounds. It was my first present from him and I felt so gorgeous in it I couldn't wait to show it off.

We went to visit Mimi. There was a delicatessen called Cooper's next door to C & A, so we bought her a cooked chicken for tea and set off, full of high spirits. But if we thought we could share our happiness with Mimi, or that she would be pleased to see us, we revised our opinion fast. When she saw the coat and heard that John had bought it for me she hit the roof. She screamed at John that he'd spent his money on a 'gangster's moll' (even with Mimi yelling at us it was funny) and hurled first the chicken, which she grabbed from me, then a hand mirror at John. 'Do you think you can butter me up with a chicken when you've spent all your money on this?' she screamed. 'Get out.'

The colour drained from John's face. 'What the fuck's the matter with you? Are you totally crazy?' he shouted.

I was rooted to the spot. Mimi's outburst was beyond my comprehension.

John pushed past his aunt, grabbed my arm and dragged me through the back door. 'Come on, let's get out of this madwoman's house,' he muttered. We ran to the bus stop. John put his arm round me and apologised for her. 'All she cares about is fucking money and cats,' he said.

I wanted to go home, so we got the bus to the station and spent the journey in silence. It wasn't the first time Mimi had attacked him violently, but it was the first time she'd done it in front of someone else, and he was ashamed as well as angry.

We didn't visit Mimi's together for a while, and I never dared to wear the coat round there again.

That day her behaviour towards John was probably a mixture of fury that John had spent money on me, rather than handing it over to her, and jealousy because he obviously cared about me. Mimi wanted and expected John's devotion, and if you got in her way you were not popular. She constantly hounded and oppressed him. He often complained that she never left him alone and found fault with everything he did. Even before his mother died she had been the closest thing he had to a parent, and he wanted to please her, but she made it impossible for him. Years later, when he was world-famous and wealthy, he was still trying to earn her approval and she was still telling him off.

Most descriptions of Mimi that have appeared in print were based on interviews with her – she outlived John by eleven years. She loved to fuel the image of the stern but loving aunt who provided the secure backdrop to John's success. But that wasn't the Mimi I knew. She battered away at John's self-confidence and left him angry and hurt.

No doubt the impossibility of pleasing her was at least part of John's drive towards success. But, as his girlfriend, I found it hard to forgive her carping, when a little kindness or encouragement would have meant so much to him. Perhaps that was why the unconditional love and support I gave him meant so much to him.

After Hamburg John didn't contact the other Beatles for a couple of weeks, but when he did they decided to get back together. Poor George had been so embarrassed at having been sent home that he didn't realise for some time that the others were back too. He'd got another job, and so had Paul, who, urged by his dad, had done a stint at the post office, then started at a delivery firm.

While the boys had been away Allan Williams had been busy. He'd sent so many groups to Hamburg that he had decided to build a new club in Liverpool where they could really show off their talents. He called it the Top Ten, after the Hamburg one, but sadly it was burnt down six days after it opened. John and the others never even saw it.

Their first date, a few weeks after their return, was at the Casbah, where Pete's mum was more than happy to offer them the chance to perform again. Neil Aspinall, who was now living with Mona Best, put posters up all over the place saying 'Return of the Fabulous Beatles' and their loyal audience was delighted to see them again.

When I watched them that night, I couldn't believe how much they'd changed since I'd last heard them play. After so many hours of performing in Germany, they'd improved beyond all recognition. They'd gone from good to fantastic and the fans screamed with delight at the raw, noisy rock and roll the boys played. It was magnetic – you couldn't listen to them without wanting to dance. At that time the

Shadows, Cliff Richard's backing band, were huge, and their instrumental 'Apache' had stormed up the charts. Everyone was imitating their clean-cut style, all smart suits and shiny shoes, so the Beatles, in leather jackets and boots they'd bought in Hamburg, with their loud, pulsating beat, were something different.

Not long after the Casbah gig, they got a date at the Litherland town hall. It was the biggest place they'd played in so far and the kids there went wild. The boys did what they'd done in Hamburg, throwing themselves about the stage and playing numbers that went on and on. Before the end of the night, they'd been signed to appear there again. Suddenly they were really making it in Liverpool. They were no longer just one of the beat groups doing the rounds but a big attraction, with their own distinctive sound and style. A local write-up at the time described the Beatles 'exploding onto a jaded scene' and 'exploding' was the right word: the boys' energy, humour and wild, pounding music packed such a punch that it left Liverpool's teenagers crazy with excitement and demanding more.

After that first night at the Litherland, John's confidence returned. He could see they were going somewhere and he couldn't wait. They got plenty of dates and they started to make more money, earning six pounds a night and sometimes more. They'd gone from being the fourth or fifth best group in Liverpool to snapping at the heels of Rory Storm and the Hurricanes, who'd been at the top for a couple of years. Of course, Liverpool wasn't the only place in Britain spawning enthusiastic beat groups:

they were springing up all over the country. But it seemed that at that time there was a special energy in Liverpool, which produced an astonishing pool of talent, energy and ambition.

The Cavern club had been a traditional jazz venue for some years, but now that jazz was on the wane and rock and roll was taking over, the management had decided they'd better move with the times. The Cavern was in Mathew Street, a cobbled alley full of warehouses, lorries and litter. It was a dark, cramped cellar in a basement with poor ventilation and bad acoustics. Condensation dripped down the walls, the stage was tiny, and when it was crowded it was so hot you could hardly breathe the thick air that reeked of sweat and smoke. The Beatles got the job there through a local disc jockey called Bob Wooler, who was in his late twenties and very involved in the local beat scene. He thought the boys were good – he'd got them the Litherland town-hall job and persuaded the Cavern's owner, Ray McFall, to give them a try.

Bob first announced the Beatles at the Cavern on 21 March 1961, and before long they were playing there several times a week, at lunchtimes and in the evenings – over the next couple of years they played there 292 times altogether.

John was excited by the way things were going, although Mimi was still urging him to get some qualifications and a proper job. John humoured her, but he had no interest in what he called 'bits of paper'. He wasn't cocky or arrogant about it – in fact, sometimes he worried about the future – but something told him deep that things would work out and that it wouldn't be through passing exams.

He put in a lot of effort on behalf of the group, nagging anyone he thought would give them an opportunity to play. He knew that the Litherland and the Cavern were the breaks they'd been hoping for. John and Paul had always been competitive. Although the other band members – and the audiences – knew that John was the group's leader, Paul wanted to be involved in all the decisions, whether they were about which venue to play or which songs to use. The two sang alternately on stage and each had his own style. John exuded pent-up energy and sexuality, strutting and pacing the stage with his head tilted back, as if he was looking down his nose at the crowd. Most people took this for arrogance, and John had plenty of that, but he did it because he was so short-sighted. He was relatively aloof from the audience – he did most of the announcements and took the mickey out of people in suits or smart clothes, taunting rockers, students, older people – whoever was there. The more outrageous he was, the more popular the Beatles became.

Paul was also energetic on stage, but his style was more seductive than John's: he wooed the crowd, made friends with them, and the girls loved his big eyes and baby face. He was the most popular with girls, but John didn't mind – all of the boys attracted plenty of attention and loved the fans coming to tell them how great they were. John and Paul always had a special link between them, a chemistry that added to the heat: they knew intuitively how to share the stage and the limelight, how to spar with each other and how to play the audience so that the girls went wild. George, who played lead guitar, was quiet and serious. If anyone asked him why, he'd say that he couldn't afford to make a mistake. Stu and Pete were quiet too, so the way was clear for John and Paul to take centre stage.

Within weeks of the boys starting to play regularly at the Cavern, girls all over Liverpool were slipping out of school at lunchtime and coming to the club, where they'd queue down the street to get in. The most ardent of these was a group John called 'the Beatlettes', who hung around the stage, fetching the boys drinks and shouting song requests. The Beatles were kind to these regular fans and, knowing they had little money, sometimes bought them a cup of tea or gave them a free ticket.

I went to all the gigs with the group and, as John's girl-friend, I was often singled out by jealous fans. I learnt to stick like glue to my friend Pauline, the girlfriend – later wife – of Gerry Marsden. His group, Gerry and the Pacemakers, were often on the same bill as the Beatles. Gerry was an ex-van driver who became good friends with John – they matched each other for wit and humour.

Pauline and I were fine while the boys were playing, because the attention was on them, but in the breaks we'd always make sure we went to the ladies' together because that was where trouble usually flared up. We learnt to keep our heads down and not to look any girl in the eye, dashing in and out without bothering to redo our lippy. Even so, we'd often get 'She's a slag, what does he see in her' spat at us as we passed. But as well as the jealous fans there were the nice ones, who would say, 'If John loves her then we do as well. Give her a break.' They'd look out for us and hold back the angry ones if they threatened to leap on us and tear our hair out.

One girl who came to listen to the Beatles was Priscilla White, later to become famous as Cilla Black. She used to help with the cloakroom at the Cavern in return for free entry. She was studying at Anfield commercial college and had just started singing with local bands in the evenings. One evening she sang with the Beatles in the Cavern – John called her Cyril, a joke she took in good part, which he kept going for years afterwards. She was a skinny waif of a girl, but even then it was obvious that she had a distinctive voice and a big personality. Years later, when Cilla and I had become close friends, she confessed that John's caustic wit and endless teasing frightened the life out of her, although she had come to love him dearly.

Most of the boys' parents popped into the Cavern from time to time to see them play. Paul's dad, Jim, used to wait in the dressing room to hand Paul the food he'd bought in his lunch-hour for their tea. Paul would take it home and put it in the oven before Jim got back. Pete and Stuart's mums came, and George's turned up all the time with friends and relatives to see her boy.

As far as I know Mimi went only once. She heard what John was up to and had to see for herself. I don't know who was more shocked, Mimi or John, when they came face to face in the club. Mimi stayed a few minutes and that was that. Of course, her abrupt departure hurt John. He'd have loved her to be proud of him. But her disapproval – she made it clear she thought it was a disgusting place – didn't surprise him and he took it in his stride.

The boys loved having such a devoted following in Liverpool. It had taken John and Paul five years but at last they'd found success in their home town. Despite that, though, they were still keen to give Hamburg another try. It had been one of the most exciting times in their lives, but because they'd been ordered to leave, they couldn't be sure they'd be allowed back in. However, other bands were still going there and they didn't want to be left out. Allan Williams came to their rescue: he wrote to the German consulate, praising the boys' musical ability and characters, and explained that they'd been exploited on their previous visit. This time, he said, they were going to work for an honourable businessman, Peter Eckhorn, owner of the Top Ten Club.

It did the trick. The boys were given work permits, and George was now eighteen so there was no problem with his age. In April 1961 they set off for Hamburg for their second visit. This time I was quite happy about it – because I was going too.

# 6

I HAD NEVER BEEN OUTSIDE BRITAIN before, but now I was going to Hamburg with Dot Rhone, Paul's girlfriend, and we were so excited and scared that neither of us could sleep the night before we set off. We left a couple of weeks after the boys. It was the Easter holidays and I had two weeks off college, while Dot had taken time off from her job as a pharmacist's assistant.

Paul and Dot had been going out together for a few months. She was seventeen, petite and blonde with an elfin face, and she was great fun. I was pleased that another of the boys had a steady girlfriend, someone I could go around with and who knew what it was like to hang about in nightclubs waiting for your boyfriend to finish playing, then chatting to the fans before he got to you. It wasn't that I minded all this – I was used to it, I knew John loved me and I was proud of him – but it was so different from other girls' lives. I had to wait all evening and often half the night while my boyfriend was the focus of attention for dozens of girls before I got to be alone with him. When Dot came along, I could share with her the hours of hanging around as well as the perks – like travelling to Hamburg.

Paul's dad and my mum saw us off on the boat train from Lime Street station. We were armed with plenty of cheese butties and flasks of tea, and waved like mad, yelling goodbye. Two naïve girls setting off into the world without a clue about what we were doing.

The train travelled through the night and in the early morning we took the ferry to Holland where we caught our next train. We slept on and off, but by the second day our real problem was hunger. The butties were long since finished and the train from Holland to Hamburg had no dining car, so the only way we could get food was from the stations we stopped at along the way. However, we were reluctant to get off and search for something to eat in case the train left without us. Neither of us could speak a word of anything but English, so we couldn't ask how long the train would stop for. We had to hang on and dream of the meal we'd eat when we arrived.

We had a romantic picture of our arrival in Hamburg

– our boyfriends waiting with open arms on the misty platform as we stepped, fresh as daisies and looking gorgeous, from the train. The reality was a bit different. The train arrived in the early morning, and once we'd battled to get the door open, we almost fell out, bedraggled, exhausted and desperate for something to eat.

On the platform we looked around. No one was there. It was an extremely long platform, and after a while we saw John and Paul running towards us. As they got closer we saw that they looked even worse than we did – exhausted and baggy-eyed, they reeked of alcohol and their clothes looked as though they hadn't been washed for a week.

Despite the state of us all, it was a wonderful reunion. When John and Paul reached us, there were hugs, kisses and shouts of joy. The boys had played in the club until two, then stayed up all night, too excited about our arrival to go to bed. They'd both drunk plenty of booze and taken pep pills and they almost overwhelmed us with their non-stop talk and excitement. It was some time before we could make them understand that we needed to eat, but eventually they took us to the Seamen's Mission on the Hamburg docks, where we all ate a massive breakfast.

The boys had arranged that I would stay with Astrid and her mother at their home, while Dot was on a houseboat belonging to Rosa, the loo attendant at the Top Ten, so after the meal we went our separate ways. I was nervous about meeting Astrid – John had made her sound so confident and glamorous and I was afraid I'd feel gauche and awkward next to her. When we got to her house, she came to the door, looking as stunning as I'd imagined. She wore jeans, a polo neck and a leather jacket – all black. Her blonde hair was cut short and layered and her makeup was understated and beautifully done. But I needn't have worried: Astrid gave me a hug and we liked one another immediately. It was the beginning of a warm friendship. Astrid was gorgeous, but she was fun too and I never felt intimidated by her.

Her room was a different matter, though. I'd never seen anything like it. The walls and ceiling were covered with silver foil and everything else was black. Her bed had black satin sheets and a black velvet cover. Sketches and paintings hung on the silver walls, highlighted by spot-lights, and a stunning dried-flower arrangement softened the starkness of the room. It took my breath away and I couldn't help comparing it with my bedroom at home, the flowered nylon bedspread, prim wallpaper and neat little dressing-table. I longed to be as bold, creative and futuristic as Astrid.

Those two weeks in Hamburg opened my eyes in so many ways. It was exciting to be away from home, with no restrictions or rules, and I felt so free. Hamburg was like a larger, more wicked version of Liverpool – a city by the sea, with a massive shipping industry along its docks and a huge number of people passing through it from what seemed like every nation on earth. In some ways, German was similar to the guttural sound of pure Scouse and I wasn't surprised that the boys felt so at home there.

It was a happy, carefree time for John and me, without

Mimi or Mum hovering over us. John spent every free moment with me and showed me around the city with pride. He pointed out the spots he'd got to know, including Hamburg's Soho where the girls sat in their windows in varying stages of undress. Much to John's satisfaction my jaw dropped. I was shocked and fascinated – it was all such a long way from the neat streets of Hoylake.

Every evening Astrid and I would doll ourselves up and head for the Top Ten. It was a basic set-up – simple metal tables and chairs in a vast room – and the dazzling lights shone down on a rich assortment of people. There were teenagers out for a good time, sailors on shore leave, portly middle-aged men on the way home from work, and gangsters, who looked so menacing I hardly dared glance at them. Fights broke out most night – flick-knives would appear with terrifying speed and we'd dive for cover until the gun-toting police stormed in and hauled out the troublemakers.

Dot, Astrid and I sat together to watch the boys play and waited for them to join us in their breaks. When they did John and I often sneaked off to the boys' room to make love before he had to dash back for the next set. The room they were in was a slight improvement on the last, but still pretty rough. It was small and drab, with a tiny window overlooking an internal staircase, two sets of bunk beds and a single, with barely a foot or two of floor space around them.

John sometimes begged me to stay the night with him rather than going home with Astrid, and I'd sneak upstairs and slip into his bottom bunk with him. The two of us

Cynthia Lennon

could barely fit into it side by side so we'd lie there wrapped round each other, trying to muffle our giggles, as George snored above and Pete, Stu and Paul lay comatose around us. Not that we ever got to bed before dawn. The boys played every night until two and kept going on pep pills. Dot, Astrid and I took them too so that we stayed awake for the boys. They gave us an incredible buzz – we felt as though we could dance all night and didn't think twice about them. It was all part of the fun.

After the club closed each night we'd all be far too high to go to bed so we'd head off to eat, the boys cavorting and shouting their way down the street, cracking jokes and fooling around, us girls hysterical with laughter at their antics. Those weeks in Hamburg were among the happiest times John and I had together. We were free and in love, life was full of promise and the sun shone.

Not all the others were as happy, though. Paul and Stuart were bickering increasingly and Pete was getting on the others' nerves. It wasn't that he did anything wrong: he was a nice guy and a good enough drummer. It was simply that his personality was different: he preferred to sit on his own rather than join in with the others' non-stop banter. They'd tease him, sometimes quite cruelly, but most of it seemed to go over his head. He was so laid-back that he just didn't mind. Although George was quieter than the others too, when it came to banter he could give as good as he got. Just when you thought he wasn't listening to one of John's wicked teases he'd shoot back a withering line that had everyone in stitches.

When it was time for Dot and me to head for home I

was heartbroken. I would have loved to stay with John in Germany, but it wasn't possible: Mum and college were waiting for me, and John had a contract to fulfil. We hugged, cried and kissed as, laden with extra food to ensure that we didn't starve on the way home, we climbed back on to the train.

Paul had bought Dot and himself gold engagement bands, like Stuart and Astrid's. She was thrilled – she was mad about Paul and hoped they'd get officially engaged when he came home. When John saw the rings he'd said to me, 'Perhaps we should get engaged too.' I was pleased, but I wasn't worried about getting engaged: it all seemed too soon and we were happy, so why change anything?

At home I regaled Mum with tales of what we'd been up to in Hamburg – editing out the pep pills, the wild nights, the knife fights and anything else that would have horrified her. She also had news for me. My cousin and her husband were emigrating to Canada with their baby. They were both training to be teachers and they wanted Mum to go with them to look after the baby while they finished their studies and got jobs.

Mum was torn. She loved the idea of seeing another part of the world and trying something new, but she felt she shouldn't leave me on my own. I was close to her and would miss her terribly, but I didn't want her to pass up an opportunity like that and I knew I could cope.

If she went, the main problem was where I would live. She would have to let the house for the income it would provide and I didn't like the idea of living there alone anyway.

There was an obvious solution: Mimi had taken in lodgers since her husband's death seven years earlier and had a vacancy. I hesitated: Mimi had made it abundantly clear that I wasn't her favourite person. On the other hand I would see plenty of John. He was still living in the room he'd shared with Stuart, but at Mimi's I'd be much closer to him. And I knew Mimi would like the arrangement because John would come home more often.

Mum agreed to go, as long as I was settled. I waited until John came home from Hamburg, a few weeks later, before I asked Mimi. This time he arrived back in a great mood. Everything had ended on a good note and they had already been asked to go back. He loved the idea of me moving in with Mimi, and when we put it to her she agreed immediately.

When the rest of the boys came home Stuart wasn't with them: he'd decided to stay with Astrid. He wanted to live with her, get married and make a new life for himself as a painter – always his first love. A Scottish professor who was visiting the Hamburg art college had taken an interest in his work, helped him into the art college and even managed to get him a grant. It was clear he would become a successful artist and I suspect he left the Beatles with little regret.

John was happy that Stuart had found Astrid and he knew they'd see each other again when the Beatles went back for their next stint in Hamburg. Still, although he didn't say much about it, he missed Stu, who was his closest friend. They stayed in touch, writing long letters to each other about anything and everything.

Astrid had had a big influence on the group's image. She had encouraged Stuart to comb his piled-up teddy-boy hair forward, in a smoother, sleeker style that framed his face. The others laughed when they saw it, but eventually they tried it too – all except Pete, who refused to change his. The new haircut – later christened the mop-top – was distinctive and made them stand out at a time when most other young men in groups were still sporting the greased-up DA. When I first saw John with it, I loved it. It was softer, sexier and classier than the DA had ever been. Astrid also influenced their clothes: they had all bought black leather trousers and jackets, which they wore with the collars turned up to give them their own distinctive look.

Meanwhile there had been a step forward for the group. During their time in Hamburg, while Dot and I were still there, the boys had made their first record. Tony Sheridan, whom they'd met on their first Hamburg trip, had a strong voice and they'd often played as his backing band, so when it was suggested that they back him on a record they leapt at the chance. The song, 'My Bonnie', was a corny old classic that the boys laughed at, and for some reason they were called the Beat Boys on the record – perhaps the German producer thought 'Beatles' didn't make sense – but, still, it was a record and the four were proud of it. Not that it sold many copies – when they got back hardly anyone had heard of it.

At the end of the summer term, not long after the boys had come home, I moved in with Mimi. I was in a little boxroom, barely big enough for the bed, but I didn't mind because, with Stuart in Germany, John had moved back to Mimi's so we were both there. With Mimi in the house we didn't dare sneak into each other's rooms – we'd have been thrown out of the house instantly – but at least we saw a lot of each other.

Living with Mimi was even harder than I'd imagined it might be. Even though I was paying rent I had to do any number of chores around the house, and she expected me to drop everything and come running to do her bidding. There was no refusing Mimi and I was treated exactly as her other two lodgers. She always had a reason why she couldn't do a job herself, or needed help. I gritted my teeth and did as she asked. I wasn't about to fall out with her and lose my room and, besides, after her display of temper over the coat John had bought me I was more than a little afraid of her.

That summer I got a job in Woolworths, working on the cosmetics counter. It gave me enough money to pay Mimi for my keep as my grant was just about finished, and I liked it: the other girls were fun. I'd work in the store all day, then head out to meet John in the evenings. The Beatles were playing somewhere most nights, so I'd meet him beforehand and we'd go together to the venue.

It was around this time that a local newspaper called *Mersey Beat* was launched, devoted to the local beat scene, with big articles about the two lead groups, Rory Storm and the Hurricanes and Gerry and the Pacemakers. John was asked to write about the origins of the Beatles, which he did in true John style, in a crazy article that made little sense but had us all laughing. It began, 'Once upon a time there were three little boys called John, George and Paul, by name christened. They decided to get together because they were the getting together type. When they were together they wondered what for after all, what for? So all of a sudden they all grew guitars and formed a noise.'

John had always written prolifically – poems, stories, bits of nonsense, often on odd scraps of paper – but this was his first published piece and although he pretended he didn't care about it he was really proud when it was printed. After that he wrote regularly for *Mersey Beat*, sending them his jottings and updates on the group's progress.

In October John turned twenty-one. His aunt Mater gave him a hundred pounds and he decided to spend it on a trip to Paris for him and Paul. I'd have loved to go along, but by then I was in my final year at art college, in training to be an art teacher and busy with teaching practice at a couple of local schools. John and Paul left on 1 October and eked out their money for two weeks, hanging around the clubs in Paris and meeting up, by chance, with a couple of Astrid's friends from Hamburg.

Also, around this time, the Beatles got their own fan club. Most of the leading groups in Liverpool had one and the boys were delighted when a group of their followers started one for them. But they felt they were standing still. Playing the Liverpool scene was fine, but they wondered whether anything more than that was ever likely to happen. Sometimes John would get quite down about it, especially when Mimi was at her worst, nagging him for wasting his time instead of getting a proper job. But he held out. They were making a bit of money and most of the time he felt sure something fantastic would happen – he just didn't know what.

John was right. Something wonderful *was* about to happen. Someone was about to change their lives and he was – literally – just round the corner. When I look back to those days, the way Brian Epstein and the boys met and the events that followed seem almost touched by magic.

It began not long after John and Paul returned from Paris. A lad walked into the NEMS record shop in the Whitechapel area of Liverpool and asked for a record called 'My Bonnie' by the Beatles. The owner of the shop was Brian Epstein, and he'd never heard of the single or the group. Most people would have left it at that, but Brian prided himself on being able to get hold of any record his customers wanted. He was puzzled that he hadn't heard of this one and decided to look into it. The customer had mentioned that the record came from Germany, so he began with agents who imported foreign records.

Although Brian was aware that there was a flourishing beat scene in Liverpool he wasn't part of it. He was running one of the city's most successful record shops and he certainly aimed to have every type of record in stock, but he preferred classical music – he loved Sibelius in particular – so he didn't go out to hear the beat groups. And since most of them had never made a record, he didn't come across them very often.

Brian wrote and advertised regularly in *Mersey Beat*,

in which the Beatles were often mentioned and pictured, but he'd never noticed them. His interest in tracking down 'My Bonnie' grew when two more customers came in and asked for it. He asked them who the Beatles were and was surprised to discover that they weren't German but a local group, who had often been into his store to spend an afternoon hanging around listening to records. He decided to go and see them for himself. It made sense for him to stock the record if they were popular locally.

On 9 November 1961, Brian stepped gingerly inside the Cavern for the lunchtime session. He wasn't impressed by the place, as he told us later. In fact, he thought it was awful – dirty, dark, damp and very loud, the kind of teenage haunt he'd usually be at pains to avoid. He knew that, in his smart business suit, he'd stand out among the teenagers on their lunch break, so he stood quietly at the back.

When the Beatles came on he was fascinated. They were scruffy and cheeky, they smoked, ate and chatted as they played, they shouted to friends, cracked jokes and even turned their backs on the audience. They acted as though they didn't give a damn. But the kids in the Cavern loved them, and it was clear to Brian that they had something special.

Over the next couple of weeks he saw the Beatles several times. He was fascinated by their energy and their talent, raw and disorganised as it was. As he watched them, an idea formed in his mind. He had joined his family retail business and made a success of both their record stores, but he was bored and needed a new challenge. Perhaps the Beatles were what he was looking for.

He talked to the boys and discovered that their record was on the Polydor label but that they had no manager and no record contract. He ordered two hundred copies of 'My Bonnie', then started asking his contacts in the record industry what managing a group would involve.

On 3 December, just over three weeks after he'd first seen the Beatles, he invited them to his office for a formal chat.

By then he had talked to them a few times, but they hadn't paid him much attention and John hadn't even mention him to me until, shortly before the meeting, he told me that they were seeing the owner of the NEMS record store who wanted to talk about managing them.

John was immediately impressed by Brian. At twenty-seven, he was only six years older, but he couldn't have been more different: he wore smart suits, he was well off and he owned a record shop. John wondered why a man like that had taken an interest in the Beatles, but he was more than willing to hear what he had to say. John took Bob Wooler, the Cavern MC, with him for support and introduced him to Brian as his dad. John thought it hilarious that Brian had believed him – and it was some time before Brian realised he'd been had.

When Brian began to talk John saw that he meant business. He was calm and to the point. He asked the boys where they wanted to go. They told him they were all certain they wanted a future in music and to be successful, but that they were frustrated because, despite their popularity locally, nothing much was happening.

A week later Brian called them to his office again and

told them he wanted to manage them. John trusted him straight away and was keen to agree, and the other three didn't take much persuading. A few days later they signed a contract, giving Brian 25 per cent of their earnings. They argued for less, but Brian pointed out that he'd probably be paying all their costs for months, if not longer, before they made any money.

So the deal was done, and immediately Brian set about getting the boys to clean up their image and finding them a record contract. It was clear from the start that he was going to put all his energy into making them a success, and his determination impressed them.

The Beatles provided Brian with the opportunity to break away from his family's business. His parents had always expected him to join his father in running the family's group of shops, but Brian hadn't been keen: he had set out to be an actor and even got into RADA, London's most prestigious drama school, but after a year or so there he decided acting wasn't for him and headed back to Liverpool to do what his father had always wanted him to.

Although he had a flair for managing shops and increased the profits, he wasn't fulfilled by the work. In the Beatles he found a project that excited him, something he could set out to achieve without his family. It was a challenge – but, more importantly, he believed in the Beatles.

Over the next few months John and Brian formed a close friendship. John thought Brian was sharp, that he knew what he was doing and could be trusted. He went to Brian's house regularly to talk to him about the direction the group should go in and what they would need to do.

I met Brian soon after he had signed the Beatles. He was charming, polite, well spoken and I liked him. He accepted that I was John's girlfriend, but he told John that it would be better if all girlfriends kept a low profile. I didn't mind because I had no interest in the limelight. As long as I could be with John, I was content to stay in the shadows.

It didn't occur to us then that Brian was gay, or that he was Jewish. Life must have been difficult for him. In the sixties if you were gay you kept it secret. Gays were called 'queers', and were disliked and distrusted by many – there was a huge amount of prejudice against them. Some accounts of that time claim that Brian was in love with John, which was why he wanted to manage the Beatles. I don't believe this for a second. They had a good relationship, but Brian cared for all the boys and he wanted success for the group because he thought they had something unique. Claims have been made since that Brian and John had a gay relationship. Nothing could be further from the truth. John was a hundred per cent heterosexual and, like most lads at that time, horrified by the idea of homosexuality.

The bond between John and Brian was one of mutual respect and friendship. They liked and admired each other. Brian could see John's intelligence and distinctive talent. John appreciated Brian's business ability and his ambition for the group. They talked for hours and planned the group's future together. They both wanted the Beatles to be the biggest thing since Elvis, and were hell bent on making it happen.

# 7

BRIAN SET ABOUT TRYING TO get the boys a record contract and groomed them for success. No more eating, chatting, drinking or shouting on stage. No more leather trousers or jeans. No more playing whatever they were in the mood for. Brian wanted them clean, punctual and well-behaved. He wanted them to play their best songs every time, in an hour-long set. The Hamburg style – stringing songs out to five times their length, larking about on stage and insulting their audience – was out.

The boys went along with all this because they could

see it made sense. Paul was keen on the changes and George was happy to accept them. But it wasn't easy for John. When Brian asked them to wear suits and ties John growled for days. That was what the Shadows – the group John most despised – did. John felt it was selling out, that it wasn't what the Beatles were about, but Brian knew it was necessary if they were ever to break out of the Liverpool scene. So John did as Brian asked. He wasn't cajoled or pushed: he did it because he understood Brian's reasoning that if the group were to have mass appeal they had to be more mainstream. But he rebelled in his own little ways, often leaving the top button of his shirt undone or his tie askew.

Brian went to see all the boys' parents to let them know what he was doing. Most were delighted that this well-off, well-spoken man was taking an interest in their sons' career. Only Mimi, who trusted no one, was unsure. She was afraid that it was a game to Brian and that he would soon lose interest, leaving John bitterly disappointed. But at least she had accepted that John wanted a career in music and if Brian was part of that, well, so be it.

Pete had always been in charge of the boys' bookings, but now Brian took over. He began to charge venues more for the boys to play and refused to let them appear unless they got the higher fee. Until then the boys' organisation was haphazard, and they'd take whatever money they were offered. They were often late and had no system for letting each other know where they were due to be. Brian sent them neatly typed lists of their dates and insisted they arrive on time. I enjoyed the new-style Beatles and, despite his grumbles, so did John. He could see that a smarter, cleaner image made sense, as did putting together a performance that included their most popular songs and lasted for a set time. He knew, too, that it was moving him closer to what he really wanted: the all-powerful record contract.

Things between us were good, but it wasn't always easy to find time to be together. I was busy during the day, with art college and teaching practice, and John was playing most evenings. But I went along to watch him play whenever I could and we'd meet between our commitments. Often we'd just sit in the pub or a café and talk. Or we'd go to the cinema – we loved films and saw everything we could.

The Christmas holidays arrived and with them great news. Brian had persuaded the A&R – artists and repertoire – manager, Mike Smith, at Decca to come and listen to the Beatles. This was unheard-of – someone from a top record company coming all the way to hear a group play – and we were bowled over. The boys played superbly that night, cover versions of rock songs plus one or two of their own, and Mike Smith told Brian he'd like them to come to London for an audition on 1 January at the Decca studios. The idea was to see how they sounded on tape, that was all, no promises beyond that, but to us it sounded like the moon and stars.

Neil Aspinall had officially become the boys' road manager: he drove them around and was paid a cut of their earnings. He took them to London on New Year's Eve, while Brian went by train. Like the rest of the boys,

Neil had never been to London. He got lost somewhere along the way and the trip took ten hours. The boys arrived in London at ten p.m. and saw in 1962 'watching the drunks jump into the fountains in Trafalgar Square', as John put it.

The next morning they arrived at Decca for the recording. John told me later that they were terribly nervous and didn't sing well. Brian had asked them to stick to familiar songs instead of their own, to which they agreed although they'd have preferred to do their own material. At the end they were told they'd done well. Brian bought them dinner and the next day they drove home.

After that there was nothing to do but wait . . . and wait. Weeks passed, and John and the others pestered Brian about Decca, but there was no answer. As each day passed, they sank a little lower.

In March Brian heard that Decca weren't interested. The chap he spoke to told him groups with guitars were on the way out and they didn't like the boys' sound. John was down about it – but he couldn't stay miserable for long. He was convinced Decca were wrong and that they'd make it.

Brian went on trying to get another record company interested. He went up and down to London and the boys would rush expectantly to see him when he got back, only to hear that yet another door had slammed. Each time they'd mooch around until their mood lifted with the next prospect.

We all felt that change was in the air. Something was

going to happen for the group and it was just a question of carrying on until it did. What we didn't know then was that, over the next few months, a series of events would change all of our lives. Nineteen sixty-two was an astonishing year, when tragedy, failure, the unexpected and the miraculous combined in ways we could never have imagined.

By April another trip to Hamburg was on the cards. This time Brian had done the negotiating and he'd secured them a better club and more money. They were opening at the Star for the equivalent of forty pounds a week – more than double what they'd been earning before. He insisted that they arrive in Hamburg in style – by plane.

Once again John and I kissed goodbye. I was used to him going now, but I didn't miss him any less. The only consolation was that I'd have plenty of time to work for my finals, which were looming alarmingly close.

John, Paul and Pete flew from Manchester to Hamburg on 10 April, with George following the next day. When their plane landed John saw Astrid waiting for them and waved to her, but as he got closer he saw her stricken face.

Stuart had died a couple of hours earlier. John was stunned and his reaction in the face of the tragedy was to laugh hysterically. He'd done the same thing when his beloved uncle George had died, and again at his mother's funeral. It was his way of coping with shock and profound grief. And after the laughter came tears. John broke down and sobbed.

From Stuart's letters, John knew that he'd been suffering

from increasingly bad headaches. X-rays and medical checks had shown nothing, but the headaches were so bad that Stuart was often unable to move for pain and lay in bed, with Astrid and her mother looking after him. His family in Liverpool were kept informed. They were terribly worried about him and begged him to come home, but Stuart wanted to stay in Germany with Astrid.

That morning he'd developed such a severe headache that Astrid's mother had phoned her at work to say she was calling an ambulance. Astrid had raced home just in time to go with Stuart to the hospital, but he died in her arms at 4.45 p.m. in the back of the ambulance. The cause of death was later given as cerebral haemorrhage, and Stuart had been twenty-two.

Stuart's mother, Millie, learnt of his death in a telegram from Astrid and flew to Germany to identify his body and authorise a post-mortem. His father was away at sea and didn't hear until two weeks later. Millie and his sisters, Joyce and Pauline, arranged for his body to be flown home and buried in the church where he had once been a chorister. The boys, contracted to play in Hamburg, couldn't go to the funeral and Astrid didn't go either, saying she was ill. I knew John would be grieving and I wanted to be there for him, but all I could do was write words of comfort.

Later, when John got back, we went round to see Stuart's family and John asked for Stuart's navy and cream striped college scarf, which they gave him and which he kept for many years.

Soon after Stuart's death John wrote to me of how angry

he was that the local press had got hold of the story. The Beatles had enough of a local following for it to be a prominent story, but John hated it being made public. He was also struggling with what to do about Astrid. He said, 'I haven't seen her since the day we arrived. I've thought of going to see her but I would be so awkward – and probably the others would come as well and it would be even worse. I won't write any more about it cause it's not much fun.' It was typical of him to appear to dismiss something so distressing with a phrase like 'it's not much fun'. He found it so hard to talk about, or show, his true feelings. Later he talked to me sometimes about Stuart and about the awful sense of loss and guilt he felt. He agonised over why he had lived and Stuart had died, and whether there was anything he could have done. But these glimpses of his real feelings were rare. Most of the time he kept it all deep inside himself. In that letter he also told me that he had lost his voice – perhaps a symptom of unexpressed grief.

During the next few weeks in Hamburg John's behaviour was wild and unpredictable – also, no doubt, a symptom of grief. He saw Astrid eventually, and comforted her in his own blunt way: he told her she must choose to live, not die, and made her come to the club to watch them play. She did, despite her deep depression, which lasted for many months, and valued John's efforts to help her carry on.

Meanwhile I'd run out of patience with Mimi – I could no longer ignore the tension between us or her bad temper – and decamped to my aunt Tess, Mum's sister, who kindly took me in. When I told her I was leaving, Mimi was tight-lipped and stalked out of the room. I wrote to tell John, who was more understanding because he knew how hard it had been.

My aunt's family were lovely, but they didn't really have room for me and, besides, the secondary school where I was doing my teaching practice was a three-bus journey away. I began to look for somewhere of my own. I scoured the accommodation pages of the *Liverpool Echo* and found a bedsit I could afford in a tatty little terraced house reasonably close to college and the two schools where I was teaching. Glad as I was to be independent at last, it was hardly the sweet little home I had visualised. For fifty shillings a week I had a grubby room, with a one-bar electric fire to heat it, a minute one-ring cooker, a single bed, an ancient chair and a moth-eaten rug. I had to put a shilling in the meter if I wanted hot water for a bath in the shared bathroom, and even then the water was barely ankle deep. Still, it was a place of my own and I was determined to cheer it up ready for John's return.

My neighbour in the larger room next door was an eccentric elderly woman who filled her room with cats and bags of coal. The smell every time she opened her door was appalling – she clearly never washed or cleaned and hardly went out. What unnerved me most, though, was that every time I opened my door she was peering at me or lurking on the landing. She was probably lonely, but at the time I found it sinister.

Much to my relief, she left soon after I arrived – I never knew why, or where she went. For a couple of weeks her room was empty, until I had a great idea. I suggested to Dot Rhone that she move in. She was keen, and the land-lady agreed, but she had to work hard on her parents before they would allow it. In those days it was still unusual to move out of the family home so young, especially for a girl. Eventually, persuaded that their daughter wasn't going into some hotbed of corruption, her parents agreed and Dot moved into my room, while I took over the larger one.

It took us several days to clean and paint my new room. It was in a truly dreadful state, murky browns and greens, and absolutely filthy, but with a gallon or two of bright paint we transformed it.

With Dot next door, I was a lot less lonely. As Mum and John were away weekends especially had been pretty miserable, but Dot and I had a great time, shopping and cooking together while we looked forward to the boys coming home.

John's letters made it clear that this time round life in Hamburg was much easier for them. They stayed in a flat above the Star Club and he was keen to tell me how much better their accommodation was: 'Did I tell you that we have a good bathroom with a shower did I? Did I tell you? Well I've had ONE whole shower, aren't I a clean little rocker? Hee! Hee!' Working conditions were better too. They only had to play three hours one night and four the next, an hour on and an hour off, which was luxury compared to the long hours and short breaks they'd endured on their previous trips. John called their boss, Manfred, 'a good skin' and wrote that they'd got a day off and the boss was taking them on a trip to the sea. Later he said all they did was 'eat and eat and eat'.

I had written to tell him about my move and he was delighted: 'I wish I was on the way to you with the Sunday papers and chocies and a throbber. I can't wait to see your room. It will be great seeing it for the first time and having chips and all and a ciggie (don't let me come home to a regular smoker please Miss Powell). I can just see you and Dot puffing away. I suppose that's the least of my worries. I love you Cyn, I miss, miss, miss you Miss Powell – I keep remembering all the parts of Hamburg that we went to together. In fact I can't get away from you – especially on the way and inside the Seamen's Mission. I love, love, love you.' John often put poignant little pleas in his letters: 'I love you, I love you, please wait for me.' Despite his apparent confidence, and his cynicism, he often urged me to wait for him and it betrayed the fear that hovered just under the surface: he had been left so painfully as a child and was terrified that it would happen again. His letters touched my heart and buoyed me up. They were always so full of love, but his pain was clear too, and I ached to hold and comfort him.

For the moment, though, I had to work for my finals and finish my teaching practice at two schools. The first was a mixed secondary-modern in Garston, a tough area of the city. The kids there were lovely and, although they weren't much younger than I was, I had no problems with them. But the other school was a nightmare. It was a private girls' college and I hated it. The girls did every-thing they could to undermine me, even going so far as to destroy a frieze I was helping them with. Things were even worse when the proper art teacher went off sick, leaving me to prepare them for their exams. Totally out of my depth, I inadvertently taught them material for the wrong exam, which resulted in chaos come exam day when they couldn't do a thing. Much to my relief they were able to resit, but not before I'd received a dressing-down from the headmistress. I wondered whether I was cut out for teaching after all.

In early May, while the boys were still in Hamburg, Brian's efforts paid off. He'd taken tapes to the big HMV record shop in Oxford Street, owned by EMI. He wanted to get them made into a disc, so he chatted up a technician he knew there. The technician recorded the disc and was so impressed that he played it to a music publisher upstairs, who liked it and said he'd speak to a producer he knew at Parlophone, George Martin. Parlophone was part of the EMI empire, which had already turned the Beatles down, but George Martin, a classically trained musician turned producer, listened to the record and agreed to audition them. He liked John singing 'Hello Little Girl', Paul singing 'Till There Was You', and George's guitar-playing. Brian immediately cabled the boys to tell them the good news and ask them to rehearse new material. After that John's letters were full of their excitement: they were writing songs, planning their future and dreaming of all the money they'd make. George's father was a bus driver and he was planning to buy him his own bus.

They came back from Hamburg in early June and a

couple of days later, on the sixth, they had their audition with George Martin at the EMI studio in St John's Wood. Neil Aspinall drove them to London where they stayed at a cheap hotel. The next morning they met up with Brian, who took them straight to the studio where George Martin was waiting. Introductions were made and the boys, all nervous, played a mixture of their own songs and well-known hits. George Martin said thanks, he'd let them know. All the way home they debated how it had gone. They knew they'd have done better if they hadn't been so nervous – John and Paul felt their voices had wobbled. They just hoped they had been good enough to earn themselves a contract.

When they got back to Liverpool they were in a state of high suspense, but there was nothing to do but wait and see. Meanwhile they had a 'welcome home' night at the Cavern and Brian had lined up plenty of bookings for the next few months.

Both John and Paul loved the new set-up with the bedsits. They had the freedom to nip round to me and Dot whenever they liked and the four of us had a great time. The only drawback was the landlady: she had an unnerving habit of popping in to empty the meter or collect the rent. Since boys in the rooms overnight were strictly forbidden – and John was staying overnight with me whenever he could persuade the eagle-eyed Mimi that he would be with a friend – we crossed our fingers and prayed we'd get away with it.

Then one morning the nightmare happened. John and I were in bed together when I heard the landlady calling up the stairs – she was coming to empty the meter, which was in my room. We panicked. I wrapped myself in a sheet and fled into Dot's room, leaving John to his fate. I heard the landlady go in and, a few minutes later, come out and go back down the stairs. Baffled, I went back into my room. No sign of John. It took me a minute or two to work out that he was under the large pile of blankets, coats and clothing on the bed. He'd grabbed whatever he could see from around the room and piled it on top of himself. Gasping for air and red-faced, he crawled out, cursing the landlady.

We thought we'd got away with it – until I spotted John's cowboy boots sitting beside the bed. We fell around laughing and thanked heaven that the landlady must have been in a good mood that day.

One evening things went horribly wrong for Dot. She and I were having a girls' night in and she'd just washed her hair. She had it up in giant rollers and was dressed in an old sweater and a pair of her mother's bloomers. We were both giggling, imagining what the boys would think if they could see us, when there was a knock at the door. I opened it and there was Paul.

Poor Dot was horrified, but there was nothing she could do and they went into her room. A short time later the door opened again and I heard footsteps running down the stairs. Then the front door slammed. A moment later Dot appeared in my room, sobbing. Paul had broken off their romance. I attempted to comfort her as she cried for the next couple of hours, convulsed with sobs, her little pixie face blotchy with tears, the rollers falling out and her hair coming down in damp strands. She couldn't believe it had happened, but Paul wanted to be free and that was that. To make things worse, she thought it was the sight of her in rollers and bloomers that had made him finish with her, although in fact he'd come round to end it and her startling appearance had made no difference.

A few days later Dot moved back to her parents. I understood that she found it too painful to see me, but I missed her so much. She left behind the gold ring Paul had given her and when I found it I kept it for years. One day, when Dot was happily married and living abroad, I saw her and gave it back.

With Dot gone, the bedsit was lonely again, although John came over as often as he could. One day he arrived with a bottle of yellow liquid that smelt disgusting. 'It's the latest thing, instant tan,' he said. 'You just slap it on and you look as though you've been sunbathing for a fortnight.' Excited by the prospect, we plastered ourselves with it and waited with bated breath. A couple of hours later the colour appeared: we were both streaked from head to toe in an appalling shade of murky yellow. We didn't know whether to laugh or cry. We'd been so enthusiastic that we hadn't thought of trying it out on an inconspicuous bit of skin first. We did our best to wash it off, but no amount of scrubbing made any difference. Yellow-streaked we remained for several days.

The next few weeks brought a series of momentous events. First, just after I sat my finals at art college, the boys' breakthrough arrived: after almost two months of agonising, they heard that George Martin wanted them to sign a contract with Parlophone Records. We were all ecstatic and John kept shouting, 'This is it, Cyn, this is it, we're going to be making records, we're famous!' We laughed, danced, drank and celebrated.

There was just one hitch: John, Paul and George had agreed that they didn't want Pete in the band. They decided not to tell him about the record contract and asked Brian to fire him. Brian liked Pete and was reluctant to do so, but he accepted that the others felt he was wrong for the group. My heart bled for Pete. I'd grown fond of him and he was a good drummer, but his face didn't fit: the boys wanted someone they could have a laugh with, someone who shared their irreverent sense of humour.

Brian hated having to tell Pete the bad news and offered to try to find him work with another band. Pete sat at home for two weeks, unable to face anyone or to understand why the boys he had thought were his friends had dumped him. It was cruel. He had believed he would share the group's success and his drumming had been part of what had got them that far, but suddenly he was ditched, with no prospects. Even worse, none of the other three had the courage to go and see him, apologise or even to stay friends. Pete was closer to John than he was to the others, they'd been friends for four years, and was particularly hurt that John avoided him. I urged him to go but, as ever, he avoided the confrontation. He thought Pete would be angry and wanted to avoid a row. It was John at his most cowardly.

Without doubt he, Paul and George could have handled

things much better, but were they wrong to do it? I don't think so – they knew that for the group to work it had to have the right combination of people, and they knew whom they wanted in Pete's place.

Ringo Starr was the drummer with Rory Storm and the Hurricanes and they'd got to know him in Hamburg. Six months earlier Ringo had sat in with the Beatles for a couple of gigs when Pete had been ill and they'd got on brilliantly with him. When they heard he'd left Rory's band, they decided to offer him a job. Another band offered him a job at the same time, but the Beatles were paying more and had a record contract, so Ringo said yes to them and that was it. He was in. The only condition was that he had to brush his hair down, like theirs. He said he would.

For the next couple of weeks the boys had to face Pete's furious fans wherever they went. His moody style had attracted a big following and the girls couldn't understand why he'd been dumped. It was front-page news in *Mersey Beat* and the boys were heckled and jeered at several times. But they stuck it out and gradually, as audiences got used to the new face, the fuss died down.

While all this was going on I had only half an eye on it because I was contemplating my own future. On the same day in July I discovered, first, that I had failed one of my final exams and, second, that I was pregnant. I had to decide whether to retake the exam or to give up on becoming an art teacher. Under any other circumstance this decision would have occupied all my thoughts for days to come. But the discovery that I was pregnant wiped everything else out of my mind.

## John

Amazing as it sounds now, John and I had never used contraception. No one had ever said anything to us about it. In those days schools and parents wouldn't have dreamed of discussing such matters with us. Of course we knew how babies were made and that pregnancy could be prevented, but the level of our ignorance was such that we honestly thought it would never happen to us. Until it did.

When I realised my period was late I didn't know who to turn to. Eventually I told Phyl, who agreed to go with me to the doctor. The female GP I saw was frosty and patronising. She examined me, confirmed my fears, then delivered a stern lecture on morals. I left the surgery feeling utterly bleak. What on earth had I done? I didn't want a baby, not yet. I wanted a career, marriage and a life before children. Now I'd messed everything up.

Phyl was sympathetic and kind, but in the end she had to go home. I sat alone in my bedsit, crying and railing against life, dreading the shame of telling my mother, who was about to visit me from Canada, and wondering how John would react. I expected him to take it badly, and before I told him I made up my mind that I'd cope alone. I knew I couldn't face an abortion, which, in any case, would be hard to come by and dangerous. I was going to face the consequences and bring up the baby on my own if I had to, although that would have made me a social outcast.

I put off telling John for several days. I would wake up feeling sick, remember the awful truth, then burst into tears. It was a nightmare that wouldn't go away.

Eventually I plucked up all my courage and told him. As the news sank in he went pale and I saw the fear in his eyes. For a couple of minutes we were both silent. I watched him as I waited for a response. Would he walk out on me? Then he spoke: 'There's only one thing for it, Cyn. We'll have to get married.'

I asked him whether he meant it. I told him he didn't have to marry me, that I was prepared to manage on my own, but he was insistent. 'Neither of us planned to have a baby, Cyn, but I love you and I'm not going to leave you now.'

I was grateful, relieved and happy. I'd have understood if he had walked away, although it would have hurt. And although I hadn't thought about marriage yet – believing we had years in which to make that kind of decision – I was certain that I loved him and wanted to be with him. So, on a summer's night in my little room, John and I decided to marry, have our baby and become a family together. We loved each other and it was what we both wanted, even if it had been forced on us far sooner than we'd have wished.

The next day John told Brian, who asked him to think carefully and told him he didn't have to go through with it. No doubt Brian was thinking of the future of the group, who'd just got their first recording contract. I'm sure he felt that it would be more appealing to the fans if the boys were all single. But John insisted he wanted to marry me and Brian agreed to help. He got an emergency licence and booked the register office for the first available date.

## John

John's next task was the one he dreaded. Mimi, predictably, was furious. She screamed, raged and threatened never to speak to him again if he went ahead with it. Julia, John's sister, was sixteen at the time and she was there. Later she told me that John had stood up to Mimi: 'You don't understand, Mimi. I love Cyn, I WANT to marry her.' Mimi certainly didn't understand. She accused me of planning the whole thing to trap John and made it clear that she wanted nothing to do with the wedding.

Mimi's reaction must have hurt John deeply. She was his closest family, had taken the role of mother in his life, yet she'd undermined and criticised him for years. Now, when he needed support and love, she was disowning him and he knew that she would stay away from the wedding and make sure no one else in his family went either. He walked out of the house, vowing not to go back.

I had a much easier time with my mother. Even then, I was so afraid that she would be disappointed in me that it was not until the last day of her visit that I found the courage to tell her. She was staying with my brother Tony and I went to see her the night before she sailed back to Canada. When I told her about my pregnancy her only concern was for me: she felt terrible that she was leaving the next day and couldn't stay to look after me. But she put her arms round me, told me that we all make mistakes and that it would turn out all right in the end.

The next day Tony, his wife Marjorie and I went with Mum to the ship. As we said our last goodbyes and she waved to us from the railing, I collapsed in tears and sobbed so hard that Tony and Marjorie had to practically carry me home. I wanted Mum to be with me so badly, but I had to manage without her.

The next day I married John.

# 8

Our wedding was an odd mixture of the comically funny and the downright bizarre. Both John and I were so stunned by the suddenness with which our lives and relationship were changing that the day felt unreal. It was as if we were watching a film, yet taking part in it at the same time.

My panic at finding myself pregnant had subsided when John said he wanted to marry me, but that was when his panic took over. He wasn't ready to marry, it hadn't been part of his plan. It wasn't part of mine at that stage either, but for John there was an additional concern: he was afraid that his marriage would damage the group's future. The boys had been told many times that fans wouldn't accept steady girlfriends and that any women in their lives had to be kept right out of the spotlight. How on earth could he get away with having a wife?

What no one around us understood was that John and I were deeply connected to each other on many levels. Not only were we passionately attracted to each other, we loved each other's company and we were each other's best friend. John needed my uncritical love and support and I needed his confident belief in himself and in me. I respected him enormously for standing by me when he knew it might ruin his career, just as he was on the brink of success: he had a streak of fundamental decency that went far beyond simply observing the convention of the day and I loved him for it.

Many commentators on John's life have said that John would never have married me if I hadn't been pregnant. In the film *Backbeat* I was portrayed as a clingy, dim little girlfriend in a headscarf. Totally wrong, of course. Quite apart from anything else I never wore a headscarf.

It's hard to see our relationship, given that untruthful twist. Both John and I believed we'd marry one day, and so did most people who knew us then – Phyl used to say we were joined at the hip and that John was besotted with me. My pregnancy changed our plans, but not our intentions or feelings for each other.

Accounts of our wedding have often portrayed it as a miserable last-minute shotgun affair that John was virtually forced into. Again, it's a long, long way from the truth. It was last-minute, which meant that we had no flowers, no reception, no beautiful dress and no photographer, but it wasn't miserable. In fact, it was the opposite: we were very happy. And John – who was never one to do things the conventional way – was the most determined of us that it would go ahead.

It was all arranged so fast that we had barely time to think. Once the decision had been made it was booked for a couple of weeks later, on 23 August. During those two weeks we got more and more excited. We giggled about being Mr and Mrs, teased each other about being boring old marrieds and pictured ourselves together in our rocking-chairs. Our map of the future had changed, but we began to like the idea and John rose to the occasion, becoming gentle and protective towards me.

The wedding was booked for late in the morning at the Mount Pleasant register office and Brian, who proved himself a true friend and far more than just John's manager, arranged to pick me up from my bedsit and drive me there.

That morning I got ready on my own. I'd have loved Phyl to be with me, but she was away on holiday. I couldn't afford a new outfit so I wore my best purple and black checked two-piece suit with a frilly high-necked white blouse that Astrid had given me. I put my hair up in a French pleat, added black shoes, a black bag, a touch of pink lipstick, and I was ready.

Brian arrived, dapper in a pin-striped suit, and escorted me to his chauffeur-driven car. He looked so smart and the glamorous car made me feel special. He was kind to me on the journey, soothing my nerves and telling me I looked lovely.

The weather was awful. It might have been August but the sky was overcast and grey, and it looked as though it would rain at any moment. I prayed it wouldn't – at least until I'd got into the register office because I didn't have an umbrella. When we arrived, John, Paul and George were pacing about in the waiting room. They were all alarmingly formal in black suits, white shirts and white ties – the only smart outfits they had. George and Paul had made a big effort to look the part and clearly felt it was their role to support John, who was sitting between them, white-faced. I was touched by the effort they'd made, although their clothes would have been more in keeping with a funeral and so would their expressions: all three were horribly nervous.

When I came in John leapt to his feet to hug and kiss me and tell me I looked beautiful. Someone made a crack about the boys' suits, which broke the ice – we all started to giggle. My brother Tony and his wife Marjorie arrived next, hurrying in at the last minute because they'd had to come in their lunch hour. Tony was there to represent our family and he played the big brother to perfection, putting an arm around me protectively. I know John would have loved someone to be there from his family, but Mimi had put paid to that.

Moments later we were ushered into the register office, where the registrar, a dour, solemn man, was waiting for us. As John and I stood in front of him, preparing for our vows, the whole thing took a ludicrously comical turn. A workman in the backyard of the building opposite started up a pneumatic drill and we stared at each other in disbelief. The noise was ear-splitting. Clearly the man wasn't going to stop so we had no choice but to carry on.

There was another comic moment when the registrar asked the groom to step forward and George did so. But the registrar saw nothing funny in either the drilling or George's joke, so we all struggled to keep our faces straight. John and I leant forward, straining to hear the registrar and shouting our responses. Paul, Tony and Marjorie signed the register as our witnesses, and a couple of minutes later we were outside the room. We all burst out laughing, overwhelmed by relief that it was over.

Tony and Marjorie hugged us, then went back to work. The rest of us looked at each other. What next? Brian suggested we go to nearby Reece's café for lunch.

Outside, the rain was bucketing down. We ran along the street, laughing at the madness of it all, and burst into Reece's, where we had to queue for the set lunch of soup, chicken and trifle. Reece's had no licence so, when we finally got a table, we toasted ourselves with water. But we didn't care: we were on a high. A full church wedding with all the extras couldn't have made me happier. And despite the anxiety I knew he had felt, John wore a look of pride and pleasure that touched me. It was as though something had changed in him: he was a married man now, soon to be a father, and he liked it.

Neither of us was aware at the time of another bizarre

aspect to our nuptials: our wedding had been a carbon copy of John's parents' wedding twenty-four years earlier. Julia and Fred had also married at Mount Pleasant, then gone to Reece's for lunch. They even had Mimi clicking her tongue disapprovingly in the background. When I learnt about it later, I hoped that their wedding had been as happy as ours was but, more importantly, that we would not part as they had.

Brian treated us all to lunch – fifteen shillings a head – then gave us his present. He announced that we couldn't possibly live in my bedsit and that he had a flat he seldom used, which we could live in for as long as we needed it. John looked at him in amazement and I was so excited by his kindness that I threw my arms round him. Never keen on public displays of emotion, even at weddings, he was embarrassed but pleased to have made us so happy. Neither of us had known that he owned a flat. Brian told us he used it from time to time to entertain clients and we didn't question it. In fact, we later realised, it was his bolthole. He still lived at home with his parents and needed somewhere to escape to, away from prying eyes, where he could take partners or just enjoy a little privacy. It was typical of his sensitivity and generosity to give it up for us.

That afternoon, we moved into the flat in Faulkener Street. Brian helped me collect my things from the bedsit and John fetched what he needed from Mimi's. This was something John's parents most definitely hadn't had: a little place in which to start married life together. We had a lovely ground-floor flat, tastefully decorated and furnished, with a bedroom, kitchen, bathroom, sitting room and even a little walled garden.

That evening John and the boys had a gig to go to in Chester, which had been booked long before the wedding. I decided to stay in our new home and sort out our things ready for when John got back. When he did we lay on the bed in each other's arms and revelled at the joy of being together. We could hardly believe our luck. We'd never had anywhere to spend time together without the fear of interruption or being caught. This was the first time we could shut the front door and relax, knowing that no one would walk in on us.

'Well, Mrs Lennon, how does it feel being married?' John asked.

It felt new and a little strange for both of us. John once said it felt like walking about with odd socks on and I knew what he meant. We'd had no time to get used to the idea and suddenly we were married.

Over the next few weeks John was preoccupied with the group, fulfilling engagements and preparing for their first recording. I had a lot of time on my hands, but I was feeling so sick that I didn't mind. Every morning I rushed to the bathroom to throw up, then groaned my way back to bed. And Brian's flat had one major drawback: the bedroom was separate from the rest of the flat. To reach it, you crossed the main hall, which was shared with the tenants of the other flats. It meant that anyone could walk in off the street and open our bedroom door, or the door to the rest of the flat, unless we kept them both locked. This was a real pain, especially when I had to throw up. A lot of the time we didn't bother locking the doors, but when I was there on my own I felt vulnerable so I had to.

When John was at home we enjoyed playing husband and wife. He was loving towards me all through my pregnancy, bringing me flowers or little bits and pieces for our home. Once he came in from a gig and said, 'Close your eyes, Mrs Lennon, I've got a surprise for you.' He insisted I feel the gift he'd brought home before I opened my eyes. It was an ancient coffee-table, but he behaved as though he'd just got it from Harrods – 'Isn't it fantastic, Cyn? Our first coffee-table. Look, it's hand-made, battered copper, only a fiver?' Despite my misgivings about it, I couldn't help but be swept along by his enthusiasm.

I cooked for us – John wouldn't have dreamt of cooking and I was fairly hopeless at it. My culinary skills extended to sausage and mash, cheese on toast or – our favourite – a packet of Vesta beef curry and rice with sliced banana on top.

Not long after we married I met Ringo. He'd joined the group five days before our wedding and a few days afterwards he came round to see John. Although I'd seen him play many times with Rory Storm's group we'd never spoken. Unfortunately we got off on the wrong foot. I invited him to stay for supper and served up curry. I didn't know that after a lot of childhood illness he couldn't eat spicy food. He refused it and was very offhand, virtually ignoring me as he chatted to John. I was put out, unable to work out what I'd done wrong. Later I decided he probably thought I was stuck-up – it was the old over-the-water thing again. He came from one of the roughest parts of Liverpool and thought I was a snob from the posh side. Perhaps I made him nervous.

Ringo and I were wary of each other for some time, but gradually we became good friends and I discovered that he was one of the most kind-hearted, easy-going and good-natured men alive. Even before I came to like him I knew he was right for the group. At twenty-two, he was three months older than John and he fitted in with the other boys as though they'd all been together for ever. They shared the same quick-witted, irreverent sense of humour. He could give back as good as he got from them and they loved that.

Ringo was also a decent drummer, even though he hadn't had any interest in music until he was seventeen and had taught himself to play. Like John, he came from a broken home. He'd been an only child, brought up by his mother, Elsie, who had parted from his father when Ringo was three. He'd been named Richard, shortened to Richie, after his father, but saw him very little after the split; eventually his father had moved away and remarried.

Ringo's childhood was plagued by illness. At the age of six his appendix burst and he developed peritonitis. He had two operations, was in a coma for ten weeks and spent a year in hospital. In those days parents weren't allowed to visit their children very much as it was thought to be too disruptive so his mother often had to settle for peeping at him when he was asleep. Nowadays we'd be horrified at the idea of keeping a six-year-old apart from his mother for a year, but then it was the norm.

When Ringo was almost thirteen his mother had married Harry Graves. Ringo and he got on from the start: Harry took him to the pictures and bought him comics. But at thirteen Ringo became ill again. A cold turned to pleurisy, which affected his lungs, and he went back to hospital, this time for almost two years. He learnt to knit and make papier-mâché to pass the time but he missed a huge amount of schooling.

When he was discharged he was fifteen and old enough to leave school. He convalesced at home for a while, then had to think about jobs. As a result of his illness he was small and weak, and couldn't do heavy work. He went to British Railways as a messenger, but failed the medical and had to leave. After that he got a job as an apprentice fitter.

It was then that the skiffle craze arrived and he helped to start a group, using some second-hand drums that Harry bought for him for ten pounds. Ringo's group did the rounds of dances and parties, and then he was asked to join Rory Storm. When the group was offered a season at Butlins he had to decide whether or not to give up work. He was twenty and had only a year of his apprenticeship to go. Everyone thought he was mad, but for Ringo there was no contest: he chose Rory Storm and Butlins.

It was then that he changed his name from Richard Starkey to Ringo Starr, 'Ringo' after the gold rings he always wore and 'Starr' because Butlins shortened his name so that they could call his solo spot 'Star Time'. After a couple of years' playing with Liverpool's top group

Ringo was thinking about emigrating to the States with a friend, just to try something different, but he hadn't filled in all the forms when the Beatles made their offer.

A couple of weeks after our wedding Brian and the boys had gone to London to sign a five-year recording deal with Parlophone. A week later they recorded their first single – 'Love Me Do'. John and Paul had written it, and I was surprised they'd chosen it because I didn't think it was one of their best – I found it a bit monotonous. But John said that was the one George Martin wanted: he liked John's harmonica playing on it. John told me they'd done so many takes of 'Love Me Do' that he never wanted to hear it again. They'd had to play it about twenty times, and the same with the record's B side, 'PS I Love You'. Worst of all, half-way through, George Martin had decided he didn't want Ringo and had brought in a session drummer. The boys were upset and Ringo was devastated, but none of them dared say anything – George had already overstepped the mark when George Martin told them, 'Let me know if there's anything you don't like,' and George quipped, 'Well, for a start I don't like your tie.' That remark, so typical of the boys' humour, has been immortalised since, but at the time they were afraid they'd gone too far, that they'd better shut up and get on with the record. It was released in early October and, to Ringo's relief, they used the version in which he was playing drums, although the other drummer played on the B side.

Once the record was out we held our breath, praying it would hit the charts, but for a while nothing much happened. It was on the radio a few times – Radio Luxembourg played it – and John whooped with excitement. They were even featured on a local TV programme, Granada's *People and Places*, singing it, but it wasn't until three weeks after it had come out that it hit the charts, and then only at number forty-nine. Better than nothing, but not quite what we'd imagined. Brian did not lose confidence. 'Give it time. It's early days yet,' he would say.

Nice as it was, Brian's flat wasn't proving an easy place to live. I felt nervous there when John was away, and neither of us liked having to keep locking and unlocking the doors. Worse, the area attracted rather a lot of shady characters. One night, just as we were going to bed, there was a knock at the door. When no one from the other flats answered, John went. Two dodgy-looking men asked for Carol. John told them they'd got the wrong house and came back to bed.

A few minutes later we heard the front door open and someone banged violently on our bedroom door. The two men were back and they were shouting, 'We know she's in there, you ponce. Hand her over or we'll tear you apart.' They thumped so violently that we were sure they'd break the door down. Then they tried to pick the lock. John and I were petrified so I gathered all my energy and shouted, 'The only bloody woman in here is me. My name is Cynthia and I'm three months pregnant.'

Thank God, they left, but that night we didn't get any sleep, and afterwards when I was alone in the flat I was never entirely happy. Over the next few weeks, as John's life became busier, mine became more lonely. I'd always had to keep a low profile when I went to watch him play. Now that I was pregnant and beginning to show we agreed that I should stay safely at home. By this time the Beatles, though still unknown elsewhere, had a huge and possessive local following. Girls queued outside the Cavern for hours to see them. Sometimes in the middle of the night the boys drove past and saw girls waiting for the next day's lunchtime session, so that they could get places right at the front. Just before the boys came on they would take turns to nip into the loo and tart themselves up so that they looked their best for the Beatles.

Ringo's new girlfriend, Maureen Cox, had already discovered that some fans hated any girl who got involved with the Beatles. Maureen had been a fan herself, though never a violent one. She was a trainee hairdresser who went to the Cavern with her friends whenever the Beatles were playing. She'd kissed Paul for a dare but really fancied Ringo and was thrilled when he asked her to dance a couple of weeks later.

When Ringo started dating Maureen, she had to pretend she wasn't seeing him. One night she was waiting for him in the car outside a gig when a girl came up, put her hand through the window and scratched her face. She managed to lock the doors and wind up the window before the girl could do anything worse, but it shook her. With incidents like this on the increase, John didn't want me to put myself at risk and I agreed with him. But it meant I saw less of him.

I did what I could to keep occupied, and saw friends, including Phyl. It was a relief to be with someone who knew I was married to John and pregnant. Most of the time I had to keep it secret. I had a gold wedding ring John had bought for ten pounds, but I couldn't wear it when I went out, in case a fan who knew my face spotted me. It was hard, not being able to announce to everyone that I was married, or enjoy talking about the baby. I had few of the joys of being a young wife looking forward to the birth of her child.

One morning I woke up to discover I was losing blood. John was away in London for a meeting with the record company so I was alone. I was afraid I was losing the baby, but there was nothing I could do. Thankfully, my brother Tony decided to drop in to say hello that day, and called a doctor, who prescribed total bed rest. Tony made me as comfortable as he could, then had to go back to work. For the next three days I lay in bed, too afraid to move, even to the bathroom. I had a bucket for a loo and a kettle by the bed to make myself tea, and that was it. I didn't phone John because I didn't want to worry him. He was due home in three days, and until then there was nothing he could do. I just had to get through it and find out whether our baby was going to stay or not. If I'd told him he would have agonised about whether to come home early and I didn't want him to feel pulled in two directions.

By the time John got home the bleeding had stopped. He was concerned, but relieved that the baby seemed all right. He put his arms round me, said I should have told him and that he wished he'd been there.

A few days later I persuaded John that we should go to see Mimi. He had been so hurt by her reaction to our marriage that he hadn't spoken to her since. Though I wasn't fond of Mimi I knew she was important to John and I hated the idea of any lasting rift – John had lost too many people already.

It took a lot of cajoling, but eventually he agreed. When we got to Mimi's, both of us dreading a repeat of the flying-chicken incident, she welcomed us with open arms. I realised she must have missed John and, too stubborn to approach him, was glad of an opportunity to put things right.

Over a meal we caught up on all the news. While she wouldn't admit to being impressed by the boys' record deal, Mimi conceded that they had 'done all right'. It was high praise. When she learnt about my near-miscarriage she was touchingly anxious and suggested we move back to live with her so I wouldn't have to face any more of the pregnancy alone. To our surprise, she offered us the ground floor of the house and said she would move upstairs. It was kind of her and we couldn't really refuse. I'd been afraid and lonely in Brian's flat. Also the Beatles were about to go back to Hamburg so John and I were in for another long separation. At least with Mimi around there would be someone to call on in an emergency. The only hitch was that Mimi still had student lodgers who wouldn't leave before Christmas, so we couldn't divide the house until then.

We packed our belongings, thanked Brian for the use of his flat, and headed back to Woolton. Soon afterwards the boys flew to Hamburg to fulfil a contract they'd agreed to months earlier. Before they left we heard that 'Love Me

Do' had reached number twenty-seven in the charts and Brian announced that he'd got them a tour in February, supporting Helen Shapiro, a sixteen-year-old girl whose powerful voice and hits like 'Walking Back To Happiness' had made her a big star.

'Love Me Do' crept up to number seventeen. I was hearing it often on the radio, which was wildly exciting.

John's letters were full of stories of their wild celebrations in Hamburg, the weeks when he was in Germany crawled by and it was hard to be back under Mimi's roof. Until Christmas, when the students were to move out and Mimi would move upstairs, the house was full of lodgers who weren't supposed to know I was married and pregnant in case the news leaked out to a wider public. I wore baggy blouses and waistcoats to cover my bump, and tried to stay out of their way.

It was impossible to stay out of Mimi's way, though. With John away, she went back to treating me like the hired help. She was frequently grumpy and never asked how I was feeling. I often felt that she had no real interest in me, other than as a means to keep John close to her. It was so rare for her to be cheerful that I actually noted in my diary when she was. One entry read, 'Mimi in a good mood (change).' On top of this when John phoned me once a week – all he could afford as international calls were expensive – Mimi always leapt to the phone before I could get there. She would chat to him for the next ten minutes, giggling girlishly, and only called me to the phone when John's money was about to run out. Often we had barely time for more than a quick hello.

## John

Occasionally I went out with Phyl or to visit my aunt Tess, a family friend I knew as Auntie Muriel or John's aunt Harrie and his sisters. Julia and Jacqui were excited about the baby and thrilled at the thought of being aunties. I was always welcome at Harrie's and spent some lovely evenings there, but mostly I passed the time in my room, writing to John, reading and knitting bootees.

John sent me enough to pay Mimi's rent and to buy what I needed, and I was grateful that I didn't have to worry about money. Since my dreams of being an art teacher had collapsed I wasn't sure what I wanted to do and I was glad to put off any decision until after the baby was born.

John came home briefly in late November, but had to head off to London almost immediately to record the group's second single. George Martin had a song lined up for them, which he thought would be perfect. It was called 'How Do You Do It?' and he was sure it would be a hit. But the boys didn't like it and balked at recording it. He told them to come up with something better so they suggested their own song, 'Please Please Me'. Reluctantly, George Martin agreed. 'How Do You Do It?' went to Gerry and the Pacemakers, and became a huge hit for them.

John and Paul always wanted to do their own songs, though, not someone else's, and they wanted it enough to stand up to the great George Martin, of whom they were in awe. When he agreed to 'Please Please Me' they were delighted – but nervous too. 'God, I hope it works,' John said, when he told me about it. 'We'll look like idiots if it doesn't.'

Soon after they'd finished the recording, they returned to Hamburg for their fifth – and final – trip, and stayed there through December.

At Christmas the lodgers left and I moved downstairs, as Mimi had promised. John and I now had the kitchen, breakfast room and two sitting rooms. Mendips was a cold, draughty house with no central-heating and only ancient electric fires in most of the rooms. Until we took them over the two downstairs sitting rooms were never used and I imagined the ghosts of Mimi's husband George and little John playing there together. When George was alive, the house was too.

Since his death Mimi had sat in the breakfast room, the closest she had to anywhere cosy. It was next to the kitchen and held a drop-leaf table, three not-so-cosy chairs and a television between the table and the fire-place. The fire was always lit and the coal bucket was being constantly filled. I knew it was a wrench for Mimi to leave her fire, and I was grateful to her for giving John and me our own space. Upstairs, there were two double bedrooms and the small boxroom I had used. Mimi turned one of the bedrooms into a sitting room and the boxroom into a makeshift kitchen with a small portable cooker. I converted one of the downstairs sitting rooms into a bedroom for me and John. The only room we had to share with Mimi was the icy cold black-and-white-tiled bathroom.

All the furniture at Mendips was old and faded. Mimi used to say, 'If my husband hadn't lost all our money and put us in debt with his gambling I would go out and replace the whole damn lot.' I never knew if George really had been a gambler or if what she said was true.

Mimi had two distinct modes of appearance. At home she was frumpy: she wore scruffy sweaters and skirts that bagged at the back because she spent so much time sitting in front of the fire. Her stockings were often laddered and she lived in slippers. She had bronchial problems and hugged her chest, as if she was protecting it. It was another matter when she went out: her wiry greying hair was scraped under a felt hat, she wore smart high heels and her best beige coat with the fur collar. Her sallow complexion was warmed with rouge and she looked like a different woman.

Her passions were money, John and her cats – in about that order. She had three cats and I'm certain she preferred them to most people. As I had the proper kitchen, I was expected to feed them, which made me feel sick. Every day, fish scraps had to be collected from the fishmonger, boiled, then boned. The kitchen stank, but it was a job worth doing in exchange for the extra space and privacy.

I was glad things were going so well for the boys, but John had been away virtually since we'd moved back to Mimi's. I missed him: we were newlyweds with almost no opportunity to enjoy being married and I wanted to be with him, sharing in the excitement. But I accepted it all: it was the way things had to be. The message I had received from my parents was that once you were married you stayed together, whatever that entailed. And John and I grew up at a time when it wasn't unusual for women to wait at home for their men. Many Liverpool men went to sea, as John's father had, and their wives had no choice but to wait. My situation was much the same; he was off earning our living and it was my role to be there for him, loving and supportive, when he came home. No one around me was saying, 'Why do you put up with it?' It was considered normal. So I waited, hoping that our time apart wouldn't last too long.

My pregnancy was advancing, becoming harder to hide, so I went out less and less frequently. Not that getting out was easy for me; that winter was one of the most severe on record, with plenty of snow that turned to hard-packed, treacherous ice. When I had to go to the hospital for check-ups I slipped and slithered around on the pavements, dreading a fall and clinging to anything or anyone in sight.

It was a difficult time. Cold, miserable weather, John away more than he was at home, Mimi resentful of my presence in the house. And I was about to become a mother – a notion that terrified me.

# 9

AFTER SUCH A TUMULTUOUS YEAR I wondered what else could possibly lie in store for me. We'd lived through a series of life-changing events in rapid succession: Stuart's death, my pregnancy, our marriage, the group's first recording contract. I longed for a few quiet months with John to take it all in, but there was to be no chance of that.

At the beginning of 1963, the Beatles were hugely popular in Liverpool but still virtually unknown anywhere else. None of us could have known what an astounding difference a year would make, or that by the end of it the entire country would be Beatle-mad, and John, Paul, George and Ringo's faces the most famous in the country.

That kind of success was still beyond our wildest fantasies when John came home from Hamburg in the New Year. He was only back for a couple of days before he went off on a short tour of Scotland, but it was wonderful to see him. He was brimming with excitement, relieved that Mimi and I hadn't killed each other, and poured out stories of what he'd been up to. When he got back from Scotland I'd got our new living quarters organised and we had a couple of weeks of relative peace together. John spent a lot of time composing, strumming his guitar and jotting down lyrics, all on the front porch because, despite his achievements, Mimi still wouldn't let him play in the house. He would call me to listen when he'd come up with a song he liked. In the evening we ate together, talked and curled up in front of the TV. Bliss.

In February he set off on the Beatles' first national tour, backing Helen Shapiro, which gave them the opportunity to introduce themselves to a wider audience. John needed a lot of reassurance before he left, but he needn't have worried: audiences everywhere loved the Beatles. They were so successful that they regularly stole the show.

John rang me from whichever hotel they were staying in. Even though he was enjoying the gigs because audiences loved them, he was missing me. The tour company spent long hours criss-crossing the country on a coach and John whiled away the time reading, composing, having impromptu jamming sessions or taking the mickey out of

Helen who, despite her star status, was completely in awe of him. Later she said that although John teased her mercilessly, he was also kind and protective.

'Please Please Me' was released in mid-January and reached number one in the charts on 16 February, while the boys were still on tour. When John phoned to tell me I screamed with delight – they'd made it to the top of the hit parade. At that point I felt things couldn't get any better.

John was due home a few days later and, keen to look as good as I could for him, I decided to get my hair done. Unfortunately I let the hairdresser talk me into having it cut to shoulder length. When I saw it I was horrified and I knew John would hate it. My hair was one of the things he loved best about me. I spent the evening wondering what to do.

That night I put my hair in rollers, so that when he came in late he couldn't see what had happened. But the next morning there was no hiding it. I had known he wouldn't like it but I wasn't prepared for just how angry he would be. He looked at me with real hatred and screamed, 'What have you done?' Then he refused to talk to me, or even look at me, for two days. It was awful. I preferred him shouting to the furious silence.

Now I can see that he was overwhelmed by everything that was happening. His whole world was changing and the last thing he wanted was for me to change too: he needed to come home to the Cyn he knew and loved. At that time I was an anchor for John, providing a secure and stable base to which he could escape and be himself.

Of all the people in his life I was the most constant, the one who didn't demand, criticise or give orders, and who loved him unconditionally. Thankfully, on the third day he thawed, put his arms round me and apologised. I never contemplated having my hair cut so radically again.

Over the next couple of months John was away for days and sometimes weeks at a time, playing at venues all over the country or recording in London. The boys made their first LP, *Please Please Me*, in a single eleven-hour recording session – amazing, when you think of the months it takes most groups now to put an album together.

Meanwhile I felt as big as a whale and was desperate for the baby to be born, but I wasn't certain of my dates. The doctors thought it would arrive in early March, but in my diary for that time the due date kept hopping back – 4 March, then the seventh, the twelfth, the fifteenth. They all passed with no sign of the baby.

On Saturday 6 April John was away on yet another trip so Phyl came over and we decided to go shopping in Penny Lane. I was sick of feeling huge and decided to buy myself something pretty to wear after the birth. Half-way through the afternoon the contractions began. 'I think this could be it,' I told Phyl. But they weren't severe and I'd had plenty of false ones before, so I decided to wait and see. We went home and I put my feet up. Phyl made us something to eat and promised to stay with me. By late evening nothing much had happened, so we went to bed.

At four in the morning I was woken by a surge of pain that took my breath away. I screamed and Phyl, shocked awake and trying not to panic, phoned for an ambulance.

In India, sitting between John and Paul after dinner with George, Patti and other guests

We hadn't realised how cold it would be in India! Wrapped in blankets because we hadn't brought enough warm clothes

The Beatles' wives –
even Mo went
blonde for this
photo shoot!
Clockwise from
top: me, Mo, Jennie
Boyd (Patti's sister)
and Patti in 1966

John and Julian at
Kenwood in 1966,
next to the shelves full
of leather-bound
books John insisted
we must have

what we said about it. It's not much
bother really, is it? when you think about
it – 'cause I'm sure Dot and Lil' and
Bennis, Tommy, wee Jackey etc can understand
something as simple as us wanting
to be alone for a day. – I don't mean
Julian tho' – I mean don't pack him
off to Dots or anywhere – I really miss him
as a person now – do you know what I mean,
– he's not so much 'The Baby' or 'my baby'
anymore he's a real living part of me now
– you know he's Julian and everything and
I can't wait to see him, I miss him more
than I've ever done before – I think its been
a slow process my feeling like a real father!
I hope all this is clear and understandable,
I spend hours in dressing rooms and things
thinking about the times We wasted not
being with him – and playing with him – you
know I keep thinking of those stupid
bastard times when I keep reading bloody
newspapers and other shit whilst he's in the
room with me and I've decided it's ALL
WRONG! He doesn't see enough of me
as it is and I really want him to

An extract from the letter John wrote me in 1965, talking about his feelings
for Julian. Paul later bought the letter as a gift for Julian

John with George
Martin, on a ski-ing
holiday in St Moritz,
January 1965

With John in
St Moritz

John and Julian at
Kenwood in 1965
when Julian was two

Our lovely boy – Julian, aged two, beside the pool at Kenwood, our new house in Weybridge

Standing in the front entrance at Kenwood

In the garden at Kenwood in 1965, fooling around for a photo shoot

Arriving for the premiere of *A Hard Day's Night*, after Mum raced across London to collect my dress in time

Arriving at the Foyles lunch to celebrate the publication of John's book, hungover and with no idea that John would have to make a speech

On holiday in Tahiti in 1964 – a chance to relax and escape the fame and the fans

Arriving home from Tahiti with George and Patti

On our belated honeymoon in Paris, collapsed into bed after our wild night out with Astrid and friend

Brian and the boys at the premiere of *A Hard Day's Night*, July 6 1964

It arrived so promptly that we were both still in our nighties and Phyl's hair was in curlers. She grabbed a coat, rolled up her nightie under it and put a scarf over her rollers. I was helped out to the ambulance and she followed with the bag I'd packed – a rather large suitcase. It was far too big for the few items I was taking with me, but it was all I had.

When we arrived at the hospital I was wheeled away and Phyl was handed back the case, minus my things. She was told that she was no longer needed and might as well go home and get some sleep. The only problem was that she was in her nightclothes with no money. She had hoped the ambulance might drop her back at Mimi's but was told firmly that it wouldn't.

Poor Phyl started to walk the couple of miles back from Sefton General Hospital to Mimi's, freezing cold and lugging my case. Luckily for her, a kind-hearted taxi-driver stopped, thinking she was a runaway. He told her not to be too hasty and offered to take her home for free. Phyl was too tired to explain but accepted gratefully.

When she arrived at Mimi's she fell into bed. Mimi must have heard the disturbance during the night, but she hadn't appeared and Phyl saw no sign of her until, at six the next morning, she began vacuuming outside Phyl's bedroom door. It was how she liked to indicate to me that she considered it time for me to get up, and Sunday mornings were no exception. Exhausted, Phyl got her things together and went home.

Throughout that Sunday I was in labour, and feeling horribly alone. John had phoned often over the last few days and I knew Mimi would tell him I'd gone into hospital when he called again, but it hadn't occurred to me that he should be at the birth: it wasn't the done thing for fathers to be present. The person I wanted was my mum, but she was still in Canada, so I had to go through the birth alone.

My labour went on for another twenty-four hours. I was put into a ward in which half of the women had already given birth and were sitting, beaming, with their babies beside them, while the rest of us writhed in agony and prayed for it all to be over. I was next to a very large girl who told me she was unmarried and kept crying out for her mother. The two of us struggled on together through that endless Sunday. Every now and then the other girl decided it was all too much and staggered out of bed, insisting that she was going home. Each time I had to talk her into staying: I pointed out that she wouldn't get far and tried to offer her some encouragement and moral support.

The worst part of the day was visiting time, when beaming husbands and relatives, bearing flowers and chocolates, turned up to sit beside every other woman in the ward. Only my neighbour and I had no visitors. I'm sure the staff thought I, too, was unmarried and we both got our fair share of disapproving looks. Phyl had assumed that, in the absence of my mother, Mimi would come to see me but Mimi didn't even send a message. I felt hurt, abandoned and frightened.

In the early hours of Monday morning, 8 April, I was taken into the delivery room and told that I was ready to give birth. I was given gas and air, but it made me feel sick and I was exhausted from the long hours in labour with no food and only a few sips of water to keep me going.

It was an Afro-Caribbean midwife who told me firmly that if I didn't get on with it my baby would die and so would I. Terrified, I rallied every last ounce of strength and pushed my baby into the world at six fifty a.m. Our son arrived with the cord round his neck and yellow with jaundice, so the midwife whisked him away. I was cleaned up and moved back to a bed in the ward. A short while later he was put into my arms and I looked down at him for the first time. He was tiny, his small face scrunched up and bright red. I thought he was absolutely perfect. He was put into a cot beside me and we both slept, exhausted from the long hours of labour.

The next day Mimi finally turned up. She had phoned and been told the baby had arrived, and came to inspect John's son. She was as stiff as ever with me, but when she looked at the baby her face softened. She had already phoned John to tell him he was a father and she offered to send my mother a telegram. I thanked her and told her I would write to her myself. I was glad John knew, but disappointed not to have told him myself.

John didn't make it to the hospital until three days after our son was born – the first opportunity he'd had to get away from the tour. As soon as he got there he arranged for me to be moved into a private room: he knew that both he and I might attract unwelcome attention if I stayed in the public ward. He came in like a whirlwind, racing through the doors in his haste to find us. He kissed me, then looked at his son, who was in my arms. There were tears in his eyes: 'Cyn, he's bloody marvellous! He's fantastic.' He sat on the bed and I put the baby into his arms. He held each tiny hand, marvelling at the miniature fingers, and a big smile spread over his face. 'Who's going to be a famous little rocker like his dad, then?' he said.

It was wonderful to see him, but privacy was impossible. My room had a window on to the corridor outside, and when word got round that he was there, dozens of patients and staff gathered with their noses pressed to the window. The room felt like a goldfish bowl and it was obvious John couldn't stay long. He hugged me and signed dozens of autographs on his way out. I was disappointed that we'd had so little time together: he had to go straight back to the tour and wouldn't be home again for a week or so.

John's visit was brief but it had a dramatic effect: the nurses were distinctly friendlier to me than they had been when I first arrived.

I was in hospital for another few days and my room was soon filled with cards and flowers, among them a bouquet from Brian and the other Beatles.

The day after John's visit I wrote to my mother:

*Hi Granny,*
*Well Mum it's (or should I say he) has arrived at long last. April 8th 1963, 6.50am and thank heaven it's all over.*
*He's beautiful Mum, only 6lbs 11 oz but just gorgeous.*

*Well I suppose you can imagine how I feel now can't you! I only wish that you could be here to see him now – he's fast asleep in the cot beside me. I'm in a room on my own by the way, 24 shillings a day but it's well worth it. John came to see him for the first time yesterday (Thursday) poor fellow he was a nervous wreck. But I've never seen him look so proud or happy. The baby looks very much like John but he also has a look of me, so he should be handsome!!!*

*I hope you received the telegram that Mimi sent to you . . . Mimi is off her nut about him (she still drives me up the pole). Am going to try for a house as soon as I can Mum – so when you come home we can have a jolly good scout around – won't it be lovely?*

A couple of days later the baby and I went home from hospital in a taxi. Both John and I wanted to give our son family names so we called him John after his father. I suggested Julian, because it was as close as we could get to John's mother's name, and we chose Charles after my father. So, John Charles Julian Lennon, to be known as Julian. John had a moment's worry about whether Julian was too cissy a name, but decided to forget it: we both liked the name and he wanted it in memory of Julia. So that was settled.

It took me a while to get used to being a mum. It was only when I got Julian home that I realised how little I knew about babies. I hadn't been offered any antenatal classes and I was having to learn through trial and error. And it seemed to be mostly error. Julian cried day and night and I was exhausted. I was also having trouble breast-feeding him, and I was constantly anxious that he was disturbing Mimi, who complained about how little sleep she was getting. It would have been much easier if I'd been in my own home, with Mum on hand to help. As it was I simply had to cope, but on more than one occasion I broke down in tears of frustration and helplessness.

A week later John came back, and was enchanted by his son, although he refused point-blank to change nappies or even to stay in the room while I did. He'd bolt out, saying, 'God, Cyn, I don't know how you do it. It makes me want to throw up.' It didn't matter to me: few men had much to do with babies in those days and I hadn't expected John to be any different.

What John did love was watching Julian at bath-time, and the smell of baby talc; he loved to cuddle him when he was warm and fragrant and ready for bed. In those early days he couldn't get over the fact that this tiny creature was ours. It was touching to see them together and I was sad that John wasn't able to deepen his bond with his child over the following months.

When Julian was three weeks old, Brian invited John to go to Spain with him. John asked if I'd mind and I said, truthfully, that I wouldn't. I was preoccupied with Julian and nowhere near ready to travel, but I knew how much John needed a break where he wouldn't be recognised and could really relax. I gave them my blessing and they went off together for twelve days. It was a holiday John came to regret because it sparked off a string of rumours about his relationship with Brian. He had to put up with sly digs, winks and innuendo that he was secretly gay. It infuriated him: all he'd wanted was a break with a friend, but it was turned into so much more.

A few weeks after he got back, Paul had his twenty-first birthday party in his aunt's back garden. The Beatles had just finished another national tour, this time with Gerry and the Pacemakers and Roy Orbison. They had also released a third single, 'From Me to You', which again went straight to number one. Two number ones in a row! We were stunned. What amazing luck!

At the party the boys' old friend Bob Wooler, the Cavern MC, made a crack to John about his holiday. John, who'd had plenty to drink, exploded. He leapt on Bob, and by the time he was dragged off Bob had a black eye and badly bruised ribs. I took John home as fast as I could, and Brian drove Bob to hospital.

I was appalled that John had lashed out again. I'd thought those days were over. But John was still livid, muttering that Bob had called him a queer. A day or two later when he had cooled down he was ashamed. He kept repeating, 'Oh, God, Cyn, what have I done?' He sent Bob a telegram saying, 'Really sorry Bob stop terribly worried to realise what I had done stop what more can I say John Lennon'. Unfortunately the local press got hold of the story and the *Daily Mirror* ran it, which didn't help John's image. He swore he'd never do anything like it again and, to my knowledge, he didn't, certainly for as long as we were together.

At the same party I'd been introduced to Paul's new girlfriend, a seventeen-year-old called Jane Asher. She was beautiful, with auburn hair and green eyes. Also, although she had been a successful actress since she was five, she was unaffected, easy to talk to and friendly. She and Paul had met the previous month when she'd been sent by the *Radio Times* as a 'celebrity writer' to do an interview with the Beatles after their performance at London's Royal Albert Hall. It had been a prestigious event for them, a BBC Light Programme live broadcast in which they shared the bill with Del Shannon and Shane Fenton (later known as Alvin Stardust). That evening, the others had gone off to a West End club, leaving Jane and Paul to finish the interview. When they returned two hours later, the pair were still talking.

Jane was different from the girls Paul had been out with previously. The daughter of a psychiatrist father and a music-teacher mother, she was highly intelligent and cultured. She had a strong inner confidence, with a maturity and grace way beyond her years. Paul, whose working-class background couldn't have been more different, was bowled over, and from the day they met they became an item.

The boys' fourth single 'She Loves You' was released in July. It was an instant number one and became one of the most famous Beatle songs of all. Written by John and Paul together on the Helen Shapiro tour bus, it captured the feel-good atmosphere of the early sixties. Its 'yeah, yeah, yeah' chorus was irresistible – the whole country was singing it – and it confirmed the Beatles as talented and original, a pop force to compete with the best of the stars from the States. I loved that song: it reminded me of John's first Christmas card to me – 'I love you yes, yes, yes'.

In those days it wasn't really done to publicly dedicate songs to those you loved, but I know, and John often told me, that many of his songs were for me. He and Paul wrote from their own experience, and I was so much a part of John's life that I also became part of the fabric of his writing. It was simply understood that his love songs were our songs. Some were extra-special, such as 'All My Loving', which came out on an EP in February 1964. John had written it for me during a time when we were often apart, and I loved its tender, romantic lyrics.

The British press was slow to grasp just how popular the Beatles were and recognise them publicly because nothing like the Beatles had happened before. Fans at their concerts all over the country went wild, but the press virtually ignored them. Before the Beatles most major stars, certainly rock and roll stars, came from the States. When the press did catch on, the boys were in demand for interviews and journalists were astonished to find out how intelligent and articulate they were. All four were sharp, funny and could talk the hind legs off a donkey on just about any subject, whether they knew anything about it or not. Until then most pop stars had been pretty monosyllabic and happy to stick to a few quick comments about their latest record. Not the Beatles. Paul, the most PR-minded and career-driven of the four, was always charming. George was bright but didn't enjoy interviews and kept them short. Ringo was happy to be typecast as the cheerful guy who'd got lucky when he was hauled aboard the Beatles bandwagon at the last minute. But it was John who confounded the press: he could talk for hours about anything and everything. He was self-critical, opinionated, witty and ruthless with journalists he felt were wasting his time. Papers and magazines soon learnt to send their brightest hacks to interview him.

John loved what was happening to the Beatles. But a part of him stood back from it and watched. He wanted fame and success, but not if it meant selling out or changing who he was. He had always said what he thought and wasn't afraid to criticise himself or anyone else. Fame didn't change that.

In Liverpool, I was proud, excited and a little frightened. It was all taking off so quickly, and the more successful the boys were, the further away from me John seemed. I was getting used to being a mum, but most of the time I felt like a single parent and it was hard not to feel frustrated with being stuck at home. I loved Julian, but I knew that if I hadn't had him I could have seen much more of John and that was hard. Although we'd have had to be discreet, I could have joined him sometimes in London or on tour. As it was, I felt shut off from the life he was living. After years at his side, I was excluded, just as it was all happening.

John got home whenever he could, but it was never for long: Brian had filled their diaries and the next recording or concert was always waiting. I missed seeing them play live, having watched them so many times in the past. And because the press was slow to pick up on their success, even I didn't grasp just how big they were becoming. I knew the hits were coming thick and fast, but John didn't say much about the concerts and the reception they were getting. He came home to get away from it all and feel normal.

Not that it was easy for us to lead any kind of normal life, even when he was at home. For the first few months after Julian's birth we were at Mimi's and the strained atmosphere made it impossible to relax. Julian continued to cry incessantly – probably picking up on the tension in the house – and I'd wrap him up in his pram and park him at the bottom of the garden, hoping that Mimi wouldn't hear him there. She treated Julian and me with disdain, and muttered frequently about having her home turned upside-down. Eventually I became so desperate with exhaustion that even Mimi couldn't ignore it any longer. One night she offered grudgingly to take Julian upstairs so that I could get some sleep. I was grateful, but I couldn't help worrying about how she was coping and in the end I hardly slept.

That was Mimi's only attempt to help. Most of the time she complained about being ill or grumbled that she'd been pushed upstairs, conveniently forgetting that she'd invited us to live there in the first place.

I was still feeding her cats, gagging daily over the smell of boiling fish. And although I asked Mimi many times to keep them away from my part of the house, she frequently let them in while I was out. I'd return to find them sleeping in Julian's carry-cot, which would be plastered with cat hairs. It was hard not to believe that she had done it on purpose.

To add to my woes, I was still a secret, and I hated it: I wanted to be acknowledged as John's wife. Of course, some of the Liverpool fans knew, but to the rest of the country he was young, free and single. Every now and then when I was out with Julian in the pram a girl would come up and ask whether I was John's wife. I had to play up to the role I'd been assigned, say no, and hurry off.

On 3 August 1963 the Beatles played for the last time at the Cavern. Sadly I was told to stay away. The fans had heard rumours about John's wife and baby, and Brian didn't want them fuelled by my appearance. So, I stayed at home with Julian, but John told me that the place was packed and the atmosphere electric. The local fans knew that the rest of the country had discovered the Beatles and that they were too big now for places like the Cavern. But for one last night they belonged to Liverpool, and it seemed as though half the city had turned out to see them and celebrate.

It hurt having to deny who I was and keep my marriage under wraps, but it was always Brian, not John, who insisted on the secrecy. John argued with him about it: he reckoned it wouldn't make any difference to the group's image if the fans knew he was married but Brian decreed that it must be kept quiet. Understandably he kept his own private life secret and this might have influenced his attitude to the Beatles. He didn't even want Paul to be linked with Jane.

Paul ignored him and was photographed with Jane not long after he met her. Within weeks he was living with her family in their five-storey Wimpole Street house whenever he was in London. Brian had no choice but to accept that Paul having a girlfriend hadn't done their image any harm, but he still drew the line at a wife and baby, so I remained hidden at home. There were times when I felt very down about it, but I kept myself going by dreaming of the future and a time when John and I could be openly together.

In August John suddenly announced that he'd got a surprise for me: we were going to have a belated honeymoon in Paris. My spirits rose. My only problem was what to do with Julian. There was no chance of Mimi having him, so I turned to John's aunt Harrie, who said she'd be glad to help out. Julia and Jacqui were excited about their little nephew coming to stay, and I knew they would all lavish love and attention on him.

I'd bought myself some new clothes, packed and got Julian and all his gear over to Harrie's when disaster struck: I went down with what appeared to be food-poisoning. I couldn't stop throwing up and John was on the point of cancelling, but I wasn't about to miss my honeymoon and I insisted we went, even though the taxi taking us to the airport had to stop every few minutes so that I could be sick.

When we landed in France the nausea disappeared miraculously and I began to enjoy myself. That three-day honeymoon was one of the happiest times we had during that period. Usually we saw so little of each other, and when we did it was under Mimi's roof, which meant we were never really alone. So to be away, free, unrecognised and together, was wonderful.

We stayed in a gorgeous hotel, the Georges V. The luxury took my breath away. I walked round the room touching the vast bed with its silken cover, the lush hangings and the antique furniture. In the huge white marble bathroom I came face to face with my first bidet. John had sent me a picture of him and Paul washing their feet in one when they'd come to Paris for his twenty-first birthday two years earlier. Although John was earning more than enough money to afford it, we both felt somehow as though we shouldn't be there – that we might be rumbled and thrown out. The opulence of Paris's grandest hotel was such a world away from anything we'd known before that it was hard to believe it was now ours for the taking. Determined to make the most of our time, we saw as much of Paris as we could cram in, going up the Eiffel Tower, gazing at the Arc de Triomphe and exploring Montmartre. John, always generous and now revelling in having money, wanted to spoil me. He bought me armloads of presents, including a gorgeous grey coat, a pretty white beret and a bottle of Chanel No. 5.

A couple of days into our stay we came back to the hotel to find a message from Astrid. She was in Paris for a few days and had heard we were there. We called her and that evening met up with her and a girlfriend for what turned out to be a night of excess. The four of us moved from one wine bar to the next, knocking back rough red wine in vast quantities. It was lovely to see her again. It had been almost a year since John was last in Hamburg and two and a half years since I'd seen her, so we had a lot to catch up on.

When dawn broke the four of us were so paralytic that we could barely walk. We stumbled back to Astrid's lodgings, where we managed to down another bottle of wine before we collapsed in a heap on Astrid's single bed. Unbelievably the four of us slept there, piled together like sardines, until morning when, with raging hangovers, John and I crawled back to our hotel.

Soon afterwards we went back to Liverpool. I was longing to see Julian, but I was sad too: almost as soon as we landed John would be heading off again.

When Julian was almost six months old, my mother came back from Canada. This time she was home to stay and I couldn't wait to introduce her to her grandson. The morning before she arrived Mimi came downstairs and told me she'd had a terrible nightmare about my mother coming to her front door. In the dream, she said, she had told my mother to go away and that she wouldn't allow her past the front door. 'Isn't that terrible, Cynthia? Fancy me dreaming such a thing.'

That was the last straw for me. The meaning of Mimi's 'dream' was crystal clear. She had disliked Mum at their first meeting and their second would be little better. It was at that point that I knew had to leave Mimi's.

When Mum arrived Mimi managed a frosty hello, then disappeared upstairs. I told Mum that life at Mendips was unbearable and she offered me a home with her, back in our house at Hoylake. The only problem was that her tenants weren't due to leave for another month. Mum had planned to stay with Tony and Marjorie or with friends until she could move back in, but when she saw how desperate I was she agreed that we should look for some-where to rent together. Within a few days we'd found a bedsit in Hoylake and moved into it.

When I explained the situation to John over the phone he was understanding. He was due to be away for the whole of that month and by the time he came back Mum, Julian and I would be settled in our old home, where John could join us. I'm sure he would have preferred us to stay at Mimi's, but he knew how difficult she could be and how much it meant for me to have Mum home again. He said, 'You do what you want to do, Cyn. It's fine by me.' It was a relief that he could be so relaxed about it because I knew he wouldn't relish being under my mum's roof. What I didn't know, but John did, was that our living arrangement in Liverpool would be only temporary. The boys were already spending most of their time in London, living in expensive hotels, so it made sense to find homes there, and John was determined that once he had his own place Julian and I would join him.

I spent a difficult month in the bedsit, a room in a large house with a number of elderly, rather cranky tenants. Mum and I spent all our time trying to keep Julian quiet. We walked him around in his pram for hours during the day and spent our evenings trying every trick we knew to get him to settle down. I was glad to have Mum to share the burden with me, but it was tough and there were times when we were so exhausted that we were hysterical with either laughter or tears.

Life was almost surreal: there was John, rapidly becoming famous and wealthy, living in luxury hotels, as the country took him to its heart, and there was I, in a grim little five-pounds-a-week bedsit, with his son. Ever since, people have speculated on how I could have put up with it. Was I cowed or afraid of John? Not in the least: I put up with it because I didn't want to do anything to harm John's career, and I had been told repeatedly that going public would do just that. I was loyal to John, and if he needed me to support him by lying low, then that was what I would do: it wouldn't last for ever and I was strong enough to do it for as long as I had to. John was always loving and reassuring when we spoke and I trusted him. I believed that, wherever he was, Julian and I came first in his life.

We were in touch as often as we could manage. Communications then were so different from today: all we had was letters or the phone, and phone calls were still expensive. John phoned me whenever he could, but that wasn't every day and neither of us had much time to write any more, so it was a question of trust and patience. And we did trust each other. Neither of us was jealous when we were apart. It never occurred to me to be suspicious of John, and he believed now that I really did love him. His early jealousy only resurfaced when he saw me talking to another man, and those difficult moments became increasingly rare.

Eventually Mum and I moved home, and from the day we got there Julian stopped crying. For the next few weeks life was calm. In Hoylake most people knew that I was married to John and were protective and discreet. I was able to push Julian's pram beside the sea or down to the shops, knowing that no one would bother us.

In November John arrived for a visit with news for me. He told me he was going to find us a home in London. He refused to be apart from me any longer and Brian would have to put up with it if the press found out. I threw my arms round him. No more loneliness, no more pretending or hiding. We would be together, a proper family in our own home. I couldn't wait.

# 10

BY SEPTEMBER 1963, THE MONTH after John and I had slipped away for our honeymoon, the Beatles had the top-selling LP, *Please Please Me*, the top-selling EP, 'Twist and Shout', and the top-selling single, 'She Loves You'. No group had ever managed this before. There were also advance orders for half a million copies of their next single, even though no one knew yet what it would be.

The story of the group from Liverpool who were storming the charts had caught on and pictures of them were everywhere. Brian and the boys were thrilled at their success and hoping that it would last for a while yet, with a few more hits. But no one was ready for what happened next. On 13 October they topped the bill at the London Palladium. Fifteen million people watched them on TV – *Sunday Night at the London Palladium* was almost everyone's favourite programme. Before the show thousands of fans besieged the theatre, in Argyll Street, off Oxford Street. Presents and telegrams for the boys blocked the stage door and the fans screamed and chanted so loudly that the people in the theatre could hardly hear the performers. Reporters arrived and the police, taken by surprise, rushed in reinforcements. Even so, they could barely control the hordes of hysterical teenagers.

The next day the story was on the front page of every paper, and 'Beatlemania' was the word used to describe the hysteria of fans who screamed, burst into tears, became totally overwrought and fainted as they attempted to get closer to the Beatles, swearing undying love for the Fab Four. Watching TV in Liverpool and looking at the photos in the papers of the crowd scenes I was a little alarmed – John and the others had almost been crushed by the mob as they tried to get to their car that night. What on earth was going on?

But this was just the start. For the next three years Beatlemania would overwhelm not just Britain but many other countries around the world. The scenes of mass hysteria and uncontained emotion that erupted then had never been seen before and have never been repeated on the same scale since. It was astonishing, even to those of us at the heart of it, as though a kind of madness had taken over – the boys couldn't go anywhere without being swamped. Although the majority of the fans were teenage girls, it seemed that everyone wanted a piece of the Beatles. World leaders paid tribute to them, politicians tried to get mileage with the public by mentioning them, and people of every age and class bought their records.

After the London Palladium performance they did several concerts in Britain, each of which generated hysterical crowds and, in many cases, casualties, as kids queued all night for tickets. Towards the end of October they went on tour for five days to Sweden, where the kids went just as mad as they had in Britain. When they arrived back at London airport the whole place was swarming with thousands of fans who'd been waiting there for hours to see them. Miss World passed through the airport unnoticed, and the prime minister, Sir Alec Douglas Home, was held up there in his car.

A few days later the boys appeared in the Royal Variety Performance, the biggest show of the year. The Queen Mother, Princess Margaret and Lord Snowdon were in the royal box. That was the evening when John famously said, 'The ones in the cheap seats clap their hands, the rest of you just rattle your jewellery.' The joke was on every front page the next day and confirmed the Beatles' image as a group of cheeky but lovable lads. Even the 'heavy' newspapers, like *The Times* and the *Telegraph*, which never normally covered pop stories, devoted as much space to them as the tabloids did. Soon afterwards their fifth single, 'I Want to Hold Your Hand', came out. There were advance orders for a million copies and, once again, it went straight to number one.

For the boys all this was, by turns, amazing, funny, bizarre, overwhelming and frightening. Suddenly they couldn't go anywhere without protection. Things they had taken for granted all their lives, like going to the pub, walking down the street or visiting a friend, were no longer possible. Just getting from a car to a door a couple of yards away was a major operation. In all their dreams of fame and success, they had never imagined this. How could they?

I'm sure the pressure they were under at this time strengthened John's desperation to have me with him. He wanted to leave the madness outside the front door and be himself, safe in the normality of home, with me.

While he was on yet another tour between 1 November and 13 December, things were hotting up in Hoylake. Reporters had heard rumours that John had a wife and child and were nosing around. Luckily I was well shielded by friends and neighbours, who put off the journalists by telling them they'd got it wrong. Every now and then one would come up to me and ask if I was married to John. I'd say, 'John who? I'm sorry but I don't know what you're talking about.' More than once I had to dodge into a local shop to avoid a journalist who was following me. Kindly assistants offered me refuge.

In December, while John was still on tour, I decided to have Julian christened at Hoylake parish church, which was across the road from our house – the same church where I had gone to Sunday school when I was small. The ceremony, performed by Canon Devereaux, was held quietly one Sunday morning. The only people there, apart from Mum, Julian and I, were Mum's friend Frances Reeves and her daughter Jacqueline, who had been a friend of mine since we were babies. The most notable thing about the event was that I didn't tell John about it – for two reasons: first, I didn't want a press circus, and second, I knew John wouldn't like the idea of a christening. To me it was important, I wasn't especially religious, but it was a family tradition. John was anti anything conventional.

A few days later I told John what I'd done. 'He doesn't need a bloody christening. I didn't want that,' he said. I said that since he hadn't been around it was my decision and, in any case, it wasn't a big deal. In the end John let it go: his hatred of confrontation was far greater than his concern about the christening. Neither of us ever carried on with a row once we'd said what we had to say – if we couldn't change something, what was the point in rowing about it?

I hadn't appointed any godparents, but when Brian heard about the christening he asked if he could be Julian's godfather. John and I liked the idea of our baby having a Jewish godfather and were happy to take him up on his offer.

Not long afterwards, I came out of the house with Julian and cameras flashed in my face. We had been caught – someone had finally found out who I was. The next day the papers were full of pictures of John's secret wife and baby. Shocked though I was at our 'outing', I was also relieved that the pretence was over. No more sneaking around, denials or hiding. I was only concerned about what it might do to John's career. But John wasn't worried: 'You should never have had to hide, Cyn,' he said. 'It's not going to make any difference to the Beatles.'

He was to be proved right, and Brian's fears about a married Beatle were shown to be unfounded. Although some fans were jealous, most accepted and even welcomed me, because they adored Julian. Certainly their passion for John wasn't in the least diminished by my existence.

There was no further obstacle to my moving in with John, so shortly before Christmas he took me to London for a weekend so that we could look for somewhere to live. While we were there we went to visit a photographer called Bob Freeman, who'd taken the photos for a couple of the boys' record covers. He and his wife, Sonny, lived in Emperor's Gate, just off Cromwell Road. They mentioned that the flat above theirs was empty and John and I leapt at the idea of living there. Bob put us in touch with the landlord and the deal was done. For fifteen pounds a week we had a three-bedroom flat. Early in January we moved in with our few bits of furniture, clothes and Julian's toys. My mother came with us to help us settle in. There was nothing but a cooker and a fridge already in the flat. It was dingy and needed furnishing and redecorating. While decorators transformed the rooms, I shopped until I dropped, pushing Julian's pram round Barker's of Kensington and Derry and Tom's, two smart department stores, choosing furniture, bed linen, pots and pans, china, cushions and curtains.

John and I were really excited to have our own home.

## John

We'd been married for over sixteen months and hadn't yet had a place we felt was ours, away from prying eyes and the demands of the world. It was bliss seeing so much more of him. Of course, he was out a lot of the time, but unless he was away on tour he could come home at night and we were together. For the first time I felt like a married woman.

However, the flat was not without drawbacks. It was at the top of three, each built over two floors. There were six flights of dark narrow stairs to climb to get to our front door – not easy with a baby, a pram and an armload of shopping. I had to leave the pram at the bottom, carry Julian up, park him somewhere safe, then go back down for the shopping.

I loved being in London. As a child I had been there on visits a couple of times a year with my parents, and as teenagers Phyl and I had come down to visit art galleries – we stayed at a bed-and-breakfast in Earls Court, and when we weren't in the galleries we hung around Soho and saw another side of London. Another time my brother Charles and his girlfriend Katie had taken me to my first West End play, *Salad Days*. For me London was the epicentre of culture and all that was attractive and exciting. I had always dreamt of living in the city and suddenly there I was. Every morning I'd be ready by nine thirty, with Julian in his pram, and we'd go out walking, shopping and looking at the sights for hours at a time. Even when John was away I was never bored; I couldn't wait to get out and explore.

Everything changed again a few weeks later when the

fans discovered our address. We woke one morning to find teenage girls, with the *de rigueur* beehive hairstyle and black eye-liner, camped on the pavement outside. After that they were always there, day and night. If any of the residents in the flats accidentally left the front door open they would grab their chance and slip in. We'd find them camped on the lino in the hallway, with sleeping-bags and Thermos flasks.

When I took Julian out in his pram fans would surround us, begging for a glimpse of him, and the pram would virtually disappear as they swarmed round it, clamouring, 'Oh, Cyn, isn't he sweet? Can I touch him? Can I cuddle him?' Or, 'Oh, Cyn where do you get your hair done? You're so lucky to have John. Where do you buy your clothes?' On and on it went. Most were well-meaning, and many were very young. The problem was that there were so many of them. I did my best to be polite, but it could be overwhelming and sometimes frightening. Outside, I'd push though them to the front door, trying to stay calm, but desperate to get in.

For John it was worse: they surrounded him whenever he came home, begging for autographs, locks of hair, a chance to touch him. He was always kind to fans. He could be intolerant of hangers-on, gold-diggers, money-men and sycophants, but he respected and cared for the fans. He believed that the group owed them a lot. After all, they were the ones who bought the records and paid to go to the concerts. So, however tired he was, he always stopped to sign autographs or say hello. Years later, when John was killed by a 'fan', who had waited for his autograph, the memory of his kindness to them stayed with me. I sometimes wondered whether, if he hadn't been so patient and generous, he might still be alive.

Inside the flat we were fine. With the fans down in the street below we felt safe and peaceful several floors up. But once the fans discovered us it was clear that we couldn't stay there for long and we wondered where we could go to escape the attention. It was around this time that the Beatles were due to go to the United States, a 'taster' visit, to find out whether the Americans would take to them. Several popular British artists had flopped spectacularly in the States, among them Cliff Richard and the Shadows.

John was both excited and apprehensive about the transatlantic tour. He'd swing from 'What if we don't make it, what if they think we're rubbish?' to 'They'll love us, who wouldn't?' He wanted me to come too: now that everyone knew he had a wife he didn't see why I should stay behind. The fans had accepted me so Brian was in no position to object. He agreed, and I was thrilled. I was the only girl to go with them – Paul's girlfriend Jane was working, Ringo's Maureen was still in Liverpool and was too young at seventeen, and George hadn't got a steady girlfriend. I persuaded Mum to come and stay in the flat to look after Julian – she was always great about dropping everything to help out – and raced out to buy some new clothes.

Brian had put a lot of work into launching the Beatles in the States. The boys' first four singles had flopped there, but Brian, as ever, was persistent. He had gone to the States and done the rounds of record companies, television stations and promoters. The breakthrough came at a meeting with Ed Sullivan, the legendary chat-show host. His researchers had heard how big the Beatles were in Britain and he agreed to have them on two of his shows, on 9 and 16 February. Brian wanted him to give them top billing and, after a bit of grumbling, he agreed.

Sidney Bernstein was an agent with GAC – the General Artists Corporation – one of the biggest agencies in America. He specialised in teenage music, had a strong interest in Britain, read the British papers and thought the Beatles had the potential to take off in the States. He tracked down Brian's home number in Liverpool, rang him and asked if he could arrange a concert for them at New York's Carnegie Hall, on 12 February. Brian happily agreed.

The boys' fifth single, 'I Want To Hold Your Hand', was at number one in Britain for two months before it was knocked off the top spot by the Dave Clark Five's 'Glad All Over'. Eager for a fresh story, the papers said the Beatles were finished. The boys were anxious about this, but Brian, cool and determined, told them to ignore it.

Just before they went to the States they did a tour of France. They were in their Paris hotel when they heard that 'I Want To Hold Your Hand' had got to number one in the US. After that, things took off. 'She Loves You' rocketed up the American charts, and so did the *Please Please Me* LP. The American press hot-footed to France to interview them. *Life* magazine did a six-page story and suddenly the Beatles were big news.

Their American record company, Capitol Records, a

subsidiary of EMI, put up five million 'The Beatles are Coming' posters all over the States and the record-company executives appeared in Beatles wigs. Fifty thousand people applied for the seven hundred tickets to *The Ed Sullivan Show* and Carnegie Hall was besieged by fans desperate for tickets.

We were blissfully unaware of all this when we drove to Heathrow airport to catch the plane. All we knew was that a couple of the records had done well in the States. We believed there was still a huge mountain to climb if the Beatles were really to make it there.

At Heathrow there was pandemonium. Thousands of fans had arrived from all over Britain and any ordinary passengers hoping to travel that day had to give up. Screaming, sobbing girls held up 'We Love You Beatles' banners and hordes of police, linking arms in long chains, held them back. We were ushered into a massive press conference, where journalists, spotting me at the side of the room, demanded a picture of John and me together. To my surprise John agreed. He was usually careful to keep Julian and me away from publicity, but this time, carried along by the momentum of the whole thing, he agreed.

Minutes later we were ushered to the plane. At the top of the steps the boys waved to the packed airport terraces as the screams crescendoed.

For me the whole experience was amazing. Until then I hadn't seen the fans on that scale, or understood how the boys felt when they faced that mass of people, all screaming for them. It was intoxicating, exciting, mind-blowing, yet somehow unreal. How could so many people be caught up in this vast wave of emotion over four young men?

With us on the plane were Brian and the boys' two road managers, Neil Aspinall and Mal Evans. They had taken on Mal a few months earlier when the job had become too much for Neil to manage alone. Mal was a big, friendly man who had worked as a bouncer at the Cavern. He'd always got on well with the boys, so when it became obvious that someone else was needed, they had approached him. He handled all the transporting of their equipment, the setting up and dismantling at concerts, while Neil looked after their personal needs for transport, food and so on. At least, that was the theory, but in practice they both did anything and everything that needed doing, from getting the Beatles safely on to a plane to finding sandwiches when they felt peckish.

Dozens of journalists and photographers were on the plane too, including a team from the *Liverpool Echo*. The atmosphere was like a party: champagne flowed, and the excitement and anticipation grew as we got closer to the States. The boys admitted to feeling sick: this was the big one – if they could make it in the States, they would have succeeded beyond anyone's wildest expectations. 'We can always turn round and go home again if no one likes us,' John joked, but any ideas about going home again were rapidly forgotten when we looked out of the windows of the plane as it taxied to a halt. 'Oh, my God, look at that!' John spoke for us all, as our jaws dropped at the sight of over ten thousand teenagers all singing, 'We love you Beatles, oh, yes, we do.' It was Beatlemania all over again, but bigger, louder and wilder. Only Brian remained calm and composed as he went over the last-minute arrangements.

When the door of the plane swung open the screaming and cheering were deafening. The boys hadn't yet set foot on US soil, but they had already won the hearts of the Americans. Inside the airport, we were hustled to a lounge where dozens of journalists and TV crews were waiting for the boys' biggest ever press conference. There was so much noise that John had to shout to get everyone to shut up. Then the questions flew thick and fast. As always the boys batted each question back with a witticism: 'What's your ambition?' 'To come to America.' 'Do you hope to take anything home with you?' 'The Rockefeller Center.' 'What do you think of Beethoven?' 'I love him, especially his poems,' said a totally straight-faced Ringo.

From the press conference we were escorted to our cars. In big plush Cadillacs we sailed into New York where we were to stay at the glamorous Plaza Hotel. As John and I had our first glimpse of the city the constant news bulletins on the car radio announced the Beatles' arrival. In the streets around the hotel madness had descended: thousands of singing, shrieking teenage girls in bobby-sox were waiting for us, waving Beatles wigs, banners, photos and T-shirts. Lines of police, red-faced with exertion, were holding back the crowds as our car inched towards the hotel entrance.

Inside the hotel, we were shown to a spectacular suite where our bedrooms all opened off a large central room.

When we looked out of the windows, the crowds reached back in every direction and it was clear that we weren't going to be doing much sightseeing. Hotel managers and security guards stood outside the door of our suite day and night. More security guards waited downstairs to catch stray fans trying to sneak through the barriers. We were virtual prisoners. The phone rang constantly with requests for appearances or interviews. Telegrams, including one from Elvis, were delivered to the door, with letters, cards and gifts from the fans. Even walking down the hotel corridors was hazardous – on one occasion a photographer who'd evaded the security net leapt out as John and I were passing. John threw his coat over both of us as we legged it back into the suite.

When we did go out we were whisked through side doors straight into waiting limousines, which then eased through the crowds. I was instructed to hold back until the boys were safely in the car, but the first time we tried it I was almost left behind. The police line holding back the fans broke and a sea of people cut me off from the boys and Brian. I had a terrifying couple of minutes until, with John screaming at them, the police realised who I was, lifted me bodily through the crowd, and almost threw me into the car. I got no sympathy from an irritated John: 'Don't be so bloody slow next time – they could have killed you.'

The boys appeared on *The Ed Sullivan Show* two days after we arrived – poor George felt so unwell that he had to be doped up with flu remedies – before a TV audience of almost 74 million, the largest in the history of television. The newspapers gave them huge coverage. America

loved everything about the Beatles, including their irreverent sense of humour. When George was asked who the
leading lady would be in their forthcoming film – they
were about to make *A Hard Day's Night* – he quipped,
'We're trying for the Queen.'

The next day we went to Washington by train for a concert
at the Coliseum. Once again, vast crowds turned out to greet
the boys. As with all their concerts on that tour, I watched
from the wings, taking in the screaming audience, the dazzling
lights and the music – which was almost impossible to hear
above the volume of the crowd. All four of the Beatles were
despondent about that – the whole point was for people to
hear their music, yet most of the time they couldn't because
they were screaming so loudly.

In Washington Brian persuaded the boys to accept an
invitation from the British ambassador to a dinner in their
honour. They were all, John in particular, wary of functions dominated by the wealthy and privileged. John felt
that most of the people at these dos were stuck-up and
hypocritical, pretending to like the Beatles when secretly
they despised them as working-class yobs. So, they were
reluctant to go, but were finally persuaded that it would
be undiplomatic to refuse.

Sadly, their fears were well founded: most of the wealthy,
elderly guests barely knew who they were and were
extremely patronising. 'And which one are you?' they
drawled, over their champagne glasses, then demanded
autographs and even locks of hair for their families. Pushy
officials tried to insist that the boys play along with this,
but for John it was too much: he and I left before the end,

and the others followed as soon as it was over. The ambassador and his wife, who had been very pleasant, apologised later, but none of the boys ever again accepted a similar invitation.

Back in New York they played at Carnegie Hall to an audience of six thousand. It was reported in the papers that even top film stars couldn't get tickets. That night Murray the K, a well-known local disc jockey who'd interviewed the boys at the airport and then latched on to our party, took us out. He was a fast-talking, wheeler-dealing steamroller of a man who dressed in skin-tight trousers and a cowboy hat. We went with him to the Peppermint Club, the famous New York nightclub where the Twist had originated. Protected from the fans, we danced and drank the night away. Ringo, always a keen dancer, held the floor for most of the evening, while the rest of us watched, laughed, and let our hair down.

From New York we flew down to Miami for the boys' second appearance on *The Ed Sullivan Show*. We were to stay at the Deauville Hotel, where our reception was similar to the one we had experienced in New York. Once again we were trapped, unable to go out and enjoy the sunshine, the beach or the sights. At one point I slipped down to the hotel lobby where there was a boutique. Although crowds of girls were camped outside the hotel, and plenty swarming about in the lobby, I felt safe because no one recognised me when I was away from John.

It was fun to browse though the racks of clothes after being stuck in our room for hours, but even more enjoyable to listen to the conversation of a couple of overweight middle-aged hotel guests. They wore multi-coloured Bermuda shorts, trowel-loads of makeup and diamanté-encrusted sunglasses, and they were laying into the Beatles with gusto. 'Aren't they just too awful? All that hair! I don't know what the kids see in them,' they grumbled. 'They look like something out of a zoo.' Tempting as it was to let them know that I was John's wife, I simply smiled and headed back to our room, giggling.

It hadn't been hard to slip out of our suite, but getting back in proved a problem. I'd come out without any identification and the security guard refused to believe that I wasn't just another fan. 'But I'm Mrs John Lennon, truly,' I pleaded. His face was a picture of resigned disbelief. 'Yeah, honey, they all try that one. Now, get lost.'

After several minutes' pleading I was close to tears, when a group of fans came over and started shouting at the guard: 'She's Cynthia – we've seen her picture. Can't you hear her English accent?' They even produced photos of me with John and eventually the guard was convinced. Relieved, I promised them all Beatles autographs and sprinted back up the stairs.

My next outing was even more eventful. Our cars were followed by press and fans every time we left the hotel, so the police came up with an original if bizarre plan: we would leave though the hotel's kitchen entrance and travel in an enclosed meat wagon. The plan was approved, amid howls of laughter, and the next day we sneaked through the hotel and into the waiting wagon. No sooner had the heavy metal doors slammed shut, leaving us in pitch darkness, than the wagon took off. I'd got in last, as usual, and hadn't had time to grab hold of anything secure. So, when the driver, who'd clearly seen too many high-speed chase films, took off, I was thrown against the doors and ended up with a lump on my head the size of an egg. I wasn't the only one – we were all bruised and battered as we lurched around, trying to hang on to the meat hooks hanging from the sides of the van. We shouted to the driver but he couldn't hear us. He got the message when we arrived, though; the boys bombarded him with expletives.

Despite my injury the break-out was worth it. We were taken to a gorgeous villa next to the sea, with its own pool, where George Martin and his wife, Judy, were waiting for us. The villa apparently belonged to an anonymous celebrity who had loaned it to us as a gift. This was my first meeting with George and Judy and I liked them instantly. George was tall, debonair, polite and charming. A classically trained musician, he had been taught the oboe at the Guildhall School of Music by Jane Asher's mother. Later he'd moved into record-producing, concentrating on classical and comedy records, until he discovered the Beatles. Judy was lovely, very old school – like a headgirl with a heart of gold. She had been George's secretary and was now his second wife and they patently adored each other.

Free of the fans and the restrictions of the hotel, we partied all day, swimming in the pool and enjoying an enormous barbecue prepared by the butler who came with the house. He looked like a member of the Mafia and scowled at us for most of the day, but his steaks were perfect.

The following day we met Muhammad Ali, then the world heavyweight boxing champion and a huge celebrity, for a photo call. I don't know whose idea it was to put him and the Beatles together, but the media turned out *en masse* to see him sparring and joking with the Beatles.

As our stay in the States drew to a close, we longed to be back home. Although the boys were delighted that they'd gone down so well in the US, two weeks of being besieged in hotels, prodded, pushed and treated like zoo exhibits was enough. Besides, they had missed their girlfriends, and home delights like bacon butties and proper tea. None of them had found their enormous celebrity truly enjoyable, and this had become even more apparent to them in America. They hated feeling trapped, unable to wander down the street or pop into a shop. They hated being constantly accosted by strangers. And they hated being shoved, squeezed and even carried by police and security men every time they had to run the gamut of the fans. At least in Britain they could retreat to the comfort and safety of their own homes.

Back in London we barely had time to draw breath before the boys started making their first film, *A Hard Day's Night*. The title, used for a hit single as well, came from a phrase Ringo had used to describe the non-stop exhaustion of touring and performing. John was excited about making a film: it was another creative medium for him to explore and he was fascinated by the whole process. But he hated having to get up at dawn to be driven to the studio. He wasn't home again until seven in the evening, by which time he was exhausted. The whole film was shot in the space of a few weeks. It went on to become the most successful pop spin-off film ever.

# 11

IT WAS ONLY JUST OVER a year since the Beatles had first hit the charts but our lives had changed beyond recognition. Launches, premières, concerts and receptions followed in such rapid succession that John and I were barely able to keep up, let alone find time to be together.

I was truly happy for him. I had always been proud of his talent and believed in him, and to see him publicly recognised was wonderful. His belief in himself grew daily, and after the self-doubt and fear that his childhood had instilled in him, it was wonderful to see him calmer and so much more confident. But the endless demands on him meant that he had little time left for Julian and me. John felt as sad and frustrated as I did that we were apart so much and that, so soon after becoming a husband and father, he simply couldn't be there for us.

I became self-sufficient, managing whatever problems arose and learning to be content with my own company. I comforted myself with the thought that all the craziness would pass and eventually John, Julian and I would have time together as a family. I missed John when he was away and never doubted that he loved us or that he wanted what I wanted: a settled family life. It was simply that for us the timing was askew. And, hard as it sometimes was, there was little we could do but go along with it.

Three weeks after the Beatles began filming *A Hard Day's Night*, John's book *In His Own Write* was published. It was a collection of witticisms, anecdotes, stories and drawings that John had put together over several months. An editor at Jonathan Cape had read some rhymes he had written and had asked him if he could come up with enough for a small book, and John was thrilled. For weeks he was jotting and drawing, totally absorbed in it. He had always been a fan of *The Goon Show*, a satirical radio programme then at the height of its popularity. He and Stuart Sutcliffe used to fool around for hours, imitating the Goons, Peter Sellers (John's favourite), Spike Milligan, Harry Secombe and Michael Bentine. He loved their offbeat humour, and it's easy to see their influence in John's own writings. He loved playing with words, turning them inside out, experimenting with their flexibility and inventing malapropisms. He read everything he could lay his hands on, from newspapers and magazines to books on a huge range of subjects, including music, design, mystery novels, biographies and history. Among his favourites were Lewis Carroll's *Alice in Wonderland* and *Alice Through the Looking Glass*, with their weird and wonderful imagery. In writing his own book, John's rich imagination was unleashed and it gave him immense pleasure to move beyond the conventional 'love story' limitations of the songs he and Paul had written.

John didn't expect much of the book: he was just glad to see it all in print. Paul wrote a wacky little introduction and John hoped it would amuse his friends. But if he thought it would go unnoticed he couldn't have been more wrong: it got rave reviews everywhere, from pop magazines to heavyweight papers. The *Melody Maker* said, 'John Lennon is a remarkably gifted writer . . . often hilarious, clever and funny'; and the *Times Literary Supplement* wrote, 'It is worth the attention of anyone who fears for the impoverishment of the English language and the British imagination'; while the *Sunday Telegraph* called it 'irresistible'.

The book was an immediate bestseller. Bookshops that had ordered only a few copies demanded more and it was reprinted twice in the week it came out. John was pleased if bemused by the attention it got, and even more so when we heard that a Foyle's Literary Luncheon had been arranged in his honour at London's Dorchester Hotel. A Foyle's luncheon was a great accolade for any author, and for John's the demand for tickets was unparalleled.

John

Unfortunately John and I had no idea how big an event a Foyle's literary luncheon was. We thought it would be just a nice meal, a bit of chat and a few compliments about the book. We weren't in the least worried about it, so on the night before we went out to dinner with friends and ended up in one of our favourite nightclubs.

The next morning, after only a couple of hours' sleep, we woke with appalling hangovers and realised the chauffeur would soon be arriving to take us to the luncheon. We did our best to make ourselves look presentable, but the bloodshot eyes and shaky hands were a bit of a giveaway. We told ourselves that the event would soon be over and we could go home to collapse.

What neither of us had realised was that the media would be there in force and that John was expected to make a speech. Doyens of the literary establishment rubbed shoulders with upmarket Lennon fans and everyone was waiting with bated breath to hear the words of the 'intelligent' Beatle.

As we were ushered through the lobby of the Dorchester, hordes of press and TV crews following us, I knew John wanted to turn and run, but we had to keep smiling. We couldn't even see what was going on properly because neither of us was wearing our glasses.

When we walked into the enormous dining room hundreds of people stood up and applauded. We fumbled our way to our places and found we were at opposite ends of the top table, denied even the reassurance of squeezing hands. I was sitting between the Earl of Arran and pop singer Marty Wilde, who was almost as nervous as I was.

I was terrified, until the earl put me at ease with a string of witty stories and friendly chat. I even began to enjoy myself – until we reached the last course and dozens of TV and press cameras were pointed in our direction. 'What's going on? I whispered to the earl.

'I believe your husband is about to give a speech,' he whispered back, and politely averted his eyes from the horror written on my face.

I looked at John and my heart went out to him. He was ashen and totally unprepared. Never lost for words in private, a public speech was beyond him – let alone to a crowd of literary top dogs, with a hangover.

As John was introduced silence fell. The weight of expectation was enormous. John, more terrified than I'd ever seen him, got to his feet. He managed eight words, 'Thank you very much, it's been a pleasure', then promptly sat down again. There was a stunned silence, followed by a few muted boos and a spatter of applause. The audience was disappointed, annoyed and indignant. Both John and I wished we were on another planet. John tried to make up for it by signing endless copies of the book afterwards.

John's Foyle's 'speech' went down in history as a typical Lennon gesture, a snub to the establishment from a pop-star rebel, when it was anything but. He had panicked.

Not that it affected his sales. The book went on selling and he soon began work on another, *A Spaniard in the Works*, which was published a year later and was almost as successful.

The day after *In His Own Write* went on sale the Beatles' sixth single, 'Can't Buy Me Love', was released. It had advance sales in Britain and the States of three million copies – a world record – and went straight to number one in both countries. Soon afterwards the Beatles held the top six places in the US singles charts and a full tour of the States was arranged for August.

Meanwhile George, who'd just turned twenty-one, had met a young model called Patti Boyd and fallen in love. Patti had been given a part in *A Hard Day's Night*, playing a schoolgirl, because she had appeared in a successful crisps advertisement – she was known as the Smith's Crisps Girl. She was blonde, beautiful and a sophisticated Londoner, like Jane Asher. But, like the rest of us Beatles girls, she was friendly, too, and easy to get on with. The two Liverpool girls, Maureen and I, and the two London girls, Jane and Patti, got on well from the beginning. We were all living through the same thing and it was wonderful to have friends to share it with. From the start it was obvious that Patti and George were serious, and we were all pleased for them. The other three Beatles were all in happy relationships and until now George had been on his own.

A few weeks after they met, Patti and George joined John and me for a weekend in Ireland. It was a chance for us to get to know Patti better, and for us all to sneak off and have some fun. We were determined not to let the press know so the boys wore false moustaches, scarves and hats for the journey out and Patti and I walked some distance behind them. We left from Manchester in a private six-seater plane and, despite some odd looks at the boys' ludicrous get-up, we got away without being recognised.

We were booked into a hotel called Dromoland Castle and it seemed perfect – miles from anywhere and utterly luxurious. President Kennedy had just checked out of our suite and the staff, used to high-level guests, were charming and discreet. Our first day was delightfully peaceful. We walked in the castle's extensive grounds and luxuriated in the sense of freedom. By the evening we were more relaxed than we'd been for months.

The next morning we woke to the sound of birdsong – and the chatter of numerous voices under our windows. We peered out through the curtains, to be confronted by a sea of journalists and photographers. We'd been rumbled. The press had turned out in force in the hope of getting the first pictures of George and his new girlfriend.

We were bitterly disappointed – and trapped. How on earth were we going to get away from the place? The hotel manager came up with a simple and hilarious way to foil the masses outside. While John and George would leave by the front entrance and head for the airport, Patti and I would be smuggled out of the back, dressed as chambermaids.

Stage one went fine. Patti and I, giggling, put on ill-fitting black dresses, frilly aprons and white caps. We were taken down to the staff entrance, where we climbed inside one of the hotel's large wicker laundry baskets. The idea was that a couple of staff would carry us, inside the basket, to the waiting laundry van outside. Once we were safely in the van we would be let out of the basket and the driver would take us to the airport.

Perfect, except that he shot off before anyone could let

us out of the basket. We travelled to the airport trapped inside it for the next nightmare hour. It was pitch dark and we were hurled against the wicker as the driver sped round bends and down narrow roads, the basket sliding from one side of the van to the other. To add to our misery, we could barely breathe and were convinced we'd suffocate. We shouted at the top of our lungs but it was soon clear that the driver couldn't hear us. There was nothing we could do but hang on and pray.

When we reached the airport, we were stiff, bruised, tearful and hoarse. But at least we had the satisfaction of knowing that our ruse had worked: the press had no idea we'd got away and were still camped outside the hotel. John and George thought the whole thing was a riot and teased us all the way home. Patti and I swore that next time they could try the clever tricks and we'd go the easy way – out of the front door. However, the four of us had got on so well that a few weeks later we went on holiday again, to Tahiti this time. To escape the press we decided to charter a yacht and flew out separately. John and I went via Hawaii, where the fans and press tracked us down and we were trapped in our hotel.

When we met up with George and Patti in Tahiti our 'yacht' turned out to be a rather elderly fishing-boat and it rained torrentially, monsoon style, for the first couple of days. I was seasick and wished we'd never set out. But once the storm had passed we had a wonderful time. Our Tahitian crew were happy and helpful and, much to our delight, had no idea who the boys were. The cook specialised in potatoes cooked a different way each night, which meant John and I went home considerably fatter. We lay on deck, swam, talked and ate and, best of all, the press never found us.

Patti and I were becoming close friends. I admired her gorgeous figure and perfect fashion sense, and I think she enjoyed the company of someone who'd been with the Beatles from the beginning and knew the ropes. John and George had an easy, comfortable relationship and they headed for the beach while Patti and I went shopping. Holidays were precious: they were the only chance John and I had to be together all the time, to slow down and enjoy each other. The rest of our time was now so frantic and full that often we just passed on the way in or out of the flat. Even when we went to functions we were hardly ever seated beside each other, and so many people wanted John's attention. So, lazing on a beach, walking, talking, making love, sharing a cuddle and splashing together in the sea were idyllic. We felt no pressure to make those holidays perfect, they just were, because we were together and away from it all. At those times John was happy. For a while he forgot about the security guards, the press, the fans and the endless demands. He laughed, fooled around and enjoyed himself.

We headed back to the whirlwind that was life at home. The royal première of *A Hard Day's Night* was to take place on 6 July, in the presence of Princess Margaret, and four days later the Beatles were to attend the Liverpool première, followed by a civic reception for the city's newest heroes. John had enjoyed making the film, and he was delighted with the finished result. We all hoped it would do well and show the public that there was yet another dimension to the Beatles.

As the day of the London première drew near, excitement ran high and I was determined to look gorgeous. I slogged round the smart West End boutiques and Knightsbridge stores for days, looking for the right dress. I finally found it, a full-length tunic-style sleeveless dress in black and beige silk in Fenwicks, one of London's classiest department stores. To go with it I bought a Mary Quant black chiffon coat, bordered with black feathers. The only problem was that the dress needed shortening. Fenwicks promised to have it done and delivered to me in time for the première.

That morning when it hadn't arrived I panicked. In the end my mother went to collect it and got back with about half an hour to spare. John always took an interest in my appearance, and he loved it. I asked him whether I should wear my hair up or down. 'Let's have a change – give us a bit of Brigitte and wear it up this time. And don't wear your specs. You look great without them. Don't worry, I'll guide you.' He was always proud of the finished result when I was dressed for a big occasion and would say, 'God, Cyn, you look fantastic.' I set off for the première feeling like a princess, with a black velvet bow in my hair to match the outfit.

John, Paul, George, Ringo, Brian and I were driven to Leicester Square, the venue for all major London film premières then and now, by chauffeur Bill Corbett. As we passed through streets lined with cheering fans, held back by rows of police, John stared out of the window and asked what was going on. 'Is it a cup final or something?' he asked. When Brian told him that they were waiting to see the Beatles, he was genuinely surprised. Despite all the screaming crowds he'd been confronted with over the past nine months, he still wasn't prepared for them.

As the flashbulbs popped and we stood on the red carpet waving and smiling I thought back to when I had been living at Mimi's, hidden away and managing on a pittance. It was hard to believe the difference a year had made. At the end of the film we were introduced to Princess Margaret. When it came to meeting royalty in the flesh John was as much in awe as the rest of us. He was so pleased and proud that the princess had come to see the film that his anti-establishment views flew out of the window and he stood red-faced, as she spoke to him. She was obviously fascinated by the boys, but not their entourage. Her questions to John were clipped and super-ficial: 'How are you coping with all the adulation?' John tried to introduce me: 'Ma'am, this is my wife, Cynthia.' A brief glance in my direction, 'Oh, how nice,' and she was gone. Still, I was bowled over. I'd met royalty, which was more than I'd ever imagined in my days back in Hoylake.

Brian had advised me and all the other Beatles women to stay away from the Liverpool première and reception because of the problems involved in getting us all through the crowds that were expected. It was the first time any of the boys had been back to their home city since Beatlemania had begun and they were bowled over by the welcome they received. All four had invited numerous relatives to the première. John's sisters, Jacqui and Julia, now sixteen and eighteen, were there with Aunt Harrie, all wearing new outfits provided by John, who had told them to go and choose whatever they wanted. He had them collected in a chauffeur-driven limousine and Julia later told me it was one of the most wonderful days of their lives. Mimi didn't go because that sort of thing wasn't her idea of fun.

Once again the streets were lined with cheering crowds. When they reached the town hall for the civic reception, the Mayor, in full regalia, was waiting to greet them. The boys were taken out on to the balcony at the front where they waved to crowds estimated at two hundred thousand, all screaming at the tops of their voices. After a sumptuous lunch they moved on to the Odeon, where the première was held. Before the film began the boys went on to the stage, to shouts and cheers from the audience, and John yelled, 'Where's me family?' Julia, Jacqui and Harrie waved and shouted and the whole audience laughed.

That night the boys were whisked back to London. It had been a triumphant return to Liverpool and they'd loved it, but it had also been overwhelming. John found it strange that although he'd grown up in Liverpool, he could no longer walk down its streets, drop in on old friends or pop into one of the clubs to hear the latest sounds. He could never again be ordinary in Liverpool, and he hated that. From now on he could return only as a big celebrity, one of the most famous men in the world. Publicly he said that he didn't miss Liverpool, he wanted to explore the rest of the world, and had no desire to go back, which seemed perverse, but I suspect it was the only way he could deal with a painful truth.

*A Hard Day's Night* had been put together around the album of the same name, the first comprised only of Lennon and McCartney original tracks. Several of John's songs on that album have themes of isolation and cries for help. Tracks like 'Tell Me Why', 'Any Time At All' and 'I'll Cry Instead' reflect the frustration he felt at that time. He was the idol of millions, but the freedom and fun of the early days had gone. Success had come at a price, and he often felt caught up in something he couldn't stop.

I missed the early days too. I felt isolated and stuck in a flat that had a permanent posse of noisy fans camped outside. Even worse, there was a student hostel across the road with a balcony that overlooked us and the fans had discovered it. Day and night they waved and called to us from it, so I had to keep the curtains permanently drawn.

An awful moment came when a large building next to us – a terminal for coaches to Heathrow airport – caught fire. I stood watching it, with Julian in my arms, and the huge flames lit up the night sky. The wind blew sparks in our direction and I thought we would have to flee. Bob, our photographer friend in the flat downstairs, came up to check that we were all right. He and his wife reassured me that the fire wouldn't reach us, but the incident added to my sense of vulnerability in the flat.

Then I received a series of obscene phone calls. It was the last straw. Now I was under siege from every direction. I couldn't go out, look out of my window or even answer the phone without having to deal with fans, some of whom were clearly crazy. It was just as bad for John. Every time he came home he had to force his way through the girls camped on our doorstep. His chauffeur, Bill Corbett, was a hefty man who strong-armed them out of the way but it was an ordeal: obsessive, desperate fans fought, bit and scratched their way to the front to grab any piece of John's clothing they could get hold of. He gave up wearing scarves because so many were ripped off him. The fans' other tactic was to stuff our keyhole with chewing gum to stop us getting in. After this had happened half a dozen times John had had enough: it was time to look for somewhere more secure to live.

Money was now no object, as we discovered when we talked to our accountant about buying a house. We were seriously rich and could go out to choose the house of our dreams. The Beatles' accountant lived in Weybridge, Surrey, and we had gone down there to see him. We liked the area and it was close to London, but so much more peaceful and pretty. We started looking at houses there, and eventually found the perfect place. Kenwood was a sixteen-room mock-Tudor house in the exclusive St George's Hill estate where Cliff Richard and Tom Jones already lived. It wasn't like any estate we'd ever seen – no back-to-back houses, just acres of woods, fields and discreet homes set at such wide intervals that you wouldn't know you had any neighbours. The house was elaborately decorated in a hideous, over-fussy style when we first saw it, but it was on the top of a hill and had a wonderful sense of space and openness, which was just what we needed after the cramped flat. It cost us nineteen thousand pounds, a huge sum in those days. We bought it on 15 July and moved into its small attic flat while the rest was refurbished. We'd been advised by all our friends to employ an interior designer and Brian recommended an acquaintance of his. Feeling totally out of our depth we gave him *carte blanche* to do as he chose. For the designer it was heaven and he set about spending our money with glee, coming up with what we felt were increasingly bizarre ideas. Meanwhile, a swimming-pool was being dug outside – John was determined to do the whole pop-star thing. We spent months walking past diggers and the army of builders and decorators inside on the way upstairs to our two rooms in the attic. All we dared do was peep into the rooms from time to time to ask how it was going, usually when yet another tea break was under way.

We liked some of the lavish designs, but they weren't what we'd have chosen, left to ourselves. The results in one room almost gave John apoplexy: when he came home to find the sunroom, a bright room with lots of windows, swathed in dark green material, like a bizarre wedding marquee, he was furious. He tore it all down, then and there.

We were still living in the attic when the Beatles set off for their first full tour of the US, in August 1964. While John was away I tried to keep myself busy to avoid the loneliness and isolation my new life forced on me. I invited my old friend Phyl to stay and we had a lovely couple of weeks together, shopping, going to the hairdresser and catching up with each other's lives. Phyl thought it was hilarious that we were living in the tiny attic at the top of such a huge house.

We watched the Beatles' progress on television. Their tour was an enormous success, the crowds even bigger than they had been previously. We went to see *A Hard Day's Night* at the local cinema. Phyl hadn't yet been and at least it gave me a chance to see John, if only on celluloid, because I missed him badly.

In the end we were in the attic for nine months. As we'd been used to small living spaces, we were actually quite happy there. It felt far more strange to move into the whole house and suddenly have lots of large rooms to use. Once our tyrannical designer had gone, we changed most of the carpets and curtains and swapped the hard red-leather sofas he'd installed for soft green velvet ones, which were a lot more comfortable. Ringo took the red ones for his London flat. We only used the main reception and dining rooms when we entertained. The rest of the time they remained pristine and untouched.

Neither of us was at ease with a grand scale of living and we ended up making a den in the small back room. It had a comfy old sofa, a TV, a table at which we ate and Julian's toys. When John wasn't working he lay on the sofa, apparently watching the TV – which he liked to keep permanently switched on – but often a million miles away, lost in a daydream. I'd talk to him and he wouldn't hear me. This was nothing new: he had always been able to 'tune out' of his surroundings and the busier his life became the more often he was 'present but absent'. I didn't mind: it was his way of coping with the stresses of his life, and these 'absent' moments were a vital part of the creative process. After an hour or two he'd often get up, go to the piano and start writing a song.

Now that we were living in such grandeur, we had staff to help us. Dorothy Jarlett, known as Dot, had ironed for the former tenants and became our housekeeper. She was a warm, competent woman in her forties who became indispensable to us. She was loyal and reliable, and was a good friend to me when John was away for weeks on end.

Too often I've seen it reported that John wouldn't allow me to have a nanny for Julian, insisting that I bring him up with no help, as a reaction to his own mother-deprived childhood. In fact, he left it to me and I chose not to have a nanny. Julian was a source of delight: I loved being with him and watching him learn new things every day. I would gladly have had another child, but although we never took precautions, it didn't happen, so I contented myself with looking after Julian. Dot was always on hand to babysit and my mother was often with us too. We also had an elderly, grumpy but extremely good gardener, who largely ignored us, and a chauffeur. The first couple of chauffeurs were a disaster – we discovered that one lived in the car he drove for us – but Les Anthony, a jovial Welsh ex-guardsman whom we knew for some reason as Anthony was, like Dot, loyal and reliable. Neither John nor I could drive when we first moved in, so Anthony was far more than just a status symbol: he was our lifeline.

Once the house was finished John insisted we get a cat – he'd grown up with them, of course, and loved them.

The first was named Mimi, after his cat-loving aunt. Two more soon followed, and eventually we had about ten. We also got a dog, a mongrel that Julian named Bernard after Dot's husband. Sadly, we only had him for a year before he died.

Not long after we moved to Kenwood, George and Patti bought a house in Esher, a few minutes down the road, followed by Ringo and Maureen, who took Sunny Heights, just five minutes away on the same estate as us. Ringo and Maureen had continued to see each other, even though she was in Liverpool and he was in London. We all knew that Ringo had had the odd fling with other girls, but his heart belonged to Maureen. In the summer of 1964 Maureen had gone on holiday with Ringo, Paul and Jane without telling her parents. How she thought she could holiday unnoticed with a Beatle I can't imagine. Of course, within a couple of days her picture was splashed all over the newspapers. When reporters knocked on her parents' door, her dad said, generously, that he would have let her go if he'd known about it.

In early December Ringo collapsed and became very ill with tonsillitis and Maureen rushed from Liverpool to be with him. The Beatles were about to embark on a tour of Holland, the Far East and Australia, so this caused a minor crisis. In the end Brian decided that they'd set off without Ringo, who got up the moment he could and caught up with the others a few days later in Australia.

In January Maureen found she was pregnant and their wedding was hastily arranged for 11 February 1966 at

London's Caxton Hall. It was a carbon copy of the situation in which John and I had found ourselves, except that this time the world's press was waiting to capture all the details. Once again Brian did all the arranging. Maureen's pregnancy was kept secret and to avoid publicity the registrar agreed to perform the ceremony at eight a.m. Paul and Jane were on holiday in Tunisia, but George and Patti, John and I went, with Maureen's mother and Ringo's mother and step-father. Once again Brian was best man and, after a touching ceremony, we all went back to Brian's house in Belgravia for a celebration breakfast. The newlyweds went on honeymoon to Hove, near Brighton, for three days, then Ringo had to get back to work.

Maureen had just turned eighteen and, to the press, appeared shy and unsophisticated. Like me, she preferred to stay in the background and give few interviews. In a brief meeting with journalists during their honeymoon she held Ringo's hand tightly and said little. One article said that one of the world's best-known bridegrooms had married one of the least-known brides. But that was the way Ringo and Maureen wanted it. Like John, Ringo believed his family should be kept out of the limelight. He wanted to protect and shelter them and that was the best way he knew of doing it.

Far from being a shy little thing, Maureen was talkative, full of laughter and great fun: we all liked her enormously and thought she was good for Ringo. John and I were delighted when they came to live close by. Initially they'd lived in Ringo's one-bedroom flat in Montagu Square, close to London's Hyde Park but, like us, they needed more space and greater privacy. All of the Beatles' women got on with each other, but Maureen, who was one of the most down-to-earth, honest people I ever knew, became my closest friend. After their son Zak was born in September, seven months after the wedding, she and I used to go up to Knightsbridge to shop. Anthony would drop us off and we'd do the rounds of Harrods, Harvey Nichols and the designer shops in between, then stop for lunch in a smart little bistro. We'd buy cute little outfits for our sons and we were always on the lookout for something different or special for the men. We loved to surprise them with a psychedelic shirt, a piece of ethnic jewellery, or I would buy John a new plectrum for his guitar. John always loved prezzies, as he called them. No matter how small they were, he'd be delighted and I loved looking for things to surprise him.

Much as Maureen and I enjoyed our outings, she always made sure she was at home for Ringo when he came in. Such was her devotion to him that she would stay up some-times until four in the morning to greet him with a home-cooked meal. She wanted him to feel loved and cared for and, like me, she had been brought up in a family where women did the caring and nurturing while men provided.

We often went over to their house and hung out with them, it was always party time at the Starkeys'. Ringo was gregarious and fun-loving, a clown and a joker with an infectious laugh. Together, he and Maureen made an irresistible double act, both extrovert and uninhibited. Ringo had installed a replica pub in their front room, which he called the Flying Cow. It had a counter and till, tankards, mirrored walls and even a pool table. He'd nip behind the bar to serve us all drinks, while Maureen supplied us with endless plates of food. It was a cosy, comfortable house with what felt like the ultimate luxury at the time: a TV – usually switched on – in every room.

They had large grounds, in which Ringo had built in a go-kart track. He and John would race the go-karts or play pool while Maureen and I chatted over a cup of tea or took Zak and Julian for a walk. Ringo's other passion was making his own short films. He had lots of equipment and loved to experiment, so after the nanny had taken over Zak and Julian we'd watch his latest movie. One was a fifteen-minute study of Maureen's face. Innovative, perhaps, but not the most riveting entertainment.

We spent time at George and Patti's too. George had bought a large bungalow in Esher, about twenty minutes from us, and Patti moved in with him not long after they met. The outside of George's house was painted in a series of vivid colours that made it look startlingly psyche-delic as you approached it. Inside, it was in complete contrast to Ringo's. George's home was stylish and tasteful, straight out of a designer magazine and without a TV in sight.

The Beatles had become incredibly close over the years and much of the time we socialised together, perhaps because no one else was in quite the same situation. It was like having a second family. Even Paul, who'd bought a townhouse in London's St John's Wood, came down regularly with Jane to see us. We would visit each other far more often than anyone or anywhere else. We almost always went on holiday with one or all of the other Beatles and we usually spent Christmas Eve together at one of our homes and swapped presents. Remarkably, perhaps, there was seldom any tension between us. Minor disagreements were quickly sorted out.

John's relationship with each of the other Beatles was different. He was at his most relaxed with Ringo, who often had him in stitches with his jokes. He treated George with the mix of fondness and disdain he might have shown a younger brother. He was closest of all to Paul, but their relationship was more complex. They spent a great deal of time composing together, one sitting at the piano, the other jotting down lyrics or strumming a guitar, both calling for vast amounts of sarnies and tea, totally immersed in what they were doing. The time they spent working together was intense, and when it was over they needed to let off steam and relax apart. John spent less of his free time with Paul than with either of the others.

The Beatles' families' closeness never really extended to Brian. Perhaps because he was 'the boss', he chose to remain at a distance. He rarely joined us socially, although he was often invited. His role was to be a friend and mentor to the Beatles, but not a pal.

# 12

BRIAN EPSTEIN HAD ALWAYS SAID that the Beatles would be bigger than Elvis, and at first everyone laughed. Elvis was securely on his throne as the king of rock and roll and no one had heard of the Beatles. But by 1965 his prediction had come true: the Beatles were the biggest pop act in the world, eclipsing Elvis and every other popular musician. Their faces were known in every country on the globe – Beatlemania was breaking out in places as far afield as Australia, the Philippines, Japan and Scandinavia.

## John

Back in Britain we were at the epicentre of what became known as the Swinging Sixties. It was an era of unfettered expression, joyousness and extravagance. Suddenly people didn't have to wait until Christmas to light the candles: they lit them every day, and frequently burned them at both ends. To us it felt as though the whole country was ready to party and we were at the top of most of the 'in' people's invitation lists. Throughout 1965, and for the next three years, we went to hip restaurants and nightclubs, we were invited to celebrity parties, we shopped in designer boutiques and our pictures were in every newspaper and magazine.

My life then was full of contrasts. I might be at a première one night, with photographers snapping and crowds screaming, and the next morning I'd be taking Julian to school like any other mum. Or I would pop down to my local shops for some groceries – I was hardly ever recognised on my own – then meet John in the evening after his recording session. I'd go along with the other Beatles women to hear the end of the session, then spend the rest of the night in one of the hip London clubs, dancing and chatting to a host of famous faces, many of whom we counted as friends.

Despite our celebrity status we were still naïve. We had no sophistication or sense of style. Back in Liverpool our idea of a classy night out had been a Scotch and Coca-cola or a Babycham in the pub – even those were saved for Christmas or birthdays. Mostly we had beer or Coke and that was it. We hardly ever dined out and on the rare occasions when we had enough money to go to a restaurant, it was a curry house or a café for chicken and chips.

When John and I had first arrived in London we'd asked Brian to show us the ropes. He was already installed in a smart London flat in upmarket Belgravia and he'd wasted no time in acquainting himself with the best people and places. Since we hadn't a clue where to start we were grateful to be swept along in his wake. To introduce us to London nightlife he had first arranged to take us out to dinner. We went to La Poule au Pot in Ebury Street, a restaurant run by a couple of French gays who were, to our eyes anyway, utterly flamboyant and decadent. The atmosphere in the restaurant was intimate – brocade drapes in deep red, green and gold hung on the walls and the candles had melted into wax sculptures over the bottles that did duty as candlesticks. We loved it, even though we couldn't understand a word of the French menu.

Brian was in his element. Because homosexuality was still illegal it was only in places like this, where it was acceptable to be openly gay, that he could let down his guard and feel less of an outcast. Here he was a member of an in-club, as well as a sophisticated older man, showing us wet-behind-the-ears kids a bit of fine living. That night we were happy to let him take the decisions: he ordered French onion soup, coq au vin and poire belle Hélène, with Pouilly Fuissé to drink. The meal was sublime, accompanied by outrageous banter from our French hosts. John and I spent most of the evening trying to pronounce Pouilly Fuissé, much to the amusement of everyone else.

After that Brian took us regularly to London's hotspots,

and by the time we moved to Weybridge we had developed a taste for London's nightlife. But although we went to the best nightclubs, restaurants and parties, we never got to like sophisticated food. John still preferred a bacon butty or a steak sandwich to anything the finest chefs could offer him.

One of our favourite haunts was the Ad Lib Club. A discreet door led to a hallway with a lift, which took us to the club on the top floor. Inside, the design was minimalist, with cushioned benches and long, low tables set round a central dance-floor. The club would be buzzing all night and was the favourite haunt of most of the pop stars of the day. We'd talk, drink and dance the night away with singing twins Paul and Barry Ryan, our old Liverpool mates Freddie and the Dreamers and Gerry and the Pacemakers, plus the Who, the Rolling Stones, the Animals and Georgie Fame. Who drummer Keith Moon and I had wonderful philosophical conversations whenever we met. Despite his mad rocker image I found him sensitive and serious-minded and I was sad when, a decade later, he died so young.

After a long night of partying, we'd head home at dawn, usually with George and Patti or Ringo and Maureen. We'd get the chauffeur to stop off at a roadside café for a meat pie and be at home in time for me to get Julian up and off to school – after which I'd collapse into bed for a few hours.

Our other favourite haunts included the Bag o'Nails, the Scotch of St James's and Samantha's. Of these, the Bag o' Nails was our favourite. In contrast to the Ad Lib it was decorated in dark colours, with shadowy candlelit corners and deep sofas. But the best thing about it was the steak and onion baguettes, served all night. The boys would be ravenous after recording all day and would munch their way through great piles of them.

Brian's Belgravia flat was elegant and classically stylish. He often gave parties, always perfectly planned. A butler would move discreetly through the room with a tray of champagne while guests, including David Jacobs (host of top television programme *Juke Box Jury* on which John had appeared), music journalist Ray Connolly and his wife Plum, washing-machine tycoon John Bloom and Jimi Hendrix all mingled.

Along with the delights of fine wines and good food, we were introduced to cannabis. Joints were often produced at parties or even in clubs, and passed among those who fancied a 'toke'. The Beatles had first tried cannabis when Bob Dylan introduced them to it, on their original visit to America. Bob had listened to the lyrics of 'I Want To Hold Your Hand' and, thinking one of the lines was 'I get high', had assumed that this was a reference to drug-taking and that the boys were seasoned users. When he mentioned this to John, he was told that the line was in fact 'I can't hide' and that none of the Beatles had tried drugs of any kind. That was when Bob had offered to initiate them. They'd all got stoned, fallen around the hotel room with the giggles and thought it was great. After that they smoked dope fairly regularly, using it recreationally in the same way that they'd have a drink to relax and unwind. I occasionally gave it a try, but I never took to it. I found it impossible to enjoy anything that made me feel out of control.

It was both daunting and flattering to find ourselves on the guest lists of the stars and be invited for dinner or drinks by celebrities we'd been watching on TV or reading about in the papers only a couple of years earlier. Singer Alma Cogan was one. She had been one of the most successful singing stars of the fifties and in the early sixties was still hugely popular. The Beatles met her on her television show and she loved them, especially John. After that we were frequently invited to parties in her opulent apartment on Kensington High Street.

John and I had thought of her as out of date and unhip. We remembered her in the old-fashioned cinched-in waists and wide skirts of the fifties. But in the flesh she was beautiful, intelligent and funny, oozing sex appeal and charm. Walking into her home for the first time was like walking into another world. It was decorated like a swish nightclub with dark, richly coloured silken fabrics and brocades everywhere. Every surface was covered with ethnic sculptures, ornaments and dozens of photographs in elaborate silver, gold and jewelled frames. In the living room there were deep sofas, floor cushions and lamps draped with chiffon and silk. Candles were everywhere and a card table had been set up between two sofas for those who fancied a game.

On one of our visits the actor Stanley Baker was playing poker with three others, whose faces I recognised from the big screen. Film director Roman Polanski and his wife Sharon Tate were sitting on a sofa, holding hands. Just a few years later Sharon died at the hands of Charles Manson and his gang. Singer Alan Price and his girlfriend were on the opposite sofa, while in the corner a mysterious-looking woman was reading palms. John and I seated ourselves on cushions on the floor while Alma, her mother and sister served us all champagne. This was my first taste of such hedonistic luxury and I was out of my league.

It was exciting to be part of the swinging London set, but it highlighted my insecurities. While John took to it with ease, I was painfully aware of my lack of sophistication. I saw myself as a naïve girl who had simply got lucky and didn't deserve to be there. Also, in the presence of so much glamour and beauty, I was acutely conscious of my imperfections.

I did my best to overcome my fears and doubts, shopping in all the top places – Biba in Kensington High Street, Mary Quant in the King's Road and other places of the moment. I bought gorgeous outfits, had my hair done in a little Greek place in Bayswater – I'd have felt out of my depth in somewhere as high profile as Vidal Sassoon – and made every effort to look good. But I still lacked confidence in my appearance. It hadn't mattered so much when John and I were in Liverpool. In those days we had only had eyes for each other and with my black Bardot outfits and newly blonde hair I felt I was the height of glamour. Now, surrounded by rake-thin models and actresses who made their living from looking gorgeous, I could see temptation for John all around us and felt I couldn't compete.

I received my share of compliments, though. I remember

someone telling me that I was more beautiful than Britt Ekland, who was married then to Peter Sellers and most people's idea of the perfect blonde. I glowed for days. And John often told me he loved me. He also became impatient if I talked about my anxieties so, for the most part, I kept them to myself and did my best to appear as cool and glamorous as the women we mixed with. It wasn't easy though, especially when I began to wonder if John was being unfaithful. Alma Cogan was one of the women I suspected he was having an affair with. I could see the sexual tension between them and how outrageously she flirted with him, but I had no real grounds for suspicion – just a strong gut feeling.

I did a pretty good job of keeping my insecurity under wraps, but there were moments when it got the better of me. On one particularly vulnerable evening we were at a housewarming party given by Cilla Black and her husband Bobby, who had just moved to London. Cilla was now in Brian's management stable and was becoming a big star – she had been the first girl to hit number one since Helen Shapiro. The party was swarming with the up-and-coming and the famous.

Late in the evening Cilla walked into the bedroom after we'd all had a lot to drink, to find Georgie Fame talking to the closed door of her wardrobe. 'Come on out, Cyn,' he begged. 'What's the matter?'

'I'll come out when John realises I've gone missing,' I told him tearfully.

Cilla, panicking at the thought of the imminent ruin my drunken state might mean for her expensive dresses hanging neatly around me, helped him coax me out and took me off for another drink. John, talking to a crowd of people in another part of the house, never did notice that I'd gone missing.

One evening, not long after we moved to Kenwood, there was a knock at the door. John answered and I stood not far behind him. Bob and Sonny, the Beatles' photographer and his wife who had lived below us in Emperor's Gate, were on the doorstep. Bob looked furious and Sonny, a stunning Swede, was in tears, cowering behind him. Bob ignored me and said he wanted to talk to John. They all disappeared into the living room. Half an hour later Bob and Sonny left. When John came back into the kitchen I asked him what had been going on, but he shrugged and disappeared upstairs to his music room. It was never mentioned between us again, but not long afterwards I heard that Bob and Sonny were divorcing. I couldn't escape the conclusion that she'd had an affair with John, although I never had any proof.

Some accounts of John's life have said that he was a womaniser from the start, that even in our art college days he was seeing other girls behind my back. All I can say is that if it was true I never knew about it. I saw John flirting at parties, but that was all, and at college he was with me for so much of the time that he had little opportunity to see anyone else. In those days I never dreamt that he might be unfaithful. We were together, we loved each other and that was all I needed. It was only after we came to London and began our new life that the doubts crept in.

Of course I knew that John might have had the odd fling with a girl when the Beatles were on tour – if any of them had had the odd lapse, well, they were only human and it meant little. All of us Beatles women knew that girls threw themselves at the boys, but we also knew that they came home to us so we ignored it. After all, we were getting fan letters every day saying things like 'I'm in love with your husband, he doesn't love you, he wants me, so leave him alone'. Some were funny, others threatening, but either way they went into the bin and we forgot about them.

So, when I began to wonder if John had had affairs with Alma, Sonny and a couple of other women who were around us at the time, I decided to let it go. I believed that John and I were strong enough together to come through anything, and that unless an infidelity was staring me in the face I wouldn't ask him or look for evidence. I knew that if I tried to confront him he would walk away and I'd end up tormenting myself. In any case, I wasn't the sort of woman to be controlling and possessive. I knew I was the bedrock of John's stability and that he loved me, and I let that be enough.

John had made an appearance on Peter Cook and Dudley Moore's satirical comedy show 'Not Only . . . But Also' in January 1965. He and Peter hit it off immediately and became good friends. They shared an outrageous sense of humour and a fierce intelligence. Soon after they met Peter and his wife, Wendy, invited us to lunch. Their home in London's Hampstead was like something out of a glossy magazine, and as we walked in John and I glanced at each other apprehensively. These people seemed so effortlessly perfect. Their enormous kitchen, full of copper and dried flowers, had a huge Aga at one end, laden with pans of wonderful-smelling food. A long oak refectory table stood in the centre of the room laid with beautiful crystal glasses, glistening cutlery, and a vase overflowing with casually but stylishly arranged garden flowers. The food was superb and we had our first taste of garlic – amazingly, it was unheard-of in the Liverpool of our childhood.

Dudley Moore was there too and as we sat round the table Pete and Dud fell into their comedy routine. John joined in, putting on his thickest Liverpool accent, and as we drank bottle after bottle of expensive red wine the afternoon descended into hilarity. At one point John nudged me under the table and caught my eye. He grinned: this was great.

When it was time to go he invited Peter, Wendy and Dudley to dinner with us the following week. I looked at him in horror. How on earth was I going to compete with the lunch we'd just had? Admittedly my cooking skills had progressed from Vesta curry to a traditional roast, but it was hardly impressive. What on earth would I cook for London's wittiest, most sophisticated people? At least we had a beautiful dining room, crystal, silver cutlery, linen tablecloths and napkins. And, thank God, with Brian's help, we had installed some very expensive wines in our cellar – although neither of us had a clue what they were or which wine went with what food.

As the day of the dinner approached I drew up the most impressive menu I could think of. Prawn cocktail as a starter – sauce out of a bottle, frozen prawns. Roast lamb for the main course. And apple crumble for pudding – crumble out of a packet, custard out of a tin. With cheese to follow.

At the last minute Dudley begged off because of work commitments. The Cooks were due at eight and John, who was never a good timekeeper, had promised to be home from the studio in plenty of time. I was a nervous wreck and Julian, sensing it, was playing up. He knew something big was happening because he'd never seen the dining-table used before, never mind laid with its full regalia of crystal, linen and silver.

With half an hour to go I panicked. Flowers, oh, God, flowers. Stumbling around the garden picking flowers in the dark wasn't easy. With fifteen minutes to go, I changed and tried to compose myself. The meat was in the oven, the vegetables prepared . . . I breathed a sigh of relief.

Peter and Wendy arrived and I poured drinks and handed them the nuts and crisps I'd thrown into bowls so that they had something to nibble until John arrived. The next two hours were probably the longest and most embarrassing of my life. With no sign of John, I did my best to keep the conversation going and the glasses full, while Julian bobbed around, creating mini-diversions of his own. But, charming as Peter and Wendy were, it was hard for me to entertain them. And the food was gently disintegrating in the oven.

It was ten when John rolled in, full of apologies with a beatific smile on his face. He was clearly stoned. He was as nervous as I was about the evening, had smoked a couple of joints to fortify himself and lost track of the time. We had all drunk quite a lot, so when John produced a joint Peter and Wendy happily accepted. When I served the meal it was so late and they were so stoned that they wolfed it down, oblivious to its pre-packed origins and hideously overcooked state. In the end the night was a great success, and we had many more dinners together, at home and out on the town.

One weekend a few months later Mike Nesmith of the Monkees and his wife arrived. John had met them on tour and invited them over, but he was quite put out when they actually turned up. I, on the other hand, was delighted. Even though I still wasn't the most confident hostess, I had bought lots of cookbooks since the Cooks had come round and my repertoire had expanded. Now I was always glad to have new people to entertain. Mike's wife, who appeared to live to please him, wasn't the easiest guest, though. She would hover over me as I cooked, saying, 'Mike doesn't like it like that,' and 'I always do it this way for Mike.' How I managed not to thump her I'll never know.

Among the other guests who came to call were Bob Dylan and Joan Baez. Bob was a good friend of John's – their influence on each other's music was powerful – and he dropped in one evening to join us for dinner, as did Joan on a different occasion – both 'I'm in town can I come over' guests.

Although John and I knew we were rich, our true spending power still hadn't dawned on us. At home we lived simply. We drank little, usually preferring a glass of milk with a meal to anything alcoholic, and our tastes weren't extravagant. John gave me an allowance of fifty pounds a week, which in those days was a handsome sum. I didn't go on many shopping sprees, but when I did it was for shoes. Every time I went into the shoe shop in Weybridge the assistants rubbed their hands with glee. An hour later I'd emerge with several carrier-bags and most of my week's money gone.

I also loved buying Julian clothes, just as much as he loved dressing up and posing. He had a little blonde Canadian girlfriend called Lorraine, and he'd excitedly show off his new things to her. Bedding was also high on my list: Dot and I would drive to Walton-on-Thames and run amok, snapping up high-quality sheets, pillowcases and bedcovers. When I lived in the bedsit I'd only had one set of sheets, so they had had to be washed, dried, ironed and put back on by evening. If the sun hadn't shone and the sheets weren't dry, I had to roll up in a blanket. So, the luxury of being able to afford as many gorgeous sheets as I wanted was glorious.

John loved shopping even more than I did. Stores would open out of hours for the boys and they'd scoop up goodies in their own version of a supermarket trolley dash. John would come home like a kid laden with gifts for me, toys for Julian and clothes for himself. He bought me a beautiful Cartier gold watch and bracelet in a stunning velvet jewellery box. And he had a thing about lingerie, not just sexy black numbers but unbelievably extravagant creations, négligés that could have been worn at a society ball. We would giggle as I paraded around in them. 'Shall we dance?' he would say, and we'd waltz round the bedroom. The fashion shows never lasted long in such close proximity to the bed.

I always had to be his critic or admirer when it came to the clothes he'd bought for himself. 'How does this look, Cyn? Does it go better with this? Do the colours look OK together? Are the trousers too baggy? What do you think, Cyn?' A trying-on session could last hours.

Sometimes, especially when he'd been shopping with Paul, George and Ringo, he'd come home with the car boot full of toys for Julian. These were usually meant for eight- or nine-year-olds, John having forgotten, in his enthusiasm, that his son was still only two. The more complicated articles would be put away, still in their packaging, for a later date. But Julian soon learnt to root them out. One day when he was just three he found a toy that required a great deal of skill to assemble. To our pride and astonishment, he had put it together in double-quick time – even though he couldn't read the instructions. 'That's my boy,' John cried. 'I couldn't have done that myself.'

One of John's biggest indulgences, though, was cars. When we moved into Kenwood, we bought a Rolls-Royce for the chauffeur to drive us around in. A couple of years later, as flower-power swept the nation, John had it painted in psychedelic colours. It looked fantastic, but it hardly helped us to travel incognito. When I passed my test John bought me a gorgeous little white Mini, which I loved driving. Dot and I used to take it on our shopping trips. Then one day John came in very excited, made me cover my eyes and led me outside. Standing on the drive was a gold Porsche for me. A few weeks later I got up to find the Porsche gone and in its place a red Ferrari. Now that he, too, was able to drive, John had part-exchanged my car for one for himself. Generous as he was, this impulsive act was typically thoughtless: he hadn't stopped to consider whether I might mind, just rushed ahead. Once he'd taken a decision everyone and everything had to fall into line with it. He had little time for negotiation, considering or planning, preferring to act on impulse and hang the consequences.

Next I was given a green Volkswagen Beetle, which I enjoyed, though not as much as the Porsche. John loved taking off in his Ferrari. Unfortunately he was an appalling driver. His passengers had to suffer a hideous rollercoaster ride as violent swerves caused the car to hit the kerb or mount the pavement, all at breathtaking speed.

We had lots of fun at Kenwood. When we had moved out of the attic, it was gutted, renovated and became our games area. Scalextric car racing was a big craze and John bought three sets, which were laid out as one enormous track. Anyone who came to the house, and especially Paul, George or Ringo, would be dragged upstairs, past Julian's life-sized rocking horse, into the Scalextric room for a no-holds-barred race. John was ferociously competitive and always chose what he thought was the fastest car. He and Julian would team up together against the opposition and the race would be played out amid whoops, screams and shouts of 'Cheat! Your car's been souped up. Mine's been fixed.'

In another room in the attic a basic recording studio had

been thrown together. It was usually in complete disarray
– records strewn among beanbags, scribbled lyrics all over
the room, recording equipment everywhere. John often disappeared there for a few hours, and later would shout downstairs, 'Cyn, what are you doing? Come and listen to this.'
If I didn't get there immediately he'd shout again: 'Come
on, Cyn, drop whatever you're doing, I need you NOW.'

'Yes, sir, OK, boss.' When I got to the attic John would
be desperate for an audience for his new song. I would
listen and comment, trying to help when he was stuck for
a lyric. I loved all of John's music, which was evolving
beyond rock and roll. He was working on tracks for their
fifth album, *Help!*, which was also to be the soundtrack
to the film. John's songs were often angry, sad or challenging, but their honesty and intensity added to their
appeal. I remember him writing 'You're Going To Lose
That Girl' late one night, then calling me to listen.

After hours locked away with his music I'd ask, 'Fancy
a bacon butty and a mug of tea?'

'Perfect, you must have read my mind,' he would say,
and beam.

John's composing never followed a pattern: he might
have an idea and head for the attic at any time of day or
night. I got used to him leaping out of bed in the middle
of the night to write lyrics or try out a line for a song on
the piano, or sitting up half the night to finish a composition. Sometimes he would play the piano for hours while
I sat dress-making, keeping him company. Then there
would be phone calls back and forth to Paul, as they played
and sang to each other over the phone.

At other times John would get me to sing along with him. Our favourite was 'Blue Moon', but our rendition was so bad that we fell about laughing. 'I know you sang solo in a choir, Cyn, but, Christ, you'll never make it as a pop star. You're too posh. Try and rock it up a bit. You're not singing in church now, loosen up and try it again.'

Despite all that was going on around us, and my occasional unease about John and other women, we were a solid unit, happy together and amazed at our good fortune.

Of course it wasn't perfect, partly because the boys were under such pressure. They were still carrying out major foreign tours every few months as well as playing regular British concerts. And on top of the live performing they were expected to produce a steady stream of three or four original singles a year and a couple of LPs – far more than any pop star today. When John was away our affection for each other intensified. He would phone as often as he could, but preferred to write – tender, funny letters, filled with anecdotes, musings and long passages telling me how much he missed me and longed to come home. In one of his letters, written on the Beatles tour of the States in August 1965, at a time when they played to a record-breaking audience of fifty-six thousand people in Shea Stadium, he wrote of his love for Julian:

*I really miss him as a person now – do you know what I mean – he's not so much 'the baby' or 'my baby' any more he's a real living part of me now – you know he's Julian and everything and I can't wait to see him, I miss* *him more than I've ever done before – I think it's been a slow process my feeling like a real <u>father!</u> I spend hours in dressing rooms and things thinking about the times I've wasted not <u>being</u> with him – and playing with him – you know I keep thinking of those stupid bastard times when I keep reading bloody newspapers and other shit whilst he's in the room with me and I've decided it's ALL WRONG! He doesn't see enough of me as it is and I really want him to know and love me, <u>and</u> miss me like I seem to be missing both of you so much.*

*I'll go now cause I'm bringing myself down thinking what a thoughtless bastard I seem to be – and it's only sort of three o'clock in the afternoon and it seems the wrong time of day to feel so emotional – I really feel like crying – it's stupid – and I'm choking up now as I'm writing – I don't know what's the matter with me – it's not the tour that's so different from other tours – I mean I'm having lots of laughs (you know the type he! he!) but in between the laughs there's such a drop – I mean there seems no in between feelings.*

*Anyway, I'm going now so that this letter doesn't get too draggy. I love you very much.*

*To Cyn from John*

In a PS he asked me to try to ring him and in another he told me to say hello to my brother Charles for him.

His letters weren't always so reflective, but that one wasn't unusual. He found it easier, in many ways, to say what he really meant in a letter and, as he had since the Hamburg days, he used them to tell me how he really felt.

A few years later, when John and I had divorced, I sold this letter, along with several others John wrote. I was touched and delighted when, some years afterwards, the owner put it up for sale again and Paul McCartney bought it. He had it framed and presented it to me and Julian as a gift. An immensely thoughtful gesture which we appreciated deeply. This particular letter evokes mixed feelings in me now. I know John loved and missed us and that he meant every word he wrote. But he didn't change. There were periods when he tried harder with Julian, spent more time with him and got to know him better, but he was too preoccupied with other things to devote much energy to the small son who longed for his attention.

John would arrive home from tours exhausted and spend the next few days more or less asleep, which meant I had constantly to keep Julian quiet and away from our bedroom. Julian missed his father when he was away and painted endless pictures for Daddy. So when he came home Julian couldn't wait to see him.

The rule was that at two in the afternoon John had a wake-up call and a cup of tea. Julian, who had waited impatiently all morning to see his daddy, would go in as soon as I gave the signal, and jump on John for cuddles and a chat.

It was usually several days before John was back to normality, but then he was like a tornado, wanting to know about everything he'd missed and wrestling with Julian, who shadowed his every move. He'd settle down to go through the fan mail, which by then had piled up by the sackload. Julian's tiny fingers would be poking into everything. 'Look, Julian,' John would say, 'these letters are very important. They're our bread and butter. See? This one is your breakfast, that one's your dinner, and this one is a new guitar for Daddy.' After a while it would be, 'OK, OK, come on, let's go for a walk in the garden and pick some flowers for Mummy,' and they'd disappear for a couple of hours.

John loved being with his son, but in short bursts. His moods could be unpredictable and at times he was intolerant and impatient with Julian. On one occasion I remember him shouting at the dinner-table because Julian was eating messily. I was livid and stormed, 'If you were here more often you'd realise that this is how little boys of three eat. Now leave him alone.' I rushed upstairs in tears: the shock on Julian's face when John had erupted at him had really upset me. But rows like that were rare. I'd learnt to keep away from John if he was edgy, and Julian had too.

In the evening, while Julian played with his toys at our feet, John and I would have dinner in the den and catch up with our news. John would say over and over again how great it was to be home. 'Oh, God, Cyn, this is fucking fantastic. How's your mum? Has she bought me any more goodies? You did tell her, didn't you, that if she sees anything really different she must buy it? It doesn't matter how much it costs. Oh, and don't forget we need loads of books for our bookcases. Tell Lil to buy old leather-bound ones – they look great.'

Our hall had a wall of bookcases, mainly empty, and John was happy to indulge my mother's love of sale rooms to fill them and to acquire antiques and ornaments for the house. In particular he loved clocks, and Mum searched them out for him. 'Oh, yeah, and before I forget can you ask Dot to buy a whole stack of Rice Krispies? We're running out.' Rice Krispies or corn flakes were his favourite snack.

When Julian had gone to bed and Dot had gone home the house was quiet and we had a chance to unwind and reconnect. We were safe from intruders now, which made a big difference: we'd had a few nasty incidents in which fans got into the grounds or even the house – once I got up in the morning to find twenty wandering round downstairs. We had enormous sliding wooden gates installed to keep out unwanted guests. The diehard fans still camped outside the gates in all weathers, carving their names into the wood for posterity, but we felt a lot more secure.

After weeks apart, me running the household, John travelling and performing, it was wonderful to be close again and to shut out the world. That was when we shared our most affectionate moments. We'd exchange roses or loving notes and cuddle up together in front of the TV. We called these special John-and-Cyn times, and they were oases of loving peace in the madness of our world. I treasured them.

# 13

WHEN THE ANNOUNCEMENT WAS MADE that the Beatles had been appointed MBEs there was astonishment on all fronts and uproar from pillars of the establishment. Newspapers debated the issue hotly, angry letters were written to *The Times*, but most of the country was delighted to see their favourite pop stars rewarded. Commonplace now, in the 1960s no pop star had ever received an honour: they were largely reserved for judges, politicians and civil servants. Once again the Beatles were breaking new ground.

The news came via a phone call from an ecstatic Brian: 'John, you won't believe this, I didn't at first, but I've just had confirmation that you and the boys are to receive the MBE. Isn't that fantastic? I have to keep pinching myself. An MBE! From the Queen! In Buckingham Palace!'

John's reply was a little more earthy: 'Fuckin' hell, Brian, you must be fucking joking. Why? Pop stars don't get MBEs, they're supposed to be for ex-army, do-gooders, the establishment. Bloody hell, wait till I tell Cyn and Mimi.' John was as pleased as Punch and couldn't wait to tell everyone. Not for a moment did he consider turning it down.

A few months later the boys and Brian went to the palace for the ceremony. We wives and girlfriends would have loved to go, but the boys, knowing that the palace would be surrounded by crowds of hysterical fans, felt they would cope better alone. We watched on TV, bursting with pride.

'Mimi will have to eat her words now, Cyn, about earning a living with a guitar,' John kept saying. Mimi's lack of pride in him still hurt. Even though he was famous and wealthy she still behaved as though he'd got lucky and that his success was nothing to do with talent and hard work. He hoped that being honoured by the Queen might impress her. And it did. She was so pleased that she asked John if she could keep the MBE for him, to which he agreed. For a few years, it had pride of place on her television.

However, Mimi was no longer happy in Liverpool and complained to us every time we phoned her. She had no friends. She was being pestered by fans. Her life was intolerable. Eventually she told us she wanted to live somewhere else. 'Where?' we asked. 'Bournemouth,' Mimi announced. She had always wanted to live by the sea, she said, and fancied going down south. We were worried that she'd miss her sisters and their children. 'Oh, that lot.' She sniffed. 'I hardly ever see them anyway. They only come round when they want something.'

We agreed to find her a home in Bournemouth. She came down to stay with us at Kenwood for a few days and we arranged for a few estate agents in Dorset to send us details of properties. Mimi sifted through them for those she fancied. With a shortlist of four, we drove to Bournemouth in the Rolls. It was not an easy day. Mimi was grumpy and the first three properties were entirely wrong. But then, by some miracle, she decided that the fourth would do. It was a pretty white luxury bungalow, built close to the beach in Poole, the next town to Bournemouth. Suddenly Mimi's mood lifted and she was happy. She chatted all the way home. 'Oh, John, thank God for that. Maybe I can have some peace now. And you can bring Julian to visit now and again. He can enjoy the sea and sand.'

John and I breathed a sigh of relief. Mimi's house was sold for six thousand pounds and we paid twenty-five thousand for the bungalow. It was worth more than our house because it was in such a sought-after position, right on the beach. Once Mimi had settled in, during the summer of 1965, John, Julian and I went to visit her. She seemed much more at ease with herself and the world. As the sun was shining we decided to go to the beach. We put a picnic together hastily, bought buckets and spades and built sand-castles, paddled and sunbathed. John wore a large sunhat, shorts and a T-shirt and not one person on that crowded beach recognised him. It was heaven. We thought then that we had all the time in the world and would often go down to see Mimi, but sadly it was our only visit. Although we intended to go back, the demands on John were such that he didn't see Mimi again for some years, although they talked regularly on the phone.

He continued to be enormously generous. Now that he had money he wanted to share it with his family and friends. First he bought a house for my mother, in nearby Esher. He knew that it meant a lot to me to have Mum close: I enjoyed her company when he was away and she was a great help with Julian, looking after him whenever I was out late or away with John. We decided to give her and Mimi a weekly allowance each of thirty pounds. Mum was delighted and spent most of hers buying John knick-knacks for the house or clocks for his collection. But all hell broke loose when Mimi found out that she wasn't John's only beneficiary. When I picked up the phone, I didn't get beyond 'Hi, Mimi', before I was hit by a tirade: 'I have just been informed that *your* mother is receiving the same allowance as me. I am disgusted. How dare she expect or even accept money from John? What has she done to deserve anything? Tell John when you speak to him that I am very, very annoyed.' Plonk. The phone went down before I could utter another word.

When I told him John was almost dismissive: 'That's

Mimi, Cyn. She's never satisfied. You should know that by now. Forget it.'

Another recipient of John's generosity was his old schoolfriend Pete Shotton. Pete had played in the original Quarrymen group, although he wasn't very musical and had dropped out early on. He and John had stayed good friends, though, and when Pete hit hard times John gave him twenty thousand pounds to open a supermarket on Hayling Island, not far from where we lived in Weybridge. He visited us regularly at Kenwood, driving me mad by turning up without letting us know in advance, then staying for a couple of days. Later on, when the Beatles formed their own company, Apple, he went to work there at John's instigation.

Not long after we'd finished the bulk of the renovations and settled into our new home, Julia and Jacqui came to stay. They were bright pretty teenagers, thrilled to be visiting their famous big brother. Julia had finished her A levels and was planning to go to Chester College to study French, while Jacqui was planning to become a hairdresser – a career I'd quite fancied myself. We'd seen little of them since we had left Liverpool eighteen months earlier, so we planned to give them a really special time.

We sent them tickets to fly down, knowing it would be their first time in an aeroplane. Then we arranged for Anthony to meet them at the airport with the Rolls. When they arrived we showed them round the house. John was clearly proud of it and the girls oohed and aahed as we took them round the huge living room with its enormous sofas, the oak-panelled dining room with its chandelier, and the six bedrooms. What impressed them most was our bedroom, with its white carpets and en-suite bathroom, complete with sunken bath, Jacuzzi and his-and-hers basins.

John spent as much time as he could with them, but he had to go to the recording studio a couple of times. When that happened the girls and I went to see the London sights and do some shopping. Anthony and I took them on a magical mystery tour of London, ending up at Harvey Nichols where I told them they could choose anything they wanted. They were amazed and couldn't quite believe I meant it, but eventually, grinning like Cheshire cats, they launched themselves into it. Clothes were tried on, discarded or drooled over.

'You can't possibly buy this for me, Cyn, it's far too expensive.'

'Julia, will you please shut up? I've told you already. Anything.' I had as much fun giving as they had choosing. I knew John would want to spoil them and it was so good to share what we had grown used to with the girls, both so good-natured, unspoilt and bowled over by it all.

Before long they had shopped themselves out and Julian was getting tired and cranky. We took him to the children's department where, with a little help from the three of us, he tried on jeans and a polo-neck, all in black. The saleswoman, the girls, our fellow shoppers and I were all in hysterics. He posed and danced around, looking like his father – a mini Beatle. The only thing missing was a tiny guitar.

At the end of the day we drove home with the boot full

of clothes for the girls. The next evening was to be their biggest treat of all: seeing their brother live on stage for the first time. The Beatles were playing at the Finsbury Park Astoria and John had arranged VIP treatment for them. They drove up to the concert in the Rolls with me and John. They were astonished by the crowds outside, the mass of screaming, fainting girls and the rows of sweating policemen holding them back. We'd come to think of all this as almost normal, but Julia and Jacqui had never seen it before and were understandably frightened, especially when we had to dash from the car, past the crowds and in through the stage door. In the dressing room the girls met the other Beatles and Mick Jagger, who'd dropped in. Soon everyone was sitting around drinking Coke and chatting, oblivious of the seething hordes outside.

John had got us seats right at the front, as close to the stage as possible – the first few rows were cordoned off so that the fans couldn't storm the stage – but even from there we couldn't hear much of the music because of the screaming. And as the fans pushed forward, climbing over the seats and heading for the stage, we were in danger of being trampled. A minder grabbed us and we watched the rest of the show from the wings. Afterwards I took the girls home to Weybridge in the car while John went to an after-show party, which, in his role as concerned big brother, he'd decided wouldn't be suitable for them.

A couple of days later the girls went back to Liverpool and I missed them: it had been fun playing the big sister.

They were to make one more visit to us, a few months later. Then John and I took them to the Abbey Road studios to see the Beatles record 'Day Tripper' and meet George Martin. I wish we'd had them to stay more often. We always thought we would, but John was so often away or recovering from being away that it was hard to find the right moment. As it was, that was to be their last stay with us and afterwards we had little contact with them. Soon Julia was at college studying French, and Jacqui had started her training, but only a few months later their father, Bobby Dykin, was killed in a car crash. They had lost him in the same way that they had lost their mother. John, of all people, would have understood their pain yet, almost unbelievably, we were not told of what had happened. We didn't hear that Bobby had died until much later. Then we discovered that Aunt Harrie had been appointed their guardian. John was very upset that his sisters had been orphaned, and suggested that we buy Harrie and Norman a house that the girls would inherit eventually. Harrie and Norman were delighted with the idea and went about finding a suitable place.

John's intention was always that this house was for his sisters, as a nest-egg for their future, but he was always vague about finances and possessions. The house was bought and Harrie and Norman moved in, but instead of signing it over to them or to the girls, John took his accountants' advice to keep it in his name. It created all kinds of problems for the family after his death. The same was true of the other houses he had bought. Mum's house in Esher and Mimi's bungalow in Poole still belonged to John's estate, which Mimi only found out, to her consternation, after John's death.

John's love for his family was never in doubt, even though he allowed a distance to open between them. He adored his sisters, and Mimi always had a place in his heart. He visited her with Yoko, and I went to see her, too, in later years. He saw Harrie and the family in Liverpool before he moved to America, but after that he had little contact with them until he and Julia began to talk regularly on the phone in the months before he died. I was always fond of John's sisters and we are still close friends today. Julia, in particular, has kept a strong connection with me over the years and I'm glad that I've been able to tell her how much John cared for her and Jacqui, and that he wanted Harrie's house to go to them.

Aunt Mater remained another close friend of mine until her death in 1976. When John and I lived in Weybridge I used to catch the overnight train to Edinburgh to see her from time to time. We'd been close since the weekend when John and I had hitchhiked up to see her, and when London got a bit much for me, or John was away on tour, her home was a peaceful refuge. She was always warm and welcoming and we'd sit up late into the night, drinking her husband Bert's whisky, and talk about everything under the sun. She didn't have Mimi's sharpness and treated me like a daughter, bringing me breakfast in bed and cosseting me.

One family member John wasn't happy to see was his father, Alf. One day we opened the newspapers to discover that he had turned up as a kitchen porter at the Greyhound

Hotel in Hampton, south London. It seemed that he had had no idea that his son was famous – he wasn't interested in pop music – until a colleague pointed at John's photo in the newspaper and asked whether Alf was related to him.

Once he realised that the son he had walked out on twenty years earlier was now a millionaire, he was suddenly very interested in resuming contact. One day when John was out he turned up at Kenwood. I opened the door to find a tiny man with lank grey hair, balding on top, outside. He looked as unkempt and down-at-heel as an tramp – but, alarmingly, with John's face.

Once I'd got over the shock of his arrival I asked him in, gave him a cup of tea and some cheese on toast and introduced him to his grandson. He told me he wanted very much to see John, so I suggested he wait: John was due back in an hour or two. I was a little anxious about John's reaction, but felt I couldn't shut the door in his father's face. And I was curious about the man who'd disappeared so early on from his son's life.

Alf and I chatted rather awkwardly for a while until, perhaps conscious of the contrast between his appearance and our home, he mentioned that his hair was a mess. I'd always enjoyed hairdressing – I used to do Mum's when we lived together – so I was itching to take the scissors to Alf's long, stringy locks. He agreed quite happily to let me cut it, so I did my best to make him look more presentable. A couple of hours later, with no sign of John, he went on his way.

John was annoyed when I told him about Alf's visit. He

told me they'd met briefly a few weeks earlier when he'd turned up in John's dressing room one day on the set of the Beatles' second film, *Help!*. He had been brought along by a newspaper reporter, hoping to run the scoop of a Beatle reuniting with his long-lost dad. Shocked, John was furious with the reporter, and had just wanted Alf to go away. I was surprised he hadn't mentioned it to me, but I knew what a sensitive issue John's father was for him. John had pushed the incident to the back of his mind.

Over the next few months he began to see his father and they became friendly, although John always had mixed feelings about their relationship. He knew that Alf was probably gold-digging, but he was his father and perhaps a small part of John still hoped for a father–son relationship. 'He's all right, Cyn. He's a bit wacky, like me,' he told me once. 'I can see where I got it.' It meant a lot to him to see parts of himself in his father and to know where he came from.

At Christmas 1965, we heard that Alf had made a record, under the trendier name of Fred Lennon. 'That's My Life (My Love And My Home)' was awful, and hugely embarrassing to John, who was furious at his father's blatant jump on to the bandwagon of his own success. He asked Brian to do anything he could to stop it. Whether Brian did or not I don't know, but the record never made it into the charts and soon disappeared.

Alf's next stunt was to turn up with a girlfriend young enough to be his granddaughter. Pauline was nineteen and a student, and he was fifty-six. They swore they were in love, and despite the opposition of Pauline's horrified mother, they intended to get married. Alf asked us if we could give Pauline a job, so we agreed that she could help with the fan mail and secretarial work. She lived with us for a few months but it was a nightmare. She was constantly in tears and arguing with her mother over Alf. She slept in the attic and we'd hear her screaming down the phone and sobbing up there. Eventually she and Alf decided to give up trying to persuade her parents and they eloped to Gretna Green. They stayed together until Alf's death, a decade later. They had two sons, John's half-brothers, but we never met them. I believe Alf turned up to see John again after we had parted, and that John told him to get lost. I do know they spoke on the phone before Alf died, so I hope that John made his peace with the father who let him down so badly.

The biggest change in our lives during this time, and the biggest single factor that led to our break-up, was John's deepening interest in drugs. All of the Beatles had been smoking cannabis for a couple of years when we were introduced to LSD – at that time most people in music circles did – and I didn't object to John using it. He'd get the giggles, the munchies and appear a bit spaced out, but it soon wore off with no ill-effects. I felt it helped him to relax after working for twelve or fifteen hours at a stretch. But the effects of LSD were very different. At a dinner party we'd gone to with George and Patti, with a friend of theirs and his wife, the host spiked our drinks with it. I never forgave him for giving us a drug without our knowledge, although no doubt he thought it hip and harmless. A lot of people then were discovering LSD and enthused about it.

Shortly after the meal we began to feel very odd. The room swam and at first I wondered if I had food poisoning. We had no idea what was going on until our laughing hosts enlightened us. We were all so frightened that we rushed straight out of the house, desperate to get home. George had driven us there in his brand-new Mini, so we piled in and he attempted to drive us home. The trouble was we were in central London, a good hour from home, and George had no idea which way up the world was.

God knows how he made it, but after we had gone round in circles for what seemed like hours we eventually arrived at George and Patti's home. John and I weren't capable of getting back to Kenwood from there, so the four of us sat up for the rest of the night as the walls moved, the plants talked, other people looked like ghouls and time stood still. It was horrific: I hated the lack of control and not knowing what was going on or what would happen next.

The next day John and I made our way home wearily and fell into bed to catch up on the sleep we'd missed. I dismissed the incident as a foolish prank that had convinced me I never wanted to try hard drugs again. But John felt differently: although he'd been as shocked and scared as the rest of us, he was also fascinated. He had enjoyed the lack of control and the weirdness. What for me had been the end was for him only the beginning. He decided to give it another try.

In the following months John took LSD regularly. He was hungry for new experiences and never afraid to experiment. George had found it fascinating too and he also took it again, as did Paul and Ringo, but John felt it gave his life a whole new dimension. The other Beatles were much more cautious, but John threw himself into it with abandon, convinced that this was the way to greater enlightenment, creativity and happiness.

When John was tripping I felt as if I was living with a stranger. He would be distant, so spaced-out that he couldn't talk to me coherently. I hated that, and I hated the fact that LSD was pulling him away from me. I wouldn't take it with him so he found others who would. Within weeks of his first trip, John was taking LSD daily and I became more and more worried. I couldn't reach him when he was tripping, but when the effects wore off he would be normal until he took it again.

Initially John's drug-taking didn't make a big impact on his work. He took the LSD after recording sessions and concerts, not during them. Later, his drug-taking filtered through into his song-writing, but at this stage his work seemed unaffected.

Soon he was bringing home a ragged assortment of people he'd met through drugs. After a clubbing session he'd pile in with anyone he'd picked up during the evening, whether he knew them or not. They were all high and littered our house for hours, sometimes days on end. They'd wander around glassy-eyed, crash out on the sofas, beds and floors, then eat whatever they could find in the kitchen. John was an essentially private man, but under the influence of drugs he was vulnerable to anyone and everyone who wanted to take advantage of him.

I knew I couldn't go on like that indefinitely. Our house

was being invaded by people I neither liked nor wanted to know. I was afraid for Julian and myself. I didn't want to hear loud music all night, or pick my way through semi-conscious bodies when I brought my son down for his breakfast. But every effort I made to put an end to it was met by a brick wall. A gulf was opening between me and John and I had no idea how to bridge it. Was this a phase I had to ride out, or was it the beginning of the end? I wasn't going to give up on my marriage without trying everything I could, but I couldn't live with a man who was constantly in another dimension.

John hated the distance it put between us too, but his solution was for me to join him. He did everything he could to persuade me to take LSD again. 'Cyn, you know how much I love you. I wouldn't let anything bad happen to you. We could both benefit from the trip. It would bring us even closer together,' he begged. 'Please, Cyn, I know if we do it with people we love and trust you'll be amazed how wonderful it is.'

Finally, I agreed. I felt I should give it another try because I wanted to understand what appealed to him so much about it. We set a date and I arranged for Julian to stay with Dot and her family over the weekend. John, who was ecstatic that I'd agreed, spent hours trying to reassure me – he knew I was terrified.

I dreaded it. John had arranged for a group of friends to come over to 'support' me, and first to arrive was Terry Doran, an old friend of ours from Liverpool who now worked for Brian. After him came George and Patti and her modelling friend Marie Lise (who was later to be film

With Julian at Kenwood, just after John had left, in the summer of 1968

POST CARD

Just a not to
say hello after
All these years
If you're in the
area stop by for
a cup of tea
fond old land
Paul + Kids

Cynthia, Julian
The Grange
Temple Sowerbury
Penrith.
Cumbria
G.B —

LUCKY SPOT LOOKING THROUGH THE WINDOW. SCOTLAND 1977
PHOTOGRAPH BY LINDA McCARTNEY

A postcard from Paul. I hadn't seen him for 17 years, since he came to comfort me, bringing me a red rose, after John left

John and Yoko with Julian and a friend, in 1968

John and Yoko's bed-in for peace, Amsterdam, March 1969. Julian was confused by pictures on TV of his dad in bed

John with six-year-old Julian in 1969, not long before John moved to New York and cut off contact with Julian for three years

John and Yoko with Julian and Yoko's daughter Kyoko, in Scotland the summer of 1969, just before the car crash

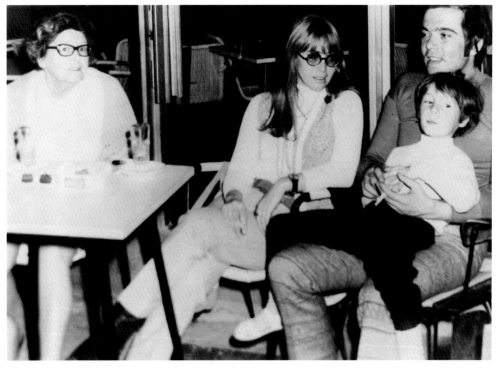

In Pesaro with Mum, Roberto
and Julian, who adored his new
step-dad

Outside the Barracuda restaurant
with Roberto and good friends
Lulu and Maurice Gibb

Taking nine-year-old Julian to school, Wimbledon 1972

With Elton John's percussionist Ray Cooper, who invited us to join his table and became a good friend

Julian, aged 13, with Sean, aged one, in New York 1976

Julian and John sharing a cup of tea in John's New York kitchen

Julian with relatives of Yoko's, in New York

Julian with Yoko and friends, New York

Julian in New York, trying out one of John's guitars

star John Hurt's great love and died after she fell off a horse). A few more of John's friends arrived, faces I only half knew. We lit candles, put on music and passed round the Scotch and Coke. This was very much 'Mission Cynthia': they were all there to ease me into the joys of LSD, but none had any idea of my panic. I wondered how I would be given the drug. Would it be pills or sugar cubes – a popular way of taking LSD? Then I realised I had already been given it, probably in my drink, so that I wouldn't refuse at the last minute.

Feeling strange, I went to the nearest loo, which was off the hall. I looked in the mirror – and saw my own skeleton. The vision kept moving, changing from a blurred prism of me to a grinning skeleton. I was transfixed. The next thing I remember was hearing John's voice: 'Cyn, are you OK? Come on, Cyn, you'll be fine, nothing will hurt you. Terry and I are here. You're safe.' His arm was round me as he guided me into the den where I began to feel safer, until Terry said something to me and immediately transformed into a snake, then an alligator. His voice emanated from a monster that kept moving towards me, every scale on its body shining, glistening and changing colour. I thought I was in hell. I couldn't look at anything without it changing its form and colour – even the carpet seemed to be breathing.

As I wandered around trying to fix on something tangible, the cat followed me. She became multi-faceted, multi-coloured. Her fur seemed to move in time to the constant beat of the music. The laughter of the people in the living room rang out like Big Ben, then echoed on and on into infinity. I felt arms round me. I heard words of love and comfort. Then I was lost again. I couldn't function on any normal level.

As the drug wore off I was aware of periods of lucidity when I was spoken to gently and my cheeks were kissed. Slowly I came back to earth. All the others were coming down, too, and they were hugging and kissing, telling each other how wonderful it all was and how much they loved each other.

Touched as I was by the warmth, my first coherent thought was that this was not for me. I didn't want false intimacy with people I didn't know well, and who didn't know me. And I didn't want drugs. To me they were terrifying and dangerous. I didn't want to tamper with my sanity. Much as I loved John, I knew I couldn't follow in that direction. I wanted calm, clear thought. I wanted to be there for my son.

After that episode I had to tell John the truth he didn't want to hear. He accepted my decision and didn't try to persuade me again, but he carried on doing drugs, as he had before, with others or on his own. And I did my best to accept it, hoping that eventually he would tire of it.

A small incident around this time, though, made me only too aware of the damaging effect that drugs had on John, and on our relationship. I told him I wanted to paint again. I'd put my career ambitions and love of art on hold since our marriage and I was longing to do something of my own. John was full of encouragement so one day I decided to paint the surround of the white TV we had in the sunroom. There was a yellow *chaise-longue* in there that Aunt Mater had given us and the walls were covered with mad posters, photos and other paraphernalia. John would lie on the *chaise-longue* daydreaming, and we often had breakfast in there.

While John was in the recording studio I spent hours painting a detailed floral design round the TV. I finished late that evening, delighted with what I'd done. I couldn't wait to show it to John.

The following morning I was up with Julian, about to give him his breakfast, when I glanced at my artwork. I could hardly believe what I saw. It was completely covered with circular stickers that read, 'Milk Is Good For You'. John had come in during the early hours, high on drugs, and destroyed my efforts. I was shaken and hurt. Did he not want me to have anything for myself? Was he so determined to have my total attention focused on him? Or was he simply so stoned that he hadn't realised what he was doing?

I chose not to say anything to him about it. I knew he'd done it under the influence of drugs, drink or both and I didn't want to hear him dismiss my efforts as unimportant or make excuses. I decided I'd rather keep my pain to myself. With hindsight, I should have spoken up but, like John, I hated confrontation or arguing. I think our mutual failure to address or resolve painful issues was a major factor in the eventual breakdown of our marriage. We both had the ability to sit on our feelings, but they inevitably resurfaced as resentment. I have no doubt that it would have made us stronger as a couple if we'd been able to deal with incidents like this more openly. But at the time I could only take what felt like the best path.

A few weeks later I tried LSD for the last time. Brian was having a party at the country house he'd bought in Sussex and John and I travelled down in the Rolls with a group of friends. On the journey everyone took LSD and I, against my better judgement but carried away by the jolly atmosphere in the car, decided to join in. Again, it was an awful mistake.

At Brian's house I followed John around, hoping he would comfort me as I went through what was, for me, a horrible experience. But he was not in a good mood: he glared at me and treated me as if I were a stranger. I felt desolate. Upstairs I found an open bedroom window and contemplated jumping out. For a few minutes, ending it all seemed like an easy solution: a chasm had opened between John and me, and I had no idea how to bring us back together.

Someone called my name, I turned back into the room and the fleeting thought passed. But I was low. For the first time I had to consider the very real possibility that my marriage might not survive.

# 14

AFTER THREE YEARS OF NON-STOP touring, recording and performing, John, Paul, George and Ringo were exhausted, both mentally and physically. During that time they had spent months on the road, doing three major foreign tours a year and performing night after night to countless thousands of fans. They had sung the same songs over and over again to screaming audiences who could barely hear them, and they'd had enough.

On top of this they'd been writing and recording several singles a year plus their LPs, and they had made two films.

*Help!*, which had its première in July 1965, had been almost as successful as *A Hard Day's Night* and the boys had hoped to follow it with a third, but their schedule was just too packed to allow for it. Faced with the prospect of continuing this constant output for another year it was clear that something had to go, so they decided to stop performing live. Instead they would concentrate on recording, filming and other new projects, and allow themselves a much-needed rest.

Before they flew to the States for what would be their final tour they went to Germany, where they received a huge welcome in Hamburg, then on to Tokyo and the Philippines.

John was happy when the decision had been taken that they would stop performing, He was worn out and had reached the point where all he wanted was to be at home without anyone bothering him. Also, the Beatles had become a magnet for the sick and disabled on their tours, which bothered John. Queues of people in wheelchairs, on crutches, or with learning difficulties, plus their helpers, formed outside the boys' dressing rooms at most of their appearances. John had always reacted badly to disability so for him this was little short of a nightmare. In our student days he'd mocked the disabled and drawn ghoulish cartoons of cripples. For some reason disability terrified him, though he could never admit it.It made him feel inadequate and guilty. The end of performing would also mean the end of this particular discomfort.

In any case, he had always been more interested in writing and recording. He much preferred working alone or with the other Beatles and a handful of talented technicians and producers. The concerts had long since lost any interest for him: he felt the Beatles were like four performing monkeys, steered on and off stage for the crowds to yell at. There was nothing creative in performing, and although he knew they would be pilloried for walking away from it, he couldn't wait to do just that.

I was optimistic too: the end of touring meant that John would be at home much more and we could be together as a family. It would make a tremendous difference to Julian to have his dad around. I also hoped that, with fewer pressures, John might cut back on his drug-taking. I longed to have the old John back, with the closeness we used to share. Before leaving on tour he confided how much he was looking forward to being at home, a proper dad to Julian, and having time for us as a couple. We agreed that in the future when he went away – he was thinking of making more films, which might involve going abroad – Julian and I would go with him as often as possible.

Happier than I'd been for months and full of plans and dreams, I decided to go on holiday with Mum and Julian while John was away. Mum suggested Pesaro in Italy: she'd been there before, to a friendly family-run hotel. 'No one will know you,' she said. 'You can have a real break.' Unfortunately the Italian press were on to us before we even reached the hotel. Magazines across Italy ran pictures of me carrying three-year-old Julian from one aeroplane to the next when we changed flights. From then on every time we went to the beach we were surrounded by crowds of people all wanting to kiss and hold the 'Beatle *bambino*'. Poor Julian was terrified – I had to whisk him away and hide in the hotel. The family who ran it, the Bassaninis, did their best to protect us, taking us out of the back entrance to cosy little restaurants where they knew we wouldn't be bothered. They were so kind that we promised to come back one day when all the Beatle madness had died down.

We got home in early July around the same time as John. He was incredibly happy to see us and we had a loving reunion, but preparations began almost immediately for the final US tour in August. A few months earlier John had given a long interview to Maureen Cleave, a journalist for the London *Evening Standard*. She was a friend of his and it was perhaps because of this that he let down his guard and said: 'Christianity will go. It will vanish and shrink. I needn't argue with that; I'm right and I will be proved right. We're more popular than Jesus now; I don't know which will go first – rock 'n' roll or Christianity. Jesus was all right but his disciples were thick and ordinary. It's them twisting it that ruins it for me.' It was a spur-of-the-moment comment and he didn't necessarily expect anyone else to agree with it, but he'd been asked for his opinion and gave it. He knew it was a controversial thing to say, but he had never shied away from saying what he thought because others might not like it.

When the interview was published in March, there was no fuss and neither John nor anyone around him thought any more of it. But at the end of July America's *Databook* magazine republished it. His comment that the Beatles were more popular than Christ was picked up by the American media and all hell broke loose.

Within days Beatles music was being banned on radio stations across the States, commentators were expressing outrage and America's Bible Belt was up in arms. From the States the outrage spread round the world. John received hate mail by the sackload, although he also had plenty of letters agreeing with him or offering support. We'd divide the letters into 'for' and 'against' piles each day and he'd ask me which was bigger. He was bemused and shaken, unable to understand why so many people thought what he'd said was so important.

The psychics' warnings were almost worse. We received dire predictions of plane crashes and other horrendous happenings, but only one made a real impact on John. Unlike the others, it wasn't hostile or angry: that John would be shot while he was in the States. When he left on the tour he was frightened and downcast. He had never meant to provoke so much anger and hatred for stating what he saw as the truth. I was frightened too, and if I could have stopped him going I would have. We parted wondering whether it was our last goodbye and clung tearfully to each other.

Throughout the tour I held my breath, watched every TV report, kept in touch with Brian's office and talked to John whenever he could get to a phone to call me. At the beginning of the tour he apologised reluctantly for his remark. Apology was never John's style, especially when he'd meant what he said, but Brian and the others persuaded him that it would be in everyone's interest if he did.

Despite the furore, the boys' concerts were all sold out and on 29 August they gave their last live performance ever, in Candlestick Park, San Francisco, to an ecstatic crowd. A day later John arrived home, relieved that it was all over, and so exhausted that he slept for two whole days.

Relieved as all the boys were to stop touring and performing, they also found themselves at a bit of a loss. Of course they would continue making records but, once they'd had a rest, what else were they going to do? Gradually they started to move in different directions and develop their own interests. For Ringo it was an opportunity to spend more time with his family. In those days he was always the most family-oriented of the boys, and liked nothing better than chilling out at home with his wife and baby. Paul and George wanted to travel. Paul decided to journey across Africa with Jane, while George and Patti headed for India. George had had a growing interest in Indian music ever since he'd picked up a sitar on the set of *A Hard Day's Night*. He wanted to find out more about it, and about Indian spiritual beliefs.

John accepted a part in a film called *How I Won the War*. He was attracted by the anti-war theme, and as he'd loved being involved in the two films the Beatles had made he wondered if his future lay in that direction. It was his first opportunity to be a straight actor and he was excited and nervous about it. Shooting was to take place in Germany and then Spain, in the seaside town of Almería. He promised that Julian and I could join him there as soon as he'd settled into filming. We flew out a couple of weeks later to the villa he was renting with the actor Michael Crawford, his wife Gabrielle, their baby and the nanny.

For the role of Private Gripweed John's long hair was cut short and he was given small, round National Health 'granny' glasses. Over the past few years he had been trying to cope with contact lenses, at least in public, but on more than one occasion fans had thrown jelly babies on to the stage – a custom that had been established after George had mentioned he liked them. One had hit John in the face and knocked out a lens, leaving him in considerable pain. Also, in those days lenses were made of thick glass and he found them uncomfortable. He loved the National Health specs and decided to stick with them and give up his lenses. Thus John's trademark look was born.

I loved being in Spain and watching the filming, but the villa we were staying in was damp and tatty. When Maureen and Ringo flew out to join us for a holiday it was the excuse we needed to find somewhere better. We searched out a vast villa with its own pool – we were told it had once been a convent. No sooner had we moved in than we discovered the place was haunted. Lights would keep going off, objects would move mysteriously and we all felt a strange presence. We planned a party to cheer the place up, but half-way through the evening the electricity was cut off and a huge storm blew up. As thunder and lightning raged outside, we lit dozens of candles in the huge main room. In the flickering candlelight the atmosphere softened and someone began to sing. Everyone joined in and the most beautiful, melodious sound filled the air. It was as though we were totally in harmony, musically and spiritually. After half an hour the lights suddenly came back on and the spell was broken, but it was easy to believe that we had been guided in our song by the spirits of the nuns who had once lived there.

The film was a success and John received good reviews. Director and producer Richard Lester told him he would make a fine actor, but although John had enjoyed the acting, he loathed learning lines and all the waiting around that is inevitable on film sets. He decided to stick with music: it was where his heart still lay and he spent long hours in the attic studio composing new songs.

I was glad he was being creative, but his drug-taking hadn't stopped and too often he was lost to me. Still, I clung to the times when he was his old self, hoping that the real, loving John would return. One morning at breakfast he pointed out an article in the newspaper to me. It was about a Japanese artist, Yoko Ono, who had made a film that consisted of close-up shots of people's bottoms. 'Cyn, you've got to look at this. It must be a joke. Christ, what next? She can't be serious!' We laughed and shook our heads. 'Mad,' John said. 'She must be off her rocker.' I had to agree. We had no understanding at all of avant-garde art or conceptualism at that point and the newspaper went into the bin. We didn't discuss Yoko Ono again until one night when we were lying in bed, reading, I asked John what his book was. It was called *Grapefruit* and looked very short. 'Oh, something that weird artist woman sent me,' he said.

'I didn't know you'd met her.'

John looked up. 'Yeah, I went to her exhibition, John Dunbar asked me. It was nutty.' John Dunbar, ex-husband of Mick Jagger's girlfriend Marianne Faithfull, was a friend of his who owned a small art gallery in central London, the Indica. It wasn't unusual for John Dunbar to invite friends to one of his exhibitions. I didn't think any more about it.

I didn't know then that Yoko was beginning a determined pursuit of John. She wrote him many letters and cards over the next few months, but I knew nothing about them at the time, or that she had even come to our house looking for him several times. On those occasions neither John nor I was at home and Dot, assuming she was just another fan, hadn't thought to tell me. Whether John knew or not, I have no idea.

What I did know was that John and I weren't as close as we had once been and I desperately wanted that closeness back. My confidence, never great, was hitting a real low. Sifting through old press pictures in an effort to work out why I so seldom felt good about myself, I decided the problem was my nose. I had a strong Roman nose, like my father's, with a pronounced bump in the middle. If it had been small and straight, like my mother's and my brothers', I reasoned, everything would have been different. Of course, some part of me knew this was silly, but I was desperate to blame my troubles on something I could resolve, which would make things better with John. If the problem was my nose, well, I could change it.

John thought I was being ridiculous when I told him I wanted to have cosmetic surgery, but said that I could go ahead. 'Don't come crying to me if they mess it up,' he added.

When the time came for the operation I was convinced I'd get a whole new personality – bubbly, confident and sexy – to go with my new nose, and checked into the London Clinic. Afterwards, as I lay there with my nose heavily bandaged, a bunch of red roses arrived with a card that read, 'A nose by any other name. Love from John and Julian.'

When the bandages came off a few days later I was terrified that my vanity would have led to disaster and I'd have a monstrously deformed nose. But when I found the courage to look at it I was overjoyed. The small, bump-free, pretty nose I'd always wanted was mine at last. I knew I'd never regret having the surgery.

I thought I looked completely different but when I went home, no one noticed. Not Julian, not my friends. 'Haven't you noticed anything different about me?' I'd ask, only to have them floundering around for an answer. 'New glasses?' ventured one, while another tried, 'Haircut?' My conclusion was that if they didn't notice it, I must have the perfect nose for my face. And John? He was fascinated by it, but still thought I was mad to have bothered.

In February 1967 the Beatles released 'Strawberry Fields Forever' (by John) and 'Penny Lane' (by Paul), the double A side about their old haunts in Liverpool. Penny Lane had been a central area, almost like a little village, close to John, Paul and George's homes. They often met, shopped or caught the bus there. Strawberry Fields was the name of a children's home in Woolton, close to Mimi's house. John had passed its red sandstone walls many times and loved the name.

The single reached number one, but it was the first to fail to go straight to the top of the charts since 'Please Please Me'. Were they slipping? Was it the fans' reaction to the end of live performing? We could only wait and see.

For the next few months the four worked hard in the studio and the result, in the summer, was a new album, *Sgt. Pepper's Lonely Hearts Club Band*. Some of the tracks on it referred to, or were written under the influence of, drugs, but its flowery, dreamy style perfectly suited the mood of the nation. It was the midst of the flower-power era, when hippies, flowers, love and peace were the themes. Psychedelia was everywhere, mini-skirts were in vogue, and everyone headed for Carnaby Street, the hippie fashion mecca, for a kaftan or a string of love beads.

*Sgt. Pepper* included one track that everyone was convinced John had written about an LSD trip – 'Lucy in the Sky with Diamonds'. In fact, the title was Julian's: he had come home from school with a painting of his friend Lucy. When John asked him what was in it, he'd said, 'It's Lucy in the sky with diamonds.' John, of course, loved this way-out and completely innocent description, straight from his son's unfettered imagination.

At this point he had been at home with us for nine months and I was having to face the fact that things between us were no better. John was still taking drugs almost daily and was unconnected, distant, moody and unpredictable. I was having to look after Julian and keep the house running on my own. John was in another world.

At the launch party for *Sgt. Pepper*, John was high and the journalist Ray Coleman, who later wrote a biography of him, was seriously worried about his health when he met him that night. Not only was John clearly drugged, he was smoking and drinking heavily, and looked haggard, old and ill; his eyes were glazed and his speech was slurred. Ray had mentioned his concern to Brian, who had replied, 'Don't worry, he's a survivor.'

I, too, was worried about John's health: the drugs had ruined his appetite and he did indeed look terrible. I feared he might kill himself. John had always had the potential to self-destruct and now he seemed hell-bent on fulfilling it.

I found it hard to understand his attraction to drugs. Was it a way of blotting out the pain of his childhood? It seemed to me that, initially, success had done that. In the first couple of years, as the Beatles soared, John had been on a high and his confidence had blossomed. But eventually the fame and idolising had become too much, and I believe he had turned to drugs to escape. He soon became addicted to them.

The chasm between us was widening. I still wanted a stable family life and a loving relationship with John, but he was restless. With the end of live performing, he was looking for something else to give direction to his life. I knew that, despite the barrier the drugs had placed between us, John still loved me. In his lucid moments he would put his arms round me and tell me so. But although he cared deeply for Julian and me, his addiction to drugs would keep him away from us. And he was most definitely addicted. I knew that he would have to make a determined effort – and needed powerful motivation – to give up now.

Almost miraculously, the motivation appeared in the form of the Maharishi Mahesh Yogi. In the midst of this summer of flower-power, George and Patti had become absorbed in Indian spiritual beliefs. Patti had been trying to teach herself to meditate but was finding it hard going until she went to a lecture about transcendental meditation at Caxton Hall, held by the Spiritual Regeneration Movement. She decided to join, and in August she heard that the movement's leader, the Maharishi, was coming from India to run a summer conference in Bangor, north Wales. A couple of days before the conference he would speak at the Hilton. George and she decided to go, and urged us all to join them. I stayed at home, but John went, with George, Patti, Paul, Jane and Ringo.

When he came home he was excited. 'It's fantastic stuff, Cyn, the meditation's so simple and it's life-changing.' Along with the others he'd been bowled over by the Maharishi's charisma and his promises of Nirvana. The Maharishi had invited the Beatles to go to his Bangor conference, which was to last ten days, beginning on the August Bank Holiday weekend. John was keen and I was happy to go along and find out what it was all about. George, Patti, her sister Jennie and Paul were all going. Ringo decided at the last minute that he'd come too. Maureen had just given birth to their second son, Jason, and was still in hospital. Also along for the ride was a young Greek, Alex Mardas: he'd been introduced to us by John Dunbar, who thought his electronics expertise might be useful to the Beatles. He soon became known as Magic Alex and joined the Beatles' inner circle, making himself indispensable both in and out of the studio.

The Maharishi was anti-drugs and had explained that through meditation you could reach a natural high as powerful as any drugs could induce. John loved this idea and was already talking about enlightenment, cosmic awareness and doing without drugs. So, I was all for the Maharishi's message. Perhaps this was the change of direction John had been looking for, and perhaps this time I could share it with him.

The lecture took place on a Thursday evening. On the Friday I arranged for Julian to go to Dot's for the weekend and packed our bags. On Saturday we set off to catch the train to Bangor from London's Euston station. As well as the Beatles party, Mick Jagger and Marianne Faithfull were coming. We were all to travel on the same train as the Maharishi's party, with the inevitable pack of photographers and reporters who'd got wind of the Beatles' wacky new interest.

It was the first time the Beatles had travelled anywhere together without Brian and their roadies, Neil and Mal, for several years. Brian knew about it and had said he might join us after the weekend. John was like an excited schoolboy, but also nervous. Going somewhere without Brian, Neil and Mal was, he said, 'like walking around without your trousers on'. Even in Spain for the film shoot he'd had Neil along to attend to his every need.

It was a bright, sunny morning when we set off. I was

ready early, but Patti, George and Ringo were coming in our car, and were late. By the time Anthony drew up at the station entrance we were cutting it fine and had five minutes to catch the train. John leapt out of the car with the others and ran for the platform – leaving me to follow with our bags. It was the result of years in which he'd taken it for granted that others would see to all the details. I followed him as fast as I could. The station was mayhem, with fans, reporters, police and passengers all milling around. I struggled to push my way through, but when I got to the platform my way was barred by a huge policeman who, unaware that I was with the Beatles party, said, 'Sorry, love, too late, the train's going,' and pushed me aside.

I shouted for someone to help. John poked his head out of the train window, saw what was happening and yelled, 'Tell him you're with us! Tell him to let you on.'

It was too late. The train was already pulling away from the platform and I was left standing with our bags, tears pouring down my cheeks. It was horribly embarrassing. Reporters were crowding round me, flashbulbs were popping and I felt a complete fool. Peter Brown, Brian's assistant, had come to see us off: he put his arm round me and said he'd take me to Bangor by car. 'We'll probably get there before the train,' he assured me, anxious to cheer me up.

But what neither he nor anyone else knew was that my tears were not simply about the missed train. I was crying because the incident seemed symbolic of what was happening to my marriage. John was on the train, speeding into the future, and I was left behind. As I stood there, watching the train disappear into the distance, I felt certain that the loneliness I was experiencing on that platform would become permanent one day.

Neil Aspinall drove me to Bangor. It took about six hours and our journey was peaceful and easy, unlike the media circus aboard the train. But as I stared out of the window at the glorious countryside my heart was full of fear. What next for me and John?

We arrived quietly, shortly after the others had been greeted by screaming crowds and even more press at the station. Bangor, a small seaside town, can't have known what hit it. I was greeted with hugs and kisses from the others and admonishments from John. 'Why are you always last, Cyn? How on earth did you manage to miss that train?'

'Perhaps if you hadn't left me to carry the bags, darling, I'd have made it,' I told him. How dare he tick me off, when he'd been so thoughtless? But I swallowed my feelings, as I had many times before. As ever, I didn't want a row, and especially not with all the others about. It was another moment when I should have been more assertive with John. I let him get away with an awful lot.

We were all staying in dormitories at a large training college, along with a couple of hundred other followers of the Maharishi. Our room was basic, with bunk beds and simple chests of drawers. John was amused. 'It's different, isn't it?' he said. It was.

At this point Mick and Marianne sauntered in, looking bewildered. 'Hey, John, what's happening? Where do we go from here?'

'Back to school.' John laughed.

The introductory seminar was held half an hour later in the main hall. It was an incongruous mix of the Maharishi's regular devotees joined by the psychedelically clad pop-star élite, all sitting cross-legged on the bare wooden floor. Marianne, next to me, whispered that she'd just started her period and did I have anything with me? Fortunately I did. I liked her: she was sweet and seemed too fragile for the world of drugs and rock.

That afternoon the Beatles held a press conference renouncing the use of drugs, in keeping with the Maharishi's teachings. Only a month earlier they, along with a string of other pop stars, had taken a full-page ad in *The Times* stating that the law on marijuana was unworkable and immoral. Paul had admitted publicly that he had tried LSD, and it was well known that the others had too. Now all that was turned on its head. The press was wildly excited.

But the boys' announcement had barely hit the news-stands when it was overtaken by news of an appalling tragedy. As we were heading back to our room, a reporter told us that Brian Epstein, who had steered the Beatles for the past six years, had been found dead.

# 15

I KNEW THAT THINGS WOULD NEVER be the same again. The disbelief and horror we felt was overwhelming. I watched the blood drain from John's face. Brian had been at the heart of the Beatles' lives for what seemed so long. The six years we had known him felt far longer because of the huge events that had taken place during that time. For the Beatles and their families, Brian had been guide, mentor, friend, big brother and father figure. He was generous and thoughtful, as well as an astute businessman. He'd orchestrated the Beatles' rise to unprecedented fame and success, and along the way he'd seen to everything – from our wedding to Julian's first pram, our house moves, social lives, birthdays and anniversaries. To John, even more than to Paul, George and Ringo, Brian had been a true friend. He and John had thought of each other with mutual respect, admiration and affection.

The details we had at that point were sketchy: it had been a suspected overdose – it was possible that Brian had killed himself. This was horrific. Not only had he died at the age of just thirty-two, but in circumstances that meant there would have to be an investigation. The press would be slavering for sordid details. As the reporters gathered at the college entrance John and I fled to our room. We put our arms round each other and wept. At that moment we were almost painfully close: shock and loss bound us tightly together. 'Christ, Cyn, what are we going to do?' John said. I was in no state to provide an answer.

The press wouldn't leave until someone came out and gave a statement. The others came to our room to talk about who should speak on their behalf. John said he'd do it, and stumbled towards the mêlée of lenses. He was hit by a barrage of questions.

'Yes, I've just heard.' He could hardly speak.

'No, I don't know.'

'Yes, it's terrible.' His face was ashen.

'Sorry, but I've got to go now.'

He came back to our room and we all went to gather in the hall, away from the clamour of the press. The silence was oppressive. None of us could speak. We all felt numb, traumatised by the news.

The sun was shining when the Maharishi's chief devotee arrived to summon us to his presence. We got to his quarters and walked, heads bowed, into an unbelievably beautiful room full of flowers and colour. Everything seemed to shimmer, including the Maharishi, who sat yoga-style in the centre. He asked us to sit down on the floor and talked to us for the next few minutes about life's journey, reincarnation, release from pain and this life being a stepping-stone to the next. He said that Brian's spirit was with us and that for him to have an easy passage we must be joyful for him, laugh and be happy, because negative feelings would hold him back.

The Maharishi's words helped us all to feel a little less bleak and we made a huge effort to be cheerful for Brian's sake. Some elements of the press, catching the Beatles smiling after Brian had died, interpreted this as callousness. But that was the opposite of the truth: John and the others were devastated, their smiles an attempt to do something loving for Brian.

That afternoon we drove home. In the car John and I held hands, trying to give each other strength. Every now and then John would mutter, 'Oh, Christ, why? Why, Brian? I just can't get it into my head.' He was as low as he had been after Stuart's death, and Brian's passing was yet another sudden loss. For the Beatles, it was the end of a hugely important chapter in their lives.

To the world Brian had had everything. Young, good-looking and successful, he had made the Beatles, he had a stable of other artists, legions of well-placed friends, two homes and every luxury money could buy. Why would he want to die? But those of us who knew him well were aware that there was far more to Brian. He was a complex and insecure man. Although he had had lovers, he had never found what he longed for: a stable, loving relationship. He flitted from one scene to another, always hovering on the outside. His loneliness was almost tangible. Even with the Beatles Brian had been an outsider. He loved them dearly, but they all had partners and families and, in many ways, looked up to him. He had remained apart from them.

We had to wait twelve days for the inquest to announce a verdict on Brian's death. When we heard it had been accidental, we wept with relief. But a picture of Brian's last days emerged that left us sad and guilty. After the Beatles had stopped touring a year earlier, they had had less day-to-day contact with Brian. A large part of his job had been organising their appearances and tours, so although he was still their manager and there was still regular contact between them, it was not at the same level. Of course, he had other acts to manage, of whom Cilla Black was the most successful, but the Beatles were always his *raison d'être*. The end of touring had evidently left a hole in his life. In addition, none of us had known how prone to depression Brian was. It seems he had suffered from depressive episodes for many years, and he had been depressed during the months before he died. Latterly he had rarely got up before lunchtime. Sometimes he went into his office in the afternoon, but more often he didn't and his work had suffered. Just over a month before his death his father, Harry, had died, which might have fuelled his depression. His mother, Queenie, had been staying with him recently and he'd made a huge effort to comfort and reassure her, getting up to have breakfast with her, then spending a normal day in the office and coming home to her in the evening. She had left on the Thursday, three days before Brian died. He had planned to spend that Bank Holiday weekend at his country house with friends, but when most of them had to cancel he had driven back to London late in the evening. His body had been found the next day, when the live-in couple who worked for him, Antonio and Maria, had become worried. His bedroom door had been locked, so they had called his secretary, then a doctor. Eventually the door was forced and he was found dead in his bed.

The police had discovered seventeen bottles of pills in Brian's house. He had been on large quantities of anti-depressants and sleeping pills for some time. He was taking more of them than he was supposed to and died not from a massive overdose but from a series of smaller, unintentional ones that had led to a build-up of drugs in his system. The conclusion was that he had been careless in taking the drugs, but had not intended to kill himself. We were certain that that was right. Apart from anything else, Brian would never have put his beloved mother through more grief so soon after the loss of his father.

His body was returned to his home in Liverpool, where his funeral took place, quietly and privately. The Beatles didn't attend because the family didn't want the attention their presence would attract. Instead, in October, we all went to his memorial service, at the New London Synagogue in St John's Wood, not far from the EMI studio. It was a memorable, moving occasion, a tribute to a man we had loved and would never forget.

Peter Brown, Brian's personal assistant, who had previously worked for him in Liverpool, took over Brian's office and did his best to deal with the business side of things, but the Beatles were unsure what to do next.

The result was that they moved in two new directions. In the absence of a manager they decided to form a company, Apple, to bring all their business affairs under one roof and to expand into new areas. The idea was that they'd have their own record label as well as all sorts of other enterprises. They bought a headquarters in Savile Row, and their first venture was a boutique in Baker Street that opened in early December 1967, managed by John's old friend Pete Shotton. The outside of the building was painted with psychedelic colours and the boys loved it, but the neighbours in the surrounding buildings complained. The press went to town and the paint came off again.

The Beatles were innocents in the business world. All four were hugely enthusiastic about Apple and spent a fortune kitting out the Savile Row building with a recording studio, luxurious furnishings and fittings, drinks cabinets in every room and so on. But no one around them had much more business acumen than they did: they employed a large staff, including several old friends, who had no idea what they were supposed to be doing. The Beatles would turn up regularly for meetings, in high spirits about their new company, but little would get done.

Meanwhile John and George were also being drawn towards the Maharishi. What had begun as a passing interest now became a life quest. It was as though, with Brian gone, the four needed someone new to give them direction, and the Maharishi was in the right place at the right time.

John and George agreed to go to his ashram, or training centre, in Rishikesh, at the foot of the Himalayas in India, to study meditation. Patti and I would go too. Paul, Jane, Ringo and Maureen were less convinced about the joys of meditation but decided to join us. The trip was planned for February 1968 and we would stay for two or three months.

John, always passionate about a new cause, was evangelical in his enthusiasm for the Maharishi, talking about spreading the message to the world, and devoting his life to meditation. I was a little more sceptical, but I liked the message and enjoyed the meditation, so I was happy to go to India and learn more. I hoped, too, that time together in such a peaceful environment and out of the spotlight would be good for John and me.

That Christmas the Beatles *Magical Mystery Tour* was shown on television. It had begun as an experiment, an idea Paul had, that they should simply get on a bus and see what happened. But it was the Beatles' first flop. The critics panned it as 'rubbish' and 'boring', and it wasn't even shown in America. It lacked plot and the songs were weak. John called it 'the most expensive home-movie ever' and privately he blamed Paul, which I always thought was unfair: they had all agreed to do it and taken part.

Just before it was shown there was a launch party at London's Royal Lancaster Hotel, on 21 December. It was an extravagant fancy-dress affair and I wore a crinoline dress, like something on a chocolate box, while John went as a greased-up, leather-clad teddy-boy. He had invited his father, Alf, to come and they got very drunk together. It wasn't a happy evening. I became more and more upset as John flirted with other women, including Patti, who was seductively attired as a belly-dancer. Lulu was outraged on my behalf and shouted at John that he should be ashamed of himself. It made me smile, despite my wounded pride, to watch her, dressed as Shirley Temple, giving teddy-boy John a dressing-down. The evening ended with John and Alf dancing drunkenly together, while I was thoroughly miserable. I had been about the only person John hadn't danced with.

Four days later, on Christmas Day, Paul and Jane got engaged. We were delighted because we loved Jane. George and Patti had married quietly, without any of us there, almost two years earlier in January 1966. Now the last Beatles couple was to tie the knot. We knew that Paul hadn't found it easy to be with such a successful girl-friend: like the other Beatles, he was essentially an old-fashioned Liverpool man, who wanted his woman tucked away at home cooking the dinner and minding the kids. Jane was never going to fit that mould: she travelled for her work almost as much as Paul did and had her own life. She wasn't willing to wait around for him and at times she was away working when we were all together.

But Paul loved her, we knew that, and we were glad they had worked things out.

Shortly before we were due to leave for India John spent the weekend with Derek Taylor, a former journalist who had become the Beatles' press spokesman and a good friend to us all. He, his wife Joan and their five children lived in a big country house where they seemed incredibly contented. When he came home after that weekend John put his arms around me and said, 'Let's have loads more kids, Cyn, and be really happy.'

Despite my increasingly strong feeling that John was slipping away from me, it seemed at moments like that as though nothing had changed. John was off drugs and seemed almost like his old self. 'We can make it work, Cyn,' he said. 'When we're in India we'll have time for us and everything will be fine.' I hoped he was right.

A few days before our departure, we had a meeting with the Maharishi's assistant at a house in London to finalise details of the trip. As we entered the main room I saw, seated in a corner armchair, dressed in black, a small Japanese woman. I guessed immediately that this was Yoko Ono, but what on earth was she doing there? Had John invited her and, if so, why?

Yoko introduced herself to the group, then sat silent and motionless throughout, taking no part in the proceedings. John chatted to the other Beatles and the Maharishi's assistant and appeared not to notice her. My mind was racing. Was he in regular contact with this woman? What on earth was going on?

At the end of the evening Anthony was waiting outside

for us. He opened the car door and, to my astonishment, Yoko climbed in ahead of us. John gave me a look that intimated he didn't know what the hell was going on, shrugging, palms upturned, nonplussed. He leant and asked if we could give her a lift somewhere. 'Oh, yes, please. Twenty-five Hanover Gate,' Yoko replied. We climbed in and not another word was said until we dropped her off, when she said, 'Goodbye. Thank you,' and got out.

'How bizarre,' I said to John. 'What was that all about?'

'Search me, Cyn.' He insisted he hadn't invited Yoko and knew nothing of her being there, but common sense dictated that it had to have been John who had asked her to come. Whatever my doubts, though, it was clear that he wasn't going to provide an explanation.

Soon after this, in a pile of fan mail, I came across a typed letter from Yoko to John. In it she talked about wanting her book to be published again right away so that she could take people into the world of surrealism to change the whole world into one big beautiful game. She said she was afraid she might flip out soon if she had to carry on holding her message by herself. She went on to apologise for always talking about herself so much and for pushing her goods to him – by which I assumed she meant being pushy about her art. She thanked him for his patience and said that when she didn't see John she was thinking very much about him. She also talked about her fantastic fear, whenever John said goodbye, that she would never see him again because she had been so selfish.

That letter made it crystal clear that they had been in contact. How well had they got to know one another? I tackled John, who told me she'd written many times, both letters and cards, but said, 'She's crackers, just a weirdo artist who wants me to sponsor her. Another nutter wanting money for all that avant-garde bullshit. It's not important.'

I had no way of knowing whether he was telling me the truth. He sounded genuine, but a sixth sense told me there was more to this than he was admitting. I tried to put it to the back of my mind. We were going to India, and I wanted that to be a special time for us.

On 16 February we flew out with George, Patti, her sister Jennie and Magic Alex. The plan was to stay for ten weeks, so Mum had moved into our house to look after Julian and, although I hated leaving him, I knew he'd be happy with her and Dot. We'd decided not to disrupt his routine by taking him and, in any case, as we would be meditating for many hours each day, there was no place for a child.

From the airport we were driven for several hours to the Maharishi's meditation training centre. It was set in a beautiful spot above the forested foothills of the Himalayas, on the banks of the Ganges, and surrounded by vibrant flowers and shrubs. The centre was built to house several dozen students. Each of its low stone cottages contained five rooms, and when we arrived dozens of people of all ages, creeds and races were gathered to take the Maharishi's path to enlightenment. Among them were actress Mia Farrow, Mike Love of the Beach Boys and, later, the singer Donovan with his friend, a burly bloke called Gypsy. Donovan was having a romance with Jennie, and wrote his hit song 'Jennifer Juniper' for her in India.

John and I had a room with a four-poster bed, a dressing-table, a couple of chairs and an electric fire. Close by were the Maharishi's house, a swimming-pool, a laundry, a post office and a lecture theatre where we would be expected to gather for regular talks from our leader.

I loved being in India, away from the stresses and pressures of our lives. Here, there were no fans, no hordes of people, deadlines, demands or flashing cameras. Just peace, quiet and sweet mountain air filled with the scent of flowers. Best of all, John and I could be together for much of the time. I hoped we would meditate together, grow in mutual understanding, talk, go for walks and rediscover our lost closeness.

Four days later Paul, Jane, Ringo and Maureen arrived to join us, looking forward to a peaceful break. Ringo, wary of the spicy Indian food he was certain would be served in the communal dining-hall and determined to take no chances, had brought a crate of baked beans and another of eggs. In fact, some of the centre's food was surprisingly ordinary: for breakfast, which was taken at long trestle tables out in the open and often shared with brazen monkeys, we had corn flakes, toast and coffee.

In the first week we settled into a routine, meditating for several hours a day and going to lectures, then spending the rest of the time on our own pursuits. I had taken pens and paper and spent hours drawing but also, for the first time in my life, writing poetry. It was crisply cold in the mornings and, having failed to pack warm clothes, we spent much of our time wrapped in the blankets from our beds. But the simplicity of life at the ashram, with few material goods, was enormously appealing and we all enjoyed slowing down and taking the time to breathe.

In the evenings we got together, occasionally breaking the no-alcohol rule with a glass of hooch, smuggled in by Alex from the village across the river and tasting remarkably like petrol. Giggling like naughty schoolchildren, we'd pass round the bottle, each taking a swig, then contorting as it scorched its way down our throats.

Every now and then the Maharishi would arrange an outing for us to a nearby town, where I would marvel at the stalls selling saris and rolls of fabric in every colour under the sun. All of us girls bought saris and learnt to wear them.

But Ringo and Maureen weren't happy: they missed their children, Ringo was soon tired of eggs and beans and Maureen had a phobia about flies, which were inescapable in India. After ten days they announced they'd had enough and were going home. 'That Maharishi's a nice man,' Ringo said, 'but he's not for me.'

Meanwhile, I was not having the second honeymoon I'd hoped for. John was becoming increasingly cold and aloof towards me. He would get up early and leave our room. He spoke to me very little, and after a week or two he announced that he wanted to move into a separate room to give himself more space. From then on he virtually ignored me, both in private and in public. If the others noticed they didn't say so.

I did my best to understand, begging him to explain what was wrong. He fobbed me off, telling me that it was just the effect of the meditation. 'I can't feel normal doing all this stuff,' he said. 'I'm trying to get myself together. It's nothing to do with you. Give me a break.'

What I didn't know was that each morning he rushed down to the post office to see if he had a letter from Yoko. She was writing to him almost daily. When I learnt this later I felt very hurt. There was I, trying to give John the space and understanding he asked for, with no idea that Yoko was drawing him away from me and further into her orbit.

After a month Paul and Jane had also had enough and decided to go home. John and I, with George and Patti, wanted to stay. Aside from my troubles with John, I got a great deal from being there. I loved the serenity of the place, and still hoped that John would work through his demons and move back to me. In the absence of the true explanation, I put his aloofness down to a bout of intense self-exploration and hoped that it would, ultimately, be good for him and for us.

Was I in denial when I clung to the belief that our relationship was not terminally damaged but simply strained? Perhaps. As I saw it, we were undergoing a whole new experience and John's behaviour towards me was part of it. I was disappointed that the closeness I'd hoped for hadn't happened, but I still saw us as a unit, shaken but ultimately solid.

A couple of weeks before we were due to leave, Magic Alex accused the Maharishi of behaving improperly with a young American girl, who was a fellow student. Without allowing the Maharishi an opportunity to defend himself, John and George chose to believe Alex and decided we must all leave.

I was upset. I had seen Alex with the girl, who was young and impressionable, and I wondered whether he – whom I had never once seen meditating – was being rather mischievous. I was surprised that John and George had both chosen to believe him. It was only when John and I talked later that he told me he had begun to feel disenchanted with the Maharishi's behaviour. He felt that, for a spiritual man, the Maharishi had too much interest in public recognition, celebrities and money.

By dawn the next morning Alex had organised taxis from the nearby village and we left on the journey back to Delhi and a plane home. After eight weeks the dream was over. I hated leaving on a note of discord and mistrust, when we had enjoyed so much kindness and goodwill from the Maharishi and his followers. I felt ashamed that we had turned our backs on him without giving him a chance. Once again John was running away, and I had little choice but to run with him.

The journey home was long and grim. I was close to tears and John was paranoid, afraid that the Maharishi would take his revenge on us in some way. Our taxi, a battered old saloon car, broke down on the way to the airport. The driver assured us he was going for help and left John and me standing at the roadside. John's solution was to thumb a lift from the next car that came along. Fortunately the driver and his friend recognised him and, with great sweetness, got us to the Delhi hotel where the others were waiting anxiously. We had planned to spend a night in Delhi, but the hotel was full and Alex had discovered that, if we hurried, we could make the night flight to London. I hated the rush, which seemed unnecessary, but with the others setting the pace we hit the road and just made the flight.

Sad as I was at the way the Indian trip had ended, it was wonderful to hold Julian again, shower him with kisses and sit in the kitchen with a cup of tea, catching up on the news with Mum and Dot. We had brought Julian back six little Indian outfits and some delicate hand-carved wooden soldiers, gifts from the Maharishi for his fifth birthday a few days earlier. He looked adorable in his Indian clothes and was thrilled to have Mum and Dad at home.

Back at Kenwood John continued to be distant towards me. Now that we were away from the others and the charms of India, I felt increasingly afraid and depressed. John and I were back in the same bed, but the warmth and passion we had shared for so long were absent. John seemed barely to notice me. He was little better with Julian and was more likely to snap at him than give him a hug.

There was just one moment of real warmth between us and that was, ironically, when John confessed to me that he had been unfaithful. We were in the kitchen when he said, out of the blue, 'There have been other women, you know, Cyn.'

I was taken aback, but touched by his honesty. 'That's OK,' I told him.

He came over to where I was standing beside the sink and put his arms round me. 'You're the only one I've ever loved, Cyn,' he said, and kissed me. 'I still love you and I always will.'

A couple of weeks later John suggested that I join Magic Alex, Jennie, Donovan and Gypsy on a two-week holiday in Greece. I told him I didn't want to go without him. Apart from those rare occasions when I had taken Mum and Julian away because he was working, we had never spent holidays apart.

'I've got a lot on at the moment and I can't go, but you should. It might cheer you up,' he said. I was uncertain, but he persisted and in the end I decided to go. John was busy writing songs for the Beatles new album, *The Beatles*, better known after its release as the *White Album*. He wrote thirteen of the tracks, including 'Julia', a tribute to his mother, and 'Goodnight', for Julian.

Cheered by the hope that John might miss me, and the prospect of a change, I left for Greece. Julian had gone to stay with Dot's family and John was lying on our bed when I left. He was in the almost trance-like state I'd seen many times before and barely turned his head to say goodbye.

Surprisingly, given my worries about the future of my marriage, it was a lovely holiday. Two weeks of Greek sun, sea, ouzo, tavernas and the laughter and companionship of the others raised my spirits. Despite my mistrust of Alex over the Maharishi episode, in Greece I put aside my doubts as he interpreted, smoothed the way and behaved in every way like a good friend. By the time we were due to head home I felt so much better. I had missed John badly and had convinced myself that we could make a fresh start. I was full of energy and plans for our future. The idea of breaking up was still inconceivable to me, perhaps because my own parents had never contemplated such a thing. I fully expected that John and I would stay together and find a way to work through our problems.

What I hadn't allowed for was that John's history, his attitude to marriage and the family were very different from mine. He had hardly ever seen his parents together: at five he had been abandoned by his father and, effectively, his mother too. His own father had suffered in the same way. Given how often and uncannily we repeat the patterns of our parents, I should, perhaps, have been more prepared for John to leave his own marriage and five-year-old son. But I was too young, too inexperienced and too determinedly optimistic to take it seriously.

On the way home our plane stopped off in Rome where we had lunch. Wouldn't it be fun to finish the day with dinner in London, after breakfast in Greece and lunch in Rome? We laughed. 'Let's get John to join us.' Alex suggested I ring him to let him know what time we would be back.

I spoke to him briefly. 'Hi, darling, I'll be home soon. Can't wait to see you.'

John's reply sounded normal: 'Fine, see you later.'

Donovan and Gypsy headed home, but Jennie and Alex came with me to Kenwood to see if John fancied dinner out. We arrived at four in the afternoon and immediately I knew something was wrong: the porch light was on, the curtains were still drawn and everything was silent. There was no Dot to greet me, no Julian bounding through the door, shouting with delight, for a hug. What was going on?

The front door was unlocked. The three of us walked in and began to look for John, Julian and Dot. 'Where are you all?' I called, still expecting them to appear from behind a door, laughing at the joke.

As I put my hand on the sunroom door I felt a sudden frisson of fear. I hesitated, for a second, then opened it. Inside, the curtains were closed and the room was dimly lit so it took me a moment to focus. When I did, I froze.

John and Yoko were sitting on the floor, cross-legged and facing each other, beside a table covered with dirty dishes. They were wearing the towelling robes we kept in the poolhouse, so I imagined they had been for a swim. John was facing me. He looked at me, expressionless, and said, 'Oh, hi.' Yoko didn't turn round.

I blurted out the only thing I could think of: 'We were all looking forward to dinner in London after lunch in Rome and breakfast in Greece. Would you like to come?'

The stupidity of that question has haunted me ever since. Confronted by my husband and his lover – wearing my dressing-gown – behaving as though I was an intruder, all I could do was carry on as if everything were normal. In fact I was in shock, operating on auto-pilot. I had no idea how to react. It was clear that they had arranged for me to find them like that and the cruelty of John's betrayal was hard to absorb. The intimacy between them was daunting. I could feel a wall round them that I could not penetrate. In my worst nightmares about Yoko I had not imagined anything like this.

As I stood in the doorway, rooted to the spot in shock and pain, John said, indifferently, 'No, thanks.'

I turned and fled.

# 16

I WAS DESPERATE TO GET AS far as possible from the scene in the sunroom. I ran towards the stairs, with no idea in my head other than to escape. Jennie, who had been hovering awkwardly in the kitchen with Alex, appeared: 'Can I stay with you?' I was holding back tears and couldn't answer.

On the landing I stumbled across a pair of small Japanese slippers, placed neatly outside the guest-bedroom door, which I opened. The bed hadn't been used. For a second, I considered hurling the slippers at their owner, but I couldn't have confronted John and Yoko again. I felt utterly humiliated, and longed to disappear.

I threw a few things into a bag and ran downstairs again. Barely twenty minutes after I had arrived for what I had hoped would be a loving reunion with John, I got back into the waiting taxi with Jennie and Alex and drove away from my home, leaving John and his lover behind.

I was in a daze. My mind seemed to be floating and I couldn't focus on anything other than the vivid image of them together. Every time I saw it I was overwhelmed by a fresh stab of pain, yet I couldn't stop myself conjuring it up as I tried to take in the betrayal. Their intimacy had been so powerful that I had felt like a stranger in my own home.

I had no memory of the journey with Alex and Jennie or of arriving that evening at the small house they shared. All I remember was that Jennie went to her room as soon as we arrived. She was shocked and embarrassed. 'Cyn, I'm so sorry but I've got to go to bed. All this has been too much for me. I'll see you tomorrow. Will you be OK?'

I was left in the tiny sitting room where beanbags and ethnic cushions were scattered all over the floor and the curtains were drawn. I collapsed on to the sofa in front of the fireplace, rousing myself only to phone Dot and ask her to look after Julian for a couple more days. Thank God he was safe and happy with her family.

I was in such a confused state that I barely knew what was real any more. My mind kept turning nightmarish somersaults. It was almost like one of the LSD trips. What was I doing here? Why had I run? What on earth was going to happen to me and Julian? The world as I knew it was disintegrating. I wondered how long John had been sleeping with Yoko. Had I been a complete fool, failing to see the obvious? Had he been lying for weeks, or months? Had he sent me on holiday to get me out of the way so that he could pursue his relationship with Yoko? Each new thought added fresh torment.

'I think you need a drink, Cyn.' Alex moved around the room, lighting candles, then brought out a bottle of red wine and two glasses. I was grateful for something that might help to numb the pain and knocked back several glasses. Then Alex produced a second bottle, and we drank that too. I was exhausted, and sank rapidly into a groggy state.

Alex had been chatting away and I was hardly registering what he was saying until I was shocked into awareness: 'Do you know, Cyn?' he said. 'I've always loved you. This is perfect. How much money do you have? Why don't you and I run away together? We could have a great life. That would show John and Yoko.'

I answered, without thinking, that I had a mere thousand pounds in the bank. No riches, no fortune, nothing. I was too dizzy, disoriented, shocked and heartbroken to take in what was going on. I could barely focus. Everything in the room was swimming around, one object merging into another. Alex was John's friend: what was he talking about? It must be some kind of joke.

I needed the bathroom. When I got there I was violently sick. Then I realised that the shock had brought on my period, several days early. I lurched out of the bathroom on wobbly legs and saw, through an open door on the landing, a bedroom. I went in, collapsed on the bed, fully clothed, pulled the covers over me and passed out.

Some time later Alex crept into the bed and was attempting to kiss and fondle me, whispering that we should be together. I pushed him away, sickened.

The next morning I woke with a hangover and the dreadful realisation that my marriage was probably over. I had no idea what to do next. After Alex's behaviour the night before I wasn't comfortable staying in his house. He'd pressurised me at a time when I was vulnerable. But I had nowhere else to go, so I spent the day huddled on the sofa, drinking tea and trying to work out what to do. Jennie was sympathetic and Alex, thankfully, kept a low profile.

After a couple of days, I knew I had to go home. I was aching to see Julian. John and I might be making a mess of our marriage, but we still had a small son who needed us.

I got myself together sufficiently to face John. There had been no word from him and I had no idea what he was planning, but I needed to find out. I took a long, hot bath, tried to make myself look better than I felt, packed up my things and took a taxi home.

Walking in was difficult: I had no idea what I might find. But the house was astonishingly normal. The sun shone, the curtains were drawn back and everything was neat as a pin. Clearly Dot had been hard at work. As I stood wondering who was at home, Julian ran to me and leapt into my arms. It was wonderful to be able to hug him.

At that moment John wandered out of the den. 'Oh, hi,' he said casually. 'Where have you been?'

I stared at him. Surely he was joking. But, no, he seemed relaxed, normal, even pleased to see me – he came over and planted a kiss on my cheek.

Had it all been a nightmare? Or was John truly capable of doing something like that, then dismissing it as unimportant? I knew him so well, but he'd surprised me before with his ability to compartmentalise his life and put aside anything that he didn't want to deal with. But ignoring the fact that I'd found him with Yoko seemed a little like stepping round an elephant in the middle of the drawing room.

I didn't want to say anything in front of Julian so, for the next few hours, I did my best to pretend everything was normal. It was good to see Dot, who looked at me a little anxiously but didn't say anything, although she must have known what had happened. She fussed over me, helping me to unpack and sort out my clothes.

It was evening before John and I had a chance to talk. I had to steel myself for the confrontation we would usually avoid to ask him what was happening with Yoko.

'Oh, her?' he said, as if surprised that I'd asked. 'Nothing, it's not important.'

'We have to talk, John,' I told him. 'Please don't pretend that nothing's happening.'

Eventually we did talk, perhaps more honestly and in more depth than we had since our student days. We talked of our failings and faults, our love for each other, our hopes and dreams. John talked again about his other women, and insisted that Yoko was no more important than they had been. 'It's you I love, Cyn,' he said. 'I love you now more than I ever have before.'

That night we went to bed and made love, and my bruised heart felt lighter. It wouldn't be easy to forget what had happened, but if it really was in the past, well, I would try. The John I had spent the evening with and in whose arms I lay was so completely different from the man who had stared at me impassively as he sat with Yoko in the sunroom that it was hard to accept that they were the same person. Had he been on drugs? I knew he was dabbling again since his disenchantment with the Maharishi. Only drugs could explain how he had behaved.

In fact, John had never completely renounced drugs. While we were in India my mother had found a stash of LSD he'd hidden at home and had flushed it down the loo. When he discovered what she'd done he was furious, but he couldn't confront her without admitting that he used it so he had to keep quiet.

For the next few days all seemed well. John was in a good mood, Julian was happy to have us around and I was daring to hope that we had got through the worst. John and I had several more honest talks. He talked about his need to explore new avenues and I told him that I knew I couldn't always share them with him. I was very much the girl I had always been, happy to be at home, a wife and mother. I had grown more independent over the years, but I was essentially the same. John had changed a great deal and was searching in new directions for answers in his life. Talking brought us closer. We agreed that we wanted to go forward together, despite our differences. After all, we had always been different and it had worked for us for ten years. Why should it not for the next ten?

I felt determined once more to make my marriage work. But this brief happy respite soon ended. John was due to go to the States with Paul on a business trip in connection with Apple. I suggested that I go with him. It seemed to me that if we were going to remain close we should spend more time together. And a trip to New York would be fun.

John's answer was a flat no. He refused to look at me or discuss it. I felt my stomach tighten: he was distancing himself from me again.

Over the next few days he was irritable and withdrawn, and I felt a rising sense of panic because I couldn't reach him.

I didn't want to be left alone in the house, waiting and wondering, while he was away, so I asked him if he would mind me taking my mother and Julian to Italy for two weeks. 'Yeah, sure,' he replied.

Organising the Italian trip gave me something to do and I threw myself into it, anxious to take my mind off John. What was going on with him now? Had he already regretted his promise that we would stick together and make a go of it? With hindsight, perhaps it wasn't wise of me to go away. Perhaps I should have stayed at home to be there when he got back from America, rather than leaving him alone in the house. Perhaps I shouldn't have gone away with my mother, who irritated John at the best of times but especially since she had destroyed his drugs.

At the time going away seemed like the best thing to do: in Italy there would be the warmth, kindness and company I needed.

By this time Mum had left her house in Esher and moved into Ringo's flat in Montagu Square, just off Baker Street, which we had rented from him for her. She'd been bored in Esher, although she was close to us, and preferred London where there was plenty for her to do and explore.

Mum drove down to Kenwood with my aunt and uncle, who were coming with us. I'd always got on with Daisy, Mum's sister, and – perhaps with an intimation of what was to come – I wanted my family around me.

In Italy we stayed at the same hotel in Pesaro where we'd been two years earlier. The Bassanini family were as hospitable as ever and this time the press had no idea that we were there. Despite my anxiety, I tried hard to make the holiday enjoyable for everyone. I played on the beach with Julian and explored the little town with Mum. Whatever lay ahead, I would deal with it when I got home.

One of the hotel waitresses was a Lancashire girl and a great laugh. We became friendly and she suggested we go out for a night. I was uncertain: I was feeling low that day and had a sore throat coming on. I certainly didn't want us to look like two single girls trying to pick up men. She suggested we ask Roberto, the son of the hotel owners, to take us out. He said he would be delighted to accompany us, and I began to look forward to it. So far I had stayed quietly in the hotel with my family: it would cheer me up to have fun.

Roberto was charming and courteous and looked after us beautifully. He was extrovert and seemed to know everyone in town, introducing us to dozens of people as we moved from bar to bar. I had a good time and, for a few hours, forgot my troubles.

When our car drew up outside the hotel at about two a.m. Roberto opened the door for me and I stepped out to see a familiar figure: Magic Alex. He was hovering outside the hotel, agitated. What on earth was he doing there? He looked at me and Roberto, and my heart sank: here I was in the small hours with a good-looking young Italian. God forbid that Alex might think I was having a holiday romance.

We went inside and found Mum sitting in the lobby, looking distressed. I asked Alex what was going on. He said, 'I've come with a message from John. He is going to divorce you, take Julian away from you and send you back to Hoylake.'

My knees gave way. I felt drained and ill. All I could think at that moment was how cowardly John was to send his lapdog because he couldn't face me. Far more than simply evasive, it was sinister and cruel.

Of one thing I was certain: no matter what, I would never let John take my beloved child. I hurried upstairs to where Julian lay asleep and leant over to kiss his cheek. 'I'll never let you go,' I whispered to him. 'Never.' I crawled into bed, desperate for oblivion, but sleep wouldn't come and I lay, unable even to cry, and watched the dawn. The pain was indescribable but at the same time I experienced a strange relief. This was it, then, after so many months of worry, fresh starts and crushed hopes. My marriage was over. I fell asleep.

I woke with a fever and such bad laryngitis that I could hardly speak. I needed to get home but I was in no state to travel. Mum said she would go back and try to find out what was going on. My aunt and uncle stayed with me to look after Julian.

For the next few days I lay in bed while Signora Bassanini fussed over me with hot drinks and cold flannels. When she brought me the newspapers, there was a picture of John and Yoko, hand in hand, on their first public outing together. They were attending the opening night of the play *In His Own Write* adapted from John's books by our friend, the actor Victor Spinetti. The papers referred to Yoko as John's 'new love'. Her persistence had paid off handsomely: after all the letters and calls, and the times she'd turned up at our door, Yoko had got her man. My man.

Yoko, it seemed, was married too. This was the first I knew of Anthony Cox, her second husband, and her daughter Kyoko, who was four months younger than Julian. How were they feeling, I wondered. Did John and Yoko spare any thought for the two families now being broken up? Did they have any idea of the price of their happiness?

The pictures of John and Yoko had been flashed round the world so everyone would know that I had been replaced. It's bad enough to be tossed aside by your husband, but to be tossed aside so publicly was humiliating as well as painful.

As soon as I was well enough to travel we headed home. Julian and I took a car from the airport to Mum's flat in London. When we arrived, Mum threw her arms round us. There was a big bunch of flowers on the table. 'Who are they from?' I asked. Mum handed me the card, which said, 'Beat you to it, Lil. John.' The flowers had been waiting for Mum when she arrived home. It was John's way of telling her that he knew exactly what our movements were and was one step ahead of us. My blood ran cold: he must have had us followed.

I tried to phone him, we had to talk, but got no answer. What was I supposed to do next? Hours after I arrived an envelope was delivered by hand. The letter inside it informed me that John was suing me for divorce on the grounds of my adultery with Roberto Bassanini. Presumably Alex had told John I had been out with him.

It was laughable. Roberto had been kind and a good friend, but I had never been unfaithful to John. It was his attempt to make himself feel better about what he was doing.

What next? Would I be accused of an affair with Alex? What was behind Alex's crude attempt to seduce me? Alex had never been interested in me – he wanted only to stay in John's favour. I felt cold with disbelief and fear: if Alex claimed we had slept together it would be his word against mine.

What had happened to John that he could do this? I had loved him passionately for so many years, but now I lost all respect for him. I had always believed he was a fundamentally decent person, but in accusing me of adultery and threatening to take Julian from me he had put aside every ounce of decency. He knew I hadn't been unfaithful to him. And he knew that to lose Julian would break my heart. Who was advising him that this was the best way to end his marriage?

Mum called my brother Charles, who flew back from Libya, where he was working, and found me a lawyer. I was obviously going to need one.

Then I called Peter Brown in the Beatles' office. He was a nice man, a good friend. He was apologetic and embarrassed. 'I'm sorry, Cyn, this shouldn't be happening,' he said. He promised to try to arrange a meeting for me with John. I was no longer hoping to save our marriage, but I felt that if John and I could sit down together and talk, we might find a way to separate amicably.

Peter called back: 'I'm sorry, Cyn, he won't see you.' He promised to keep trying.

A few days later he called to say that Mum, Julian and I could move back to Kenwood, and John and Yoko would have the London flat.

Julian and I had nothing with us but our holiday luggage: at least we could go home. Mum was losing her flat, but she could live with us. I helped Mum pack all her things, and Anthony came to drive us to Weybridge. He'd been ferrying John and Yoko around since we'd gone away and, always diplomatic, he said not a word against John but made it clear that he was delighted to see us. It was good to be home, among our own familiar things, and Dot was waiting with hugs and kisses. Unlike Anthony, she couldn't wait to tell us what she thought. 'Disgraceful!' she said.

'Carrying on with that woman under your roof. I don't know how he could do it.' It felt good to have someone rooting for me. She also told me that John had presented Alex with a brand-new white Mercedes. I couldn't help suspecting that this extravagant gift might have been a thank-you.

The atmosphere at home had changed. John had gone, and although he'd taken little, the house felt empty. I missed him, and so did Julian, who kept asking when he was coming home. I didn't know what to tell him. I tried to explain, gently, that Mummy and Daddy wouldn't be living together any more, but it didn't make a lot of sense to him. 'Why not?' he wanted to know, and at that point my resources fizzled out. All I could offer him was, 'We weren't getting on very well.' How could I say that Daddy had decided he preferred someone else? 'You'll see him soon,' I promised, praying that it would be true.

I was still calling Peter Brown, asking repeatedly for a meeting with John. Peter's answer was always the same. 'Cyn darling, I'm so, so sorry. I've given them the message. I know how desperate it is for you but they're not responding. I'll keep trying. Keep your chin up.'

Eventually he called to say that John had agreed to a meeting. He would come to Kenwood to see me the following Tuesday at three thirty.

As the time for his arrival approached, Mum, Dot and I were in the kitchen with Julian. We were feeling the strain and, drained by the effort to appear cheerful and normal, our conversation went round and round in circles as we downed cup after cup of tea. I felt terribly nervous:

despite the way he had treated me I still loved John. If I had to lose him, at least I wanted to try to part as friends.

When the front-door bell rang, we jumped up in unison. I looked out of the window. John and Yoko were standing outside together, dressed from head to toe in black. I felt panicky. God! He'd brought her with him.

Dot showed them into the den to give me a minute to adjust to this new development. It hadn't occurred to me that Yoko would be there too.

Mum was fuming: 'How dare that woman come? What's the matter with John? I'm coming with you, love. There's no way you're going to face them alone.' She was ready to annihilate them both and Dot had to physically restrain her until she burst into tears.

There was no calming her but I did my best. 'I'll be OK, Mum, I promise. I'm a big girl now.' Julian was jumping up and down, longing to see his daddy. I took his hand and headed for the den.

When we walked in I barely recognised John. It had been only a few weeks since we'd last met, but he was thinner, almost gaunt. His face was deadly serious. There was no hint of a smile, even when Julian ran up to him. He was, quite simply, not the John I knew. It was as if he'd taken on a different persona. He was sitting on the footstool. Yoko was beside him in the armchair, shrouded by her hair, her face set in an expressionless mask, the fingers of one hand picking at the nails of the other. God, why couldn't he have chosen someone easier, someone I might have related to? What power did she have over him? The thought of her looking after my son was ghastly. Did she feel no remorse or embarrassment? It seemed not.

John spoke first. 'What did you want to see me for?' he said coldly. He was defiant, ready for battle, and I was clearly the enemy.

I sat down on the sofa. I had hoped that having Julian there would smooth things, but after an initial hug John had ignored him and the atmosphere was frigid. I sent Julian to see Dot in the kitchen, then took a deep breath. 'Look, John, can't we find a better way to do this? I haven't been unfaithful to you, I'm sure you know that.'

John could barely look me in the eye. 'Forget all that bullshit, Cyn. You're no innocent little flower.'

I tried again: 'John, I was with my family in Italy, I was with Julian. Do you honestly think I would do that to them or to you?'

At this point Yoko disappeared briefly. Dot told me later that she'd gone into the kitchen to ask my mother for a glass of water, Mum told her in no uncertain terms how she felt: 'Get out of this house. Get out of our lives. Haven't you any shame? Look at this little boy. Don't you care about what you're doing? You're breaking everyone's hearts.'

Back in the den John shot another broadside at me. 'What about that Yankee cowboy?

'What Yankee cowboy?'

'In India.'

I could hardly believe what I was hearing. The Yankee cowboy was an American actor called Tom Simcox, whom we'd met at the Maharishi's centre. I'd got on well with him, but the idea that we'd had an affair was preposterous. Who would be next in line as my lover? The elderly gardener?

Tom had never even hinted at anything untoward between us, apart from once asking me, very sweetly, if I had a twin sister. Other than that we'd talked about art and poetry, mostly around the table with everyone else. Hardly a crime, especially when John was largely ignoring me. Was the old jealousy still rearing its head? Or was he simply looking for a get-out clause, with me as the villain?

There was more. 'You know that when he left Rishikesh that cowboy gave George a letter to pass on to you, but instead he gave it to me. He was being loyal.'

I was bemused and curious. 'So, what was in this mysterious letter, John?'

He refused to tell me. I suspected he was disappointed that I hadn't looked more guilty.

Yoko came back in. John announced, in the same cold, clipped tone he had used throughout, that talking was pointless.

I was feeling more and more distressed. I tried one last time: 'John, please, let's discuss things.'

'We can do that through the lawyers.' Then, to Yoko, 'Come on, let's go.' He called ''Bye', to Julian, and marched out of the house, Yoko at his heels.

I went back into the den and collapsed in tears on to the sofa. This wasn't going to be a friendly parting. I'd given it my best effort and got nowhere. Deep down I knew that cutting himself off emotionally from me and Julian was probably the only way he could go through with the separation, but I wondered if he had any idea of how much he was hurting us, and if behaving like that was hurting him too. I didn't even have the comfort of thinking that he had been on drugs. If he had been, he might have been softer. He must have needed all his wits about him to maintain that icy-cool front.

That night I lay awake. I had to survive this for Julian. I couldn't afford to crumble: I had to be strong, do what was best for him. I could fight the divorce, but that would get horribly messy and in the circumstances as clean a break as possible seemed best. By dawn I had made my decision. I would counter-sue for divorce, citing his adultery with Yoko.

For the next few weeks I lived like a hermit at Kenwood, concentrating on my son. I went to meetings with lawyers and came home drained and sad, but apart from that I barely went out. John's clothes were still in our wardrobe and I hated lying alone in our bed. I cried myself to sleep most nights, then woke in the early hours and lay awake, trying to understand what had gone so wrong. I did my best to keep cheerful when I was with Julian, but every now and then he would catch me crying. He'd throw his arms round me, saying, 'Don't cry, Mummy, please don't cry.' I lost a lot of weight, unable to face eating. Mum and Dot were there, keeping an eye on me, and I was grateful not to be alone. But apart from them I saw or heard from hardly another soul. It seemed that John had cut me off not just from him but from the whole Beatles family.

The only person who came to see me was Paul. He arrived one sunny afternoon, bearing a red rose, and said,

'I'm so sorry, Cyn, I don't know what's come over him. This isn't right.' On the way down to see us he had written a song for Julian. It began as 'Hey Jules' and later became 'Hey Jude', which sounded better. Ironically John thought it was about him when he first heard it. It went on to become one of the Beatles' most successful singles ever, spending nine weeks at number one in the US and two weeks in the UK.

Paul stayed for a while. He told me that John was bringing Yoko to recording sessions, which he, George and Ringo hated. Paul had broken up with Jane Asher a couple of weeks after John had left me. I was sorry because I'd really liked Jane. In a scenario bizarrely like ours, Jane had come home a few days early from a theatre tour and had caught him in their home with another girl. Understandably she had walked out. But that was where their story parted from ours. Paul blamed himself and was heartbroken.

He joked about us getting married – 'How about it, Cyn?' – and I was grateful to him for cheering me up and caring enough to come. He was the only member of the Beatles family who'd had the courage to defy John – who had apparently made it quite clear that he expected everyone to follow his lead in cutting me off. But Paul was his own man and not afraid of John. In fact, musically and personally, the two were beginning to go in separate directions so perhaps Paul's visit to me was also a statement to John.

He drove off, promising to keep in touch, but a month or two later he got together with American photographer

Linda Eastman and his life began a new phase. It was many years before we met again.

Apart from Paul I heard from no one. Ringo, Maureen, George, Pattie and all the Beatles' friends and followers kept away. They didn't want to bring John's fury on themselves and probably didn't know what to say to me anyway, shocked and embarrassed by what had happened. In time their marriages would unravel too, but ours was the first to go, and it must have shocked everyone.

I could have phoned them, but I didn't have the heart. I talked to one or two close friends, such as Phyl, with whom I was still in regular contact, but that was all. I felt ashamed, hurt and discarded. I dreaded awkward conversations, pity, embarrassed silences. I refused to give interviews and eventually the press left me alone. At that stage I had neither the energy nor the resources to deal with anything more than putting one foot in front of the other and getting through each day.

I felt as though John had cut me off like a gangrenous limb – total amputation from all that I had been or known. As I waited for his next move I felt increasingly vulnerable and afraid.

We were progressing towards divorce but we still had to sort out the three biggest obstacles: who would divorce whom, money and, most important of all, Julian.

In October John and Yoko's London flat was raided by the police, who found some cannabis. They were charged with possession of the drug and with obstructing the police. They were later found guilty of possession, but not obstruction.

A week later they announced that Yoko was pregnant, with a baby due in February. My humiliation was complete, although I could see the comical side too. That meant she had got pregnant in May, when John was determined to accuse me of adultery. Despite this, when I heard in late November that Yoko had lost the baby at six months I was genuinely sorry. But John could hardly deny his adultery. To my enormous relief he agreed not to contest my divorce petition. The fear that Magic Alex would suddenly decide to claim I had committed adultery with him, and provide John with ammunition against me, had lurked at the back of my mind. Thank goodness I could now let that prospect go.

I saw John and Yoko once more, in the lawyers' office to which John and I had been summoned to sign the divorce papers. John barely looked at or spoke to me throughout the meeting, and when he left the room to go to the loo, Yoko followed him. A few minutes later I went into the corridor with my lawyer for a private discussion and saw Yoko waiting outside the door to the gents'. At that moment John emerged and as they passed me he said, 'See? I don't go anywhere without Yoko.'

He had agreed that I should have custody of Julian – I imagine the lawyers had told him that he would be highly unlikely to get it – but asked for regular visits and I accepted this. However, Julian hadn't seen John since the acrimonious meeting at the house, several months earlier, and I had no idea when, if ever, John would ask to see his son.

Money was perhaps the toughest issue. My lawyers had

told me that on no account should I make contact with
John as that would be collusion. I was put in touch with
a top lawyer, a QC, who told me I could take John to the
cleaners and fight for half his fortune. But I was finding
it hard to cope with the cold legal process. I wanted to
talk to John, to tell him how much Julian missed him, to
sort things out amicably. So I phoned him.

'What do you want?' he snapped.

'John, I can't bear all the animosity. It's as though we
never loved each other. They want me to take half your
money, but I'd rather we talked and sorted it out between
us, without any of the legal jargon.'

'There's nothing to talk about. My final offer is seventy-
five thousand pounds. That's like winning the pools, so
what are you moaning about? You're not worth any more.'
On that definitive note, he hung up.

I told my lawyer that I didn't want to go for half John's
fortune: I just wanted a fair and reasonable settlement. I
couldn't bear the thought of a long-drawn-out battle, or
of exposing our marriage to public scrutiny. I just longed
to have the whole thing settled. I was asked to draw up
an estimate of my weekly expenditure. It came to seventy-
six pounds for clothes, food, entertainment and holidays
for Julian and me – just under four thousand a year. My
assets included a thousand pounds in the bank, my clothes
and a Mercedes car. I had no jewellery of any more than
sentimental value.

John's assets were submitted at £750,000, although he
was undoubtedly worth far more. He raised his offer to
£100,000. This was broken down into £25,000 for a house

and £75,000 to support me and Julian until he was twenty-one. Allowing for inflation, it would not be enough to cover even the modest annual expenses I had listed.

A further £100,000 was put into a trust fund for Julian. I would be allowed to draw on the interest from this to pay school fees, but withdrawals would have to be approved by John and Yoko, who were the co-trustees with me of the fund. It was also agreed that should John have more children the fund would be shared equally with them.

While I realised that I was lucky compared to most women divorcing at that time, it still hurt to be dismissed so lightly. John was being meaner than I'd ever known him, which baffled me: typically he was generous to those around him. Why not now, to his wife and son?

Whatever his reasons, I had no energy for a fight. In so many ways he was no longer the kind-hearted, passionate and witty man I had fallen in love with. Worn down by the miserable business of negotiating through lawyers, and still smarting from John's withering remark about winning the pools, I accepted his offer.

Our decree nisi was granted on 8 November 1968. I was summoned to the divorce court and went alone, driven there by loyal Les Anthony who, although he was now working solely for John, was still a good friend to me. Walking into court beside my lawyer was terrifying. The place was packed with the press and I had to swear in front of them under oath that my marriage had broken down irretrievably, that my husband had publicly admitted adultery and that Yoko was pregnant by him.

Cynthia Lennon

Throughout this awful, surreal experience I felt humiliated and painfully aware that I was alone. Afterwards I
fled home and collapsed, sick with apprehension about
the future. I had no idea how I would cope and still found
it hard to believe that, after ten years together, I had been
severed from John's life with a few brief words from a
judge in a public court. I should have hated John for what
he had put me through. I was certainly angry with him
and bitterly hurt. But I couldn't hate him. Despite everything, I loved him still.

# 17

WHILE THE DIVORCE PROCEEDINGS WERE in progress I received a call from Roberto Bassanini. He was in London, had heard about my split from John and wondered how I was. Could he come and see me? He and his family had been very kind to us in Italy so I invited him down for the evening. Julian was delighted to see him, leaping into his arms and dragging him off to play in the garden. Roberto had a lot of the child in him and they ran around outside, Julian shrieking and giggling, while I made some supper. After we'd eaten and put Julian to bed we talked.

Roberto was horrified at the way I'd been treated and told me John was mad to give me up. It was lovely to have male company and feel feminine and attractive again. Roberto listened, he was caring and he even made me laugh, which felt wonderful after so many months.

After that he came to see us regularly, and it wasn't long before he and I became lovers. He was good for me – fun, warm and full of laughter. He was concerned for me and Julian and he made me feel wanted and loved. After so long inside what felt like a black hole, Roberto was just what I needed.

As part of the divorce settlement, I had to move out of Kenwood. Of course, it was much too big for me, Mum and Julian, but I hated having to uproot Julian from his home and his school so soon after his dad had left. Saying goodbye to dear Dot and her family was even worse. She had been such a good friend, solid, loyal and reliable for more than four years. We shed a lot of tears and shared a lot of hugs and promised to keep in touch.

I bought a house in Pembroke Gardens, round the corner from Kensington High Street, and enrolled Julian at a small private school close by. He wasn't happy: he didn't make friends and never looked forward to going there in the mornings. After a few weeks I enrolled him at a lovely state school nearby where he settled in well.

Soon after we moved in Roberto came to live with us. I asked myself whether I was ready for a new relationship. I knew that I wasn't over John; I was still hurt, frightened and angry, and probably should have spent more time on my own rather than turning to someone else to heal the wounds. But at the time I was drawn to Roberto's warmth and sense of fun. With him life turned from the black and white of my depressed, lonely days after John to technicolour again. And if I had doubts for myself, I had none for Julian. Roberto's obvious affection for him was a huge attraction.

So, the three of us settled down to family life – Mum had moved home to Hoylake. Now that we had no help in the house, we took on an au pair, a Spanish girl called Mariquilla, who fitted in beautifully and was like a big sister to Julian. She stayed with us for the next four years and became part of the family; she left to train as an air-traffic controller in Mallorca.

For the next few months I had no contact with John. He and Yoko were involved in a series of what were – to me anyway – increasingly odd stunts. First they issued their *Two Virgins* album, three weeks after our divorce. The furore about the cover picture, which featured them naked, overshadowed the music, which was a critical and commercial flop.

Next they appeared at the Royal Albert Hall, singing inside a large white bag during the underground art movement's Christmas party. A bit of fun, perhaps, and the precursor to many more 'bag' events. They even gave a press conference inside a bag. Funny, wacky, attention-grabbing, but I couldn't get over how deathly serious they were about it. Not a glimmer of amusement, seldom a smile. What had happened to John's sense of humour?

In February Yoko's divorce came through and in March 1969 they flew from Paris to Gibraltar and married, spending just seventy minutes on the island before they flew back to Paris. It was a shock to me to hear that they'd married so soon and worse because I only learnt of it through the press. But, then, after all the publicity surrounding their getting together and the end of my marriage to John, perhaps they felt it was important to make a public statement about their relationship.

A few days later they moved to Amsterdam and spent a week in bed at the Hilton hotel. This was their famous 'bed-in' for peace, during which they were filmed and gave hundreds of interviews.

Julian was fascinated. 'What's Dad doing in bed on the telly?' he asked.

'Telling everyone it's very important to have peace,' I answered, through gritted teeth, all too aware that John had found it impossible to allow for peace between us and that the small boy asking the question was paying the price.

That glimpse on television was the first that Julian had seen of his father since the acrimonious meeting at Kenwood eight months earlier. But that was about to change. In May John and Yoko bought Tittenhurst Park, a twenty-six room mansion in Sunningdale, Ascot, for £150,000. I knew the house, because John and I had been to look at it with the other Beatles couples a year or two earlier. It was beautiful, with extensive grounds including its own market garden. For a crazy moment we'd considered buying it and all moving in together, in a kind of Beatles commune. How strange that now it was John and Yoko's home. Soon after they moved in I received a call from Peter Brown, the Beatles' assistant. 'John would like to see Julian, Cyn. If it's OK with you, Anthony will call for him on Friday and take him to Tittenhurst for the weekend.'

It didn't feel very OK, but I agreed. John had been awarded reasonable access to Julian and I felt it important that Julian saw his father. When I told him about the visit Julian was excited and a little scared. He'd just turned six and hadn't seen his dad in what, to a six-year-old, must have seemed a very long time.

That Friday Anthony arrived and came in for a cup of tea. It was good to see him – like Dot he'd been a real friend. He told me it was just as well that Julian hadn't gone to the Montagu Square flat while John and Yoko were there. 'It was a complete tip,' he said. 'They were doing heroin and other drugs and neither of them knew whether it was day or night. The floor was littered with rubbish. Couldn't have had a little one there.'

Perhaps that explained why John hadn't asked to see Julian sooner. He'd obviously been in no state to care for him. But I was alarmed. I'd had no idea that John had graduated to heroin. 'How are they now?' I asked Anthony.

'They're OK at the new place,' he assured me. 'They've sorted themselves out and it's all fine there. Julian will love it.'

With these assurances, but many nagging doubts, I sent Julian off, a tiny figure waving to me from his seat next to Anthony in John's enormous car, his little face unsure. I had told him brightly to have a lovely time with Daddy, smiled and waved until the car had rounded the corner, then cried my eyes out. One of the hardest aspects of letting Julian visit John was accepting that Yoko would presumably look after him when he was there. I didn't know what he would make of her cool manner. But in fact he seemed fine with it, and perhaps it was better for him to have a rather distant step-mother than one who was all over him. He never told me that she was unkind in any way, which was a relief.

After that first weekend Peter rang regularly to arrange visits for Julian. Much to my amusement Julian had started calling Yoko Hokey-Cokey.

Julian often came home with bizarre descriptions of life at Tittenhurst. Dad and Hokey walked around the house naked, long hair hanging over their faces, which made them look like witches. They ate funny food, all rice and seeds (Yoko had apparently got John on to a macrobiotic diet). Julian was matter-of-fact about these things, as six-year-olds tend to be, and I was careful never to say anything critical. I didn't want him to feel guilty or torn or that he had to defend me. A certain amount of division is inevitable anyway with a child of divorced parents, but I did my best to keep things stable for him, to minimise his distress and to be positive about his father.

Weekends without Julian were lonely. Because John had always been away so much he and I were incredibly close: he was an articulate and funny little companion. Although I had Roberto, I missed Julian's constant chatter. Inevitably there were times when Julian didn't want to go and see John. It's not always easy for children of separated parents to leave home and be shipped off to the other parent. If

Julian had to miss a friend's party he'd grumble, but I stuck to the agreement, feeling it was better to be consistent than making last-minute changes.

One particular weekend Julian told me he didn't want to go to his dad's because he got scared in the night. It seemed that he was in a large bedroom by himself, in a separate wing of the house, a long way from John and Yoko's room. I rang Peter Brown and asked him to talk to John about it: 'Let Julian leave the light on or move nearer to where John is.' Peter said he would talk to John and I promised Julian that Dad would change his room.

I wondered why Julian hadn't told John about this. Surely he could tell his dad that he was afraid in the night. Over the next few weeks I began to question him indirectly about his relationship with John. It seemed that John sometimes had angry outbursts towards him, shouting at him for the way he ate or being too slow, which had made Julian nervous. He was afraid of provoking John, who switched very quickly from playful to furious.

I knew of old this side of John but in the past he'd seldom displayed it around Julian – perhaps because he wasn't with him very much. I worried that these explosions of anger might have become more frequent and that they would alienate Julian from his father.

I wished that John and I could talk, but he was still refusing to speak to me. It was puzzling and frustrating. I was no threat to him, so why couldn't he talk to me, for our son's sake? Infuriating as it was, there was nothing to be done about it. I focused instead on giving Julian as much support as I could and hoping that things would improve as John's new relationship became more settled.

Meanwhile John and Yoko's stunts continued. They staged another bed-in for peace in the Bahamas, but left after a day or so because apparently it was too hot. They flew on to Canada, where they staged a seven-day bed-in in Montreal, inviting Canada's prime minister, Pierre Trudeau, to join them, and recording 'Give Peace A Chance' from their bed. At the same time their single, 'The Ballad of John and Yoko', was banned in Australia for being blasphemous – 'Christ, you know it ain't easy, you know how hard it can be, the way things are going, they're going to crucify me' wasn't to everyone's taste. I remembered the hate mail John had received after his previous remarks about himself and Christ and wondered why he'd risked public ire again. But, then, he had changed in so many ways that perhaps, encouraged by his new love, he cared far less what anyone thought.

When John was back in London after several weeks away, Peter Brown arranged once again for Julian to go to Tittenhurst for a visit. As they hadn't seen each other for a while John wanted him to stay for a few days. I sent him off with Anthony and settled down to a peaceful weekend at home.

Two days later I was watching television when a newsreader announced that John Lennon and Yoko Ono had been in a car crash in Scotland, with their children Julian and Kyoko. No one was badly hurt and they were all in hospital. I was horrified. What on earth was John doing taking Julian to Scotland without telling me? I wasn't surprised he had crashed the car – his driving had always been terrible – but he had had my son with him, and another child. I was furious that John and Yoko had taken Julian all that way without my knowledge, and even more so that I'd had to learn of the accident from TV. I felt the least John owed me was a phone call to tell me what had happened.

I rang Peter Brown, who arranged for me to fly to Scotland to collect Julian. He contacted the hospital and assured me that Julian wasn't injured and neither was Kyoko, although both John and Yoko had had stitches. To add to my misery, when I was ushered on to the plane at Heathrow it turned out to be a flight to Belfast. In my anxiety I hadn't noticed and, to his embarrassment, neither had Peter, who'd been with me at the airport. I had to fly to Belfast, and then wait for a flight to Edinburgh, so I didn't get there until the next day.

When I landed I took a taxi to the hospital, which was close to where the accident had happened, in Golspie. It was a long journey and I sat in the back, impatient to see my son. But when I reached the hospital and asked for him I was told he wasn't there. 'He's gone with his aunt,' was all they could tell me. Once again John refused to see me. I was exasperated and, to make matters worse, I was passed a note from John, saying that Yoko found Julian badly behaved. He was still a small child, having to cope with a major upheaval in his life, and if he was a little difficult, it was only to be expected. Yet instead of being concerned or constructive, John was critical.

All I wanted was to find out where Julian was and take him home. Then I realised that the aunt they had mentioned must be Mater. I rang her and, to my relief, she told me she'd collected Julian and taken him to her little holiday house, up in the Highlands. Once again the taxi set off, and a couple of interminable hours later we arrived at Mater's, where I found Julian tucked up in a blanket drinking hot chocolate.

It seemed that John had decided on the spur of the moment to take Yoko and the children up to Scotland to meet Mater, Bert and Stanley in Edinburgh. Mater regaled me with hilarious tales of Yoko refusing the roast dinner she had prepared and taking over the kitchen to steam beansprouts for herself and John. 'She looked like a witch hanging over a cauldron with all that hair,' Mater said. Not a great admirer of Yoko and concerned at the changes in John, she had cornered him and demanded to know what he thought he was doing with his life. At some point he'd said, 'I'm going to take Julian away from Cyn.' Mater was incredulous and told him, in no uncertain terms, that there was no way he'd get Julian and that he should put all thoughts of taking him from his mother out of his mind.

With Mater's words ringing in his ears, John had taken off in the car with Yoko and the children – and soon afterwards landed it in a ditch.

I was grateful to Mater for speaking up for me and glad that, despite their row, John had sent Julian to her. But his refusal to inform me about what was going on had left me trekking across the Scottish countryside for hours, before I discovered the whereabouts and condition of my son.

The next day Julian and I flew home. I was still fuming about John's behaviour, but I had grasped that nothing was going to change. I contemplated refusing to allow Julian to visit him, but I knew that I would feel awful if I severed the already tenuous relationship Julian had with his father. So, for the next few months he continued to visit John and Yoko for occasional weekends.

Meanwhile he thrived on the attention Roberto gave him. For the first time in his life he had a man around every day, someone who was interested and involved with him, who took him to school in the mornings, helped with his homework and played football with him in the park. Julian adored Roberto, and the two were like best friends.

Roberto and I were happy too, although one or two dark clouds hovered on the horizon. I worried about money: the allowance I had from the divorce settlement for me and Julian didn't go far. We needed another income, but Roberto's earnings were unpredictable. He had an allowance from his father – who'd had a soft spot for him ever since Roberto had had polio as a child – and he also managed his father's London restaurant in Victoria. But all too often his money disappeared on the good things in life before it could be used to pay bills. Roberto loved to party and was very generous. He would invite people to the restaurant and treat them all, or bring crowds back to our house and ply them with food and drink until the early hours.

Mum came to see us regularly and on one of her visits,

when the three of us were in the back of a taxi, Roberto went down on one knee before her and asked her permission to marry me. Mum loved Roberto and had no hesitation in saying yes, so I put aside my niggling doubts, reminded myself of how much he was doing for Julian and said yes too.

Roberto was a kind, loving man and I was sure that any problems we had could be sorted out. We married on 31 July 1970 at Kensington register office. We invited our families plus a few close friends, including Ringo and Maureen, and Twiggy, with her manager boyfriend Justin de Villeneuve. I'd have loved my old friend Phyl to be there, but she was living in Cumbria and had two small children. Julian was our page: he hopped up and down with excitement throughout the ceremony as my mother tried to restrain him.

It was a happy day. I was thirty and Roberto two years younger. I hoped it would be the beginning of a new and happier phase of my life. We had a wedding party at the Meridiana restaurant on the Fulham Road – the owners, Walter Maritti and Enzo Apicella were friends of ours. The only blot on the day was when Roberto's father drew me aside at the lunch and told me that he was stopping Roberto's allowance as he wouldn't need it with me. He assumed, as so many people did, that I must have received a huge settlement. My heart sank: he was expecting me to support his son.

We went on honeymoon to Italy's Adriatic coast and one of the Bassanini hotels, taking Julian with us and my brother Charles's ex-girlfriend Katie to look after him.

A couple of months after the wedding we had a party at home, as a belated housewarming. After a lot of thought I agreed with Roberto's suggestion that we invite John and Yoko, in an effort to build a better relationship with them. Roberto had always been in awe of John and even though he disapproved of the way John had treated me, he was still impressed that I had been married to a Beatle and keen to meet him.

I was surprised when John and Yoko agreed to come, and even more so when they turned up on the night. It was the first time I had seen them since the ghastly meeting at the lawyer's office two years earlier and I was nervous.

Maureen and Ringo were also at the party, with Twiggy, Justin, Lulu and her new husband Maurice Gibb, Roger Moore and his wife Luisa. John and Yoko arrived late and an awkward couple of hours followed. After a cool 'Hi' to me, John went to sit with Ringo. Roberto joined them and the three were soon engaged in animated conversation, punctuated by guffaws of laughter. Meanwhile Julian, wildly excited to see his mother and father in the same room again, ran round in frenzied circles.

Yoko, left out on a limb, made a beeline for me. I had sat down for a break after a couple of hours' circulating among the guests, only to find her on the floor at my feet. I was almost speechless as she chatted about Julian and how much they loved having him to visit. It was hard to reconcile this cheerful outpouring with the silent stonewalling I had received from her at our two previous meetings. For me it felt like a performance put on for the benefit of the other guests, John in particular, and I couldn't stomach it. I got up and left the room, and a few minutes later, as they departed, John spoke his second word to me that evening: ''Bye.'

After they'd gone I was sad but relieved. It had been strange and painful to see John again. I had hoped we might talk about Julian, but he had avoided me altogether and Yoko's exaggerated attempt at being friendly had been insufferable. In the end nothing had been gained by inviting them, except that Roberto's wish to meet the great John Lennon had been granted. Like so many people I was to encounter in the future, he couldn't see beyond John's fame to the real man. It was as though, for him, the ex-husband who had behaved so appallingly towards me was another man entirely from the superstar. A few months earlier, John had been nominated in a television documentary as one of *the* three men of the sixties, along-side President Kennedy and Chairman Mao. It was an accolade that put him in a category above mere celebrity and meant that virtually no one could see him any more as an ordinary human being. Not that Roberto ever stopped disapproving of John and Yoko's behaviour. He despised Yoko, and never forgave her for the hurt he believed she had caused Julian.

The Beatles had continued to be a phenomenon to the end of the decade, with vast record sales and twenty number-one hits – a total unrivalled at the time and only rivalled since, ironically, by Cliff Richard, whose music they had once deplored. They had also released what was to be their final album, *Abbey Road*. Despite their continued success, though, the rumblings of discontent within the band – of which I'd been aware before John and I split – were reaching the press and the music grapevine, and by mid-1969 it was clear that the end was in sight.

John and Paul, who once made music together with such passion, excitement and brilliance, had found it harder and harder to get on together as the sixties drew to a close. Their musical ideas and tastes were diverging as Paul went on writing ballads, like 'Hey Jude', while John wrote raw and challenging songs like 'Hey Jude''s B side, 'Revolution', commemorating the international student riots of 1968. The *Magical Mystery Tour* film had been their first project after Brian's death and Paul had organised it; John had backed off. Although they'd gone on to write several more songs together and even make another film, *Let It Be*, the pleasure had gone out of it for them and it was clear to me from 1967 that they would eventually go their own ways.

It was still a shock when I heard that John had told the other Beatles he was breaking up the group, prompted, apparently, by the chaos at Apple. The Beatles had no business knowledge or experience and Apple was being drained of funds with no one to run it properly. Mick Jagger had told John about a smooth-talking American showbusiness lawyer called Allen Klein. John, George and Ringo met him and decided to appoint him to sort Apple out. Paul didn't trust Klein and refused to sign with him, turning instead to his new father-in-law, Lee Eastman, who was also a successful showbusiness lawyer. This had led to enormous tension between John and Paul.

There was one more factor in the the Beatles' break-up, in my view the most important. John's relationship with Yoko had infuriated and alienated the other three. From the beginning there had been an unwritten agreement that wives and girlfriends would never be allowed to interfere with the Beatles' work. We often turned up at the end of recording sessions to hear the finished version but we knew the boys worked best when they were left to get on with it. It was the way they'd always preferred things. Few people had ever been involved in recording sessions apart from the boys and George Martin. Even Brian was encouraged to stay away and wait for the finished product. Making music was work, and they worked best without anyone hanging around.

When John got together with Yoko this changed. He refused to do anything without her – as I'd seen when she tagged along on his visit to the loo at the lawyer's office. He'd already announced that their names would be joined and they'd be known as JohnandYoko, then changed his middle name from Winston to Ono, all of which the other Beatles regarded with tolerant good humour. 'He was wacky, but we loved him anyway,' was how Ringo once put it to me. But when John began to bring Yoko to recording sessions and consult her about everything he did there, even allowing her to criticise what they were doing, it was too much for the others. They hated it, and it was clear that the arrangement wouldn't last amicably for long.

Ultimately, of course, it was John who broke up the Beatles, just as he had formed them in the first place. He had moved on in his life, not just from me but from Paul, the other person who was closest to him throughout the sixties. Yoko had replaced me as John's wife and Paul as his artistic collaborator. Of course, change is inevitable and both John and Paul went on to be creative and successful separately, but the way it happened was sad, and it was sadder still that their relationship never really healed. After six years of unparalleled success, the greatest band of all splintered amid a welter of acrimony and accusations.

While all this was going on George and Ringo couldn't do much more than watch from the sidelines. George was still deeply into Eastern spiritual beliefs and was probably relieved in some ways that his Beatles days were over. Ringo, always philosophical and funny, would say later, 'It was nice while it lasted.' But I'm sure they were both deeply affected by the end of what had felt, to all of us, like a family.

Certainly I was sorry to think that it was all over. Even though I was no longer a part of the inner circle, I had known and loved them all, and it was sad to think of them so disaffected with one another.

Meanwhile John and Yoko continued to make headlines in their quest for world peace and, I couldn't help feeling, media attention. They launched their Plastic Ono Band, whose first single, 'Give Peace A Chance', was a huge hit. They put up posters in eleven cities around the world reading 'War is Over if You Want it, Happy Christmas from John and Yoko', then cut off their hair and auctioned it for peace. John had also handed back his MBE, in an anti-war, anti-establishment gesture. I felt sorry for Mimi, who had been so proud of it.

I know that to many people John's behaviour during that period seemed brave, honest and innovative, but as the mother of his confused small son, it was hard to see it like that. Along with many other people who loved John – his family and friends – I looked on with dismay as his actions appeared increasingly self-obsessed and he spared little thought for the feelings of those who had once been closest to him. He seemed so intense and took himself deadly seriously too, which wasn't the John I had known, and many of his old friends and family felt the same.

Another aspect of John and Yoko's behaviour alarmed me more deeply and in a more personal way. As far as I knew, they had a good relationship with Tony Cox, Yoko's ex-husband. He had remarried and they had spent several weeks with Kyoko, Tony and his new partner at their home in Denmark over the New Year period in 1970. But the relationship took an abrupt downturn when John and Yoko were arrested in April 1971 and accused of trying to abduct Kyoko. It seemed that Yoko and Tony were locked in a custody battle over Kyoko. He was now living in Mallorca, where John and Yoko had taken the little girl from her school to their hotel, presumably intending to leave the island with her. They were held by the police for fourteen hours and eventually released without charge. Kyoko was returned to her father who, apparently terrified of John and Yoko's wealth and power, fled with her to America. John and Yoko followed and spent several fruitless weeks looking for them, then returned to Britain.

Yoko had left Kyoko with her father, then changed her mind and decided to sue for custody. Might John do the same with Julian? Had JohnandYoko decided they wanted their children with them, now that they were more settled? If they had, I would fight to the death for Julian.

I spent a few weeks consumed with worry, but far from asking for his son or wanting to see more of him, John abandoned him. In September 1971, I learnt from newspaper reports that John and Yoko had gone to live in the States: they had had enough of the British press and public's hatred of Yoko. This was staggering. It was only a few weeks since John had seen Julian, yet now he was leaving the country indefinitely without a word to me or to his son. I had no idea whether he planned to see Julian again or how we would make contact. Until now arrangements had been made through Peter Brown in the office, but when I spoke to him he said he, too, had no idea what was going on. It seemed that John was willing to sacrifice his son to protect Yoko.

I asked Peter Brown to get in touch with John and tell him that I wanted him to continue to see Julian and to be able to let him know about Julian's progress. A couple of weeks later I picked up the phone to hear Yoko's voice: 'Hello, Cynthia,' she said. 'John and I have decided that if you wish to make contact about Julian, you should talk to me. I will be the one to speak to you from now on and John is going to be the one to speak to Tony about Kyoko.'

This was more than I could bear. I had put up with a great deal from John and Yoko, but now they had pushed me too far. I was not willing to deal exclusively with Yoko and told her so. If John wanted to see Julian, I said, he could call me himself. Then I hung up.

I was livid. John had just released 'Imagine', the song that would become an international anthem for peace, telling the world to 'live as one', yet he couldn't pick up the phone, make peace with me and arrange to see his own son. Surely, I reasoned, Julian meant more to him than some foolish agreement with Yoko about dealing with each other's ex-partner.

I was wrong. It was three years before John saw Julian again.

# 18

WITH JOHN AND YOKO IN America, Julian and I in London and impasse between us over any potential arrangements for access, I could only get on with my life and hope that things would change. If John had, at any point, picked up the phone or written to me, asking to see Julian, I would have been delighted. In fact, I often prayed that he would. But there was no word from him between 1971 and 1974, apart from birthday and Christmas presents for Julian each year, sent by his London office with no personal note or card. It must have felt to Julian as though his dad had disappeared from the planet. He had been eight when he last saw John. After that, all he had was newspaper cuttings to tell him where his dad was and what he was up to.

John was battling with the US authorities to be allowed to stay in the States: his drug conviction meant that they were reluctant to give him the coveted green card. I explained to Julian that John couldn't leave America until he got permission to live there, which didn't mean much to a small boy. 'If Dad can't come and see me then why can't I go and see him?' he asked. What could I say? 'I think he's very busy getting settled in America. I'm sure he'll be in touch soon.' But as time passed with no word Julian drew his own conclusions. 'Dad's always telling people to love each other,' he said to me one day, 'but how come he doesn't love me?'

Privately it was hard not to agree, and I cursed John for making his child suffer, although I did my best to re-assure Julian that, of course, John loved him.

It wasn't just Julian who appeared to have been consigned to John's past. John seemed to be cutting all his ties with family and friends in Britain. He seldom, if ever, contacted his sisters, aunts and the friends we had kept in touch with all through the Beatles years.

It was especially hard for me that while John was ignoring his son he put a great deal of effort into helping Yoko find her daughter. In December 1971 he and Yoko flew to Texas when they heard that Tony Cox and Kyoko were there, and Tony was jailed for five days for refusing Yoko access to her daughter. After that he went on the run again with Kyoko and this time John and Yoko couldn't track them down. I heard that they cruised the streets, hoping to spot Tony or Kyoko, and put advertisements in newspapers, appealing for them to come forward. I felt sorry for Yoko, who must have longed for her daughter, but at the same time I wondered whether John had made some kind of odd pact with her – I won't see my child until you see yours. It was the only explanation I could think of for his neglect of Julian. On what was to be one of the Beatles' last albums, called *The Beatles* but forever after known as the 'White Album', was a little lullaby John had written for Julian. The album came out just before our divorce and I knew the song was there; I had heard it when John composed it and had loved it. But I was disappointed that John had given it to Ringo to sing on the album and for a long time after our split I couldn't bring myself to listen to it or to play it to Julian. Eventually I did, though, telling him that his Dad had written the song for him, in the hope that it might provide him with some comfort.

Soon after John left Britain we moved house again. Roberto's non-stop partying had continued and the bills were running up. I could no longer afford to spend whatever I wanted and I was keenly aware that what money I had must last until Julian grew up. In a vain bid to persuade Roberto to forgo his visits to restaurants and nightclubs, I sold the house in Kensington and we moved to Wimbledon. It meant another new school for Julian, but I hoped that a more settled family lifestyle would make up for that.

I was now just an ordinary mum doing the school run and shopping in the local supermarket and I liked it. Nothing could have compared with the life I'd had with John, and I didn't want to try to re-create it. I preferred to be back in the normal world. Most people I met had no idea who I was. I had been in the background for a lot of the time when I was with John, and when I did appear in public it had been at his side, so without him people couldn't place me. And who would expect to meet a Beatle's ex-wife at the checkout in Sainsbury's? If someone said I looked familiar, I'd smile and say I had no idea why. It usually worked. They'd apologise and hurry off.

Of course, I still had celebrity friends and went to exciting parties with Roberto, but I kept that part of my life separate from the everyday world I lived in, and even the parents of Julian's schoolfriends, who brought their children to tea, seldom commented on who we were.

Just before Christmas 1972, a year or so after our move, John's Aunt Harrie died of cirrhosis of the liver. She'd never been a drinker, so it was unusual, and sad: she was only fifty-six. I went back to Liverpool with Mum for the funeral. It was wonderful to see so many of John's family again. Julia, now twenty-five, was married with a son and a daughter. Jacqui had become a hairdresser, and David, Harrie's son, was with his wife, who was expecting her second child; his older sister, Leila, was a medical consultant and mother of three. Aunt Mimi was there, over seventy now but still, astonishingly, without a single grey hair. At the wake after the funeral she cornered me:

'What on earth were you thinking of, Cynthia, you silly girl, divorcing John and allowing him to go off with that woman?' I pointed out that I hadn't had a lot of choice in the matter, but Mimi would have none of it. 'You should have stopped him. Now look what an idiot he's making of himself.' I had to chuckle. After resenting me for years, Mimi was now ticking me off for not having stayed married to John. It was comforting, in a way, that his irascible aunt hadn't changed. I promised to take Julian to visit her in Poole and, mollified, she headed off to launch a verbal missile at her next target.

Later Julia told me that when Mimi met John and Yoko after their *Two Virgins* album was released she'd said to them, 'Why didn't you get somebody attractive on the cover if you've got to have somebody naked?' As forthright as ever.

That visit was my first trip to Liverpool in a long time and made me realise how much I missed it. I promised myself I'd move back, if I got the chance. Wimbledon was pleasant but I had no ties there, and life with Roberto was trying. Far from cutting back on his partying, he had opened a taxi account and was running up even more bills by taking taxis into the West End every evening. He still 'worked' at the restaurant his father had bought, Da Bassano, but every night he filled it with his friends and let them eat without paying. He loved to be loved and the price of his popularity was our financial burden.

Roberto had always been spoiled by his parents and had never had to deal with the harsh realities of life. His playfulness was a delight that became too much for me to handle. I didn't want to party all the time when I was spending hours each month poring over the accounts and lying awake worrying. Begging, pleading and demanding had little effect on him: he would make promises he didn't mean, then tell me how much he loved me, buying me roses and looking at me with puppy-dog eyes until I had to see the funny side.

But by 1973 I knew I couldn't stay in the marriage. The only thing that had stopped me walking out sooner was Julian. I felt terrible to be taking him away from someone who was now so important to him – Roberto had done so much to fill the gap that John had left in Julian's life – but I had to. I was no longer willing to live with someone who wouldn't grow up, take a job or share the responsibilities with me. So, with a heavy heart, I told Roberto I wanted us to separate.

There were many tears, from the three of us, and it took all my determination to see it through. My only comfort was that Roberto, even though he was deeply hurt when I left him, never deserted Julian. Most unusually for a step-parent, and particularly after such a short marriage, he kept in touch and regularly invited him to Italy, where in later years Roberto began, ironically, to make a lot of money.

He and Julian were friends until Roberto died of a massive heart-attack twenty years later. He left Julian a letter saying that there was £100,000 for him in a Swiss bank account. But when it was investigated the bank insisted that no such account existed. We never did get to the bottom of that mystery, but in 1998 Julian dedicated his album *Photograph Smile* to Roberto, in loving memory of the step-father he felt had been more of a father to him than John ever was.

In 1974, after my divorce from Roberto, Julian and I headed north. I bought a bungalow in Meols, a village in the Wirral, close to Hoylake. It was good to be home again. I was now just down the road from Mum and it was lovely to be able to see her as often as I liked. I enrolled Julian at Kingsmead School, which was close to home. I was conscious that this was his fifth school and he was not yet eleven – an awful lot of change for any child to cope with – but he was happy there, and immediately made a lifelong friend in Justin Clayton.

Our bungalow, Rosebank, was very ordinary and I was aware that it was a stopgap, somewhere safe to take stock of two broken marriages and decide what I wanted for the future. I missed Roberto and wondered whether I'd done the right thing in ending our marriage. Julian missed him too. How had I managed to fail at marriage twice? And with two men who had little in common other than the exhausting emotional toll they took from me. I was on my own for the first time since I was eighteen and felt very low. I knew I could easily fall apart, but I realised I couldn't afford to do that: I had a son to care for.

Perhaps as a result of all that he had been through, Julian was old for his age and often very serious. I always felt that his face had a medieval quality to it, like carvings I'd seen on ancient gravestones. And I was convinced he'd probably been here before, many times, because he had such innate wisdom. There were many times when he looked after me as much as I did him. He would listen, comfort and do his best to protect me. We shared the particular closeness unique to a single child and single parent.

While I was picking up the pieces after my second marriage I was stunned to read in the papers of a major development in John's life: he and Yoko had parted. Apparently they had agreed on a trial separation late in 1973 and John had moved to Los Angeles with May Pang, a young Chinese employee.

What might this mean for Julian? Would John be willing now to have contact with me and see his son? I wondered about getting in touch with him, or whether I should just wait and see. I decided, for the moment, to wait.

It seemed that John was going through a strange phase. Over the next few months he hit the headlines several times for being drunk and brawling but he was also creative for the first time in ages, working on a new album and getting together with other musicians, including friends like singer Harry Nilsson and Ringo, to record an album of his favourite songs by other musicians. I hoped that this would prove the start of a more positive period in his life that might include his son. But would he get in touch or should I contact him?

The decision was made for me when I went to London on a business trip in the summer of 1974. I'd always loved doing interior decoration and when I met someone who suggested I did some work for him I couldn't wait. That evening, after our meeting, I checked into my hotel and I discovered it was the base for the local hookers. Swarms of girls, done up to the nines, and shifty-looking men were traipsing in and out. I felt a bit out of place and in need of company so I rang an old friend and got a crossed line. After a few seconds I recognised the voice I was hearing – it was Patti Harrison's. We were delighted to be in touch again and she asked me out to dinner that evening.

She was now separated from George, but we caught up on each other's lives and she introduced me to a group of her friends, one of whom was a record producer: he was about to sail to New York on the SS *France* with Elton John, then the most successful rock star in the world, who was due to give a series of concerts in the US. I mentioned to him that I wanted John to see Julian and he said, 'Why not come with us?' I was startled, but could see that it just might work. Patti chipped in and said that her sister Jennie – whom I hadn't seen since our ill-fated trip to Greece six years earlier – was now living on the outskirts of New York and would gladly put me up while Julian was with John.

The next day I phoned Peter Brown, got John's number and rang him. He was surprised to hear from me, but not as surprised as I was to get through to him on the first try, after so many years of non-communication. Our conversation was awkward and neither of us wanted to make small-talk, so I got straight to the point and said I'd like to bring Julian over to see him. John sounded pleased with the idea and agreed. I told him I had no money, so he said he'd pay for first-class tickets for us. I was thrilled.

A few days later we sailed from Southampton. The *France* was a beautiful old ship, and this was to be its final voyage. Julian and I shared a cabin and on the first evening we sat on our own at a large round table for dinner. Elton's group were nearby and we got chatting to his percussionist, Ray Cooper, and his wife. When Elton, who was a great friend of John's, realised who we were, he invited us to visit him in his cabin. It was immense, and he proudly showed us an enormous trunk devoted to hundreds of pairs of his trademark glasses. He was charming and I reminded him of an earlier occasion when we had spotted him at Heathrow and he had given Julian his autograph.

Julian was excited at the thought of seeing his dad but nervous too. He was a bright, lovable but shy eleven-year-old and I hoped that John would be proud of him. I knew that after three years they would have a lot to learn about each other and I prayed that they would get on.

Ray Cooper and his wife were very kind to us: they listened as I poured out my fears about our reunion with John. Before we parted Ray said, 'Just call me if there's ever anything I can do for you.' It proved a valuable offer after John's death, when Julian was in need of help and support.

When the ship docked John was waiting for us on the dock, with May beside him. It was both wonderful and painful to see him again. He was pale and gaunt and clearly nervous. He pecked me on the cheek, then scooped Julian into his arms and hugged him. Then we were introduced to May, a sweet but slightly lost-looking girl of twenty-three.

## John

We were ushered into the limousine, where Julian sat between John and May and I was on the seat behind them. John chatted exclusively to Julian, leaving me an awkward, silent passenger. I was glad I was going to stay with Jennie and could leave them to it. They dropped me at the Pierre Hotel in central New York, where John had booked me a room for the night.

When I got to my room I was still finding it hard to walk straight – after days at sea everything was swaying. I ordered something to eat, then phoned Jennie. Her answering machine picked up, so I called Patti in London. 'I'm sure she's there,' she told me. 'I let her know you were coming.'

I tried her numerous times but when there was still no reply I called John, dreading an icy put-down. To my relief he was sympathetic. He told me he, May and Julian were flying to Los Angeles the next morning and that I could go with them and stay with friends of his, drummer Jim Keltner and his wife, another Cyn.

The next morning I met them at the airport. We got on to the plane together, but my seat was at the back of the first-class section while they sat at the front. It hurt, but I knew John was more comfortable keeping me at a distance. He had always hated reminders of painful episodes in his past and I realised, sadly, that that was all I meant to him now. He liked to make a clean break and move on, but because of Julian he couldn't do that with me. I could see how hard it must be for him and it wasn't easy for me, but I was grateful that he hadn't left me alone in New York.

In Los Angeles we were driven to the Beverly Hills Hotel, where John and May left us, promising to return the next morning and take Julian out for the day. We had a fun evening, flicking through the endless channels on the TV, but when John appeared the next morning Julian burst into tears and kept repeating, 'I want Mummy to come.' I did my best to persuade him to go – I knew my presence would make everyone feel awkward – but he hid behind the sofa, cried and refused point-blank to budge unless I went too.

In the end we all went to Disneyland. For me it was an excruciating day. John marched on ahead with Julian, while I tagged along at a distance, feeling redundant. May saved the day: she chatted easily with me and was kind and sensitive towards Julian, who visibly relaxed as time passed.

After a tense lunch in a burger bar and a long, hot afternoon trailing round the rides, I was glad when it was all over. I would rather not have been there at all and John's clipped manner reinforced my sadness. I still cared for him deeply and had never given up hope that we could be friends and learn to be comfortable with each other. I wondered what he was afraid of – that I would take him to task? That I wanted him back? That Yoko would find out I was there? Perhaps, in the end, it was all of these. And there was nothing I could do about Yoko, who was still in contact with John many times a day, monitoring him. As for wanting him back, or even speaking my mind about what he'd done – I'd moved on. I loved him, but I didn't want to be with him again, or even to make him

The motorbike John bought Julian for his sixteenth birthday

Julian's sixteenth birthday, on Long Island with John and Yoko

The silent vigil for John, Central Park, New York, December 14 1980

With Julian in Ruthin, north Wales in 1981. John had died and my third marriage had broken up

Going on with our lives; with Julian at the house in Cumbria in 1984

Interviewing Julian for Granada Television's *Weekend* programme

Julian's platinum disc for his first album, *Valotte*, in 1985

Happy at last: Marriage to Noel, June 7 2002

The party in Liverpool when The Beatles Story bought the cartoons I had drawn of the boys in the early days. Among the guests were Helen Anderson, Pete Best, his brother Roag, John's sister Julia, and Phyl

Still together; with Phyl in 1999, still my closest friend after
48 years

Feeling the strain: smiling for the camera with Yoko, Julian and Sean in New York, August 1989

Julian with best friend Justin Clayton, who went to New York with him after John died

With Julia
Baird, John's
half-sister, in
the house in
Normandy Jim
and I had
bought

With John's former assistant and partner, May Pang, in New York

feel bad. All I wanted was to be friends, and for him to be there for our son.

After that first day Julian was happy to go out with John and May, and I went to stay with the Keltners, who were lovely people. Amused by my Englishness, they called me Mary Poppins, and when John was working, Julian came over and played with their children. Jim and Cyn confided in me that they didn't like what had happened to John. According to them, he seemed dominated by Yoko. She called him all the time and he was constantly concerned that he'd be in trouble with her. His nickname for her was 'Mother', and they said he seemed to have a love-hate relationship with her, unable to tear himself away, yet constantly angry and resentful towards her. Apparently, Yoko had engineered his relationship with May and sent them away from the New York apartment in the Dakota building that she and John shared. For whatever reason – boredom with John, irritation, the need for space, perhaps – she had decided that they should be apart but that she would choose the woman he was to be with rather than letting him loose.

I was stunned. If it was true, John had given Yoko a vast amount of power over him. He was playing the naughty child to her controlling parent. It made me shudder. What had happened to the free-spirited, independent John I'd known? Why would he want to put himself so firmly under the thumb of a woman who, by the sound of it, so often didn't appear to be showing him love and affection?

For me there was an obvious parallel: Aunt Mimi. John had grown up in the shadow of a domineering woman –

it was what he knew and was most familiar with. While I had offered the devotion and loving acceptance he had needed after his mother's death, Yoko offered the security of a mother figure who always knew best. When, in later years, I read comments from Yoko comparing herself to Aunt Mimi I had to smile. She'd got it dead right.

Their relationship may have been set up by Yoko, but John and May seemed genuinely in love. I had noticed their affectionate gestures and the warmth between them, and the Keltners confirmed that May was good for him and made him happy.

In the first few months they'd been together John had hit the headlines a few times for drunken, rowdy behaviour in nightclubs – on one famous evening he spent the entire time with a sanitary towel stuck to his forehead, heckling the stage act they had gone to see. Jim and Cyn thought he'd been a bit like a teenager let out of boarding-school; after the restrictions of life with Yoko he was indulging in as much excess as he could. Apparently he had been so cut off with her that he no longer knew how to use a bank or shop in a supermarket. But those riotous early months had passed, and John had become calmer and enjoyed doing the everyday things that the rest of us took for granted with a woman who offered uncomplicated affection.

While Julian was with John and May, I spent time with Mal Evans, the Beatles former roadie: John had asked him to keep me company and show me around. It was good to see him again. A big bear of a man, he had always been gentle and kind. I'd known his wife, Lil, and their two children – they used to come to Kenwood – so I was sad to hear that he'd left them and was living in LA with a new girlfriend.

Mal made sure I had a good time. He took me to a great Mexican restaurant, introduced me to Tequila Sunrises and drove me all round Los Angeles. He, too, was sad about the change in John, and shook his head when Yoko's name was mentioned. On the last evening of our two-week stay he asked us all over to his house. I had told him how hard I found John's awkwardness, and Mal, ever the supportive Mr Fix-it, wanted to help. It seemed to work. While Mal poured the drinks John and May sat down with me and we chatted. At last I saw a glimmer of hope that things would ease: for the first and only time since our divorce, John seemed to put aside his guilt and embarrassment and relax with me. 'How's Roberto?' he asked. I told him we'd parted, and he said he was sorry: 'Are you OK?' I told him I was, and filled him in on the life Julian and I were leading, back in Meols. John reminisced about Liverpool and old friends, and asked me to give them his love.

Half-way through the evening Yoko called. My heart sank. Would John revert to being the sulky boy with his tail between his legs? Surprisingly, he didn't. He came back to join us, still smiling. 'Julian's a lovely boy,' he told me. 'I can't believe how grown-up he is, Cyn. He's not a little boy any more, I can really talk to him now.'

By the end of the evening I was optimistic. The thaw had begun. John and Julian had spent time together, John and I were talking, and he was with a woman who was supportive of his relationship with his son.

'We'll try to get Julian over again soon,' May told me, when I said I hoped John wouldn't leave it for another three years. I liked her and hoped their relationship would last.

I flew back to England with Julian who was happy and full of stories about John – he'd recorded Julian playing drums in the studio, then used the result on one of the tracks for his new album, *Walls and Bridges*. When it came out a few months later, Julian was credited, which thrilled him.

After our visit Julian talked to his father on the phone every few weeks. Mostly it was John who phoned, but Julian plucked up the courage to call him sometimes too. He was still in awe of his father, perhaps unsurprisingly since John was not only very distant but also a globally famous rock star. But I had real hope now that, in time, Julian would feel more at ease with him.

Towards Christmas John invited Julian over for the festivities. He was delighted and so was I. This time there was no need for me to go: I could put him on a plane, in the care of a flight attendant, and John would meet him at the other end. Everything went as planned and Julian took off, a little nervous but very excited.

By this time John and May had moved into an apartment overlooking the Hudson river in New York. Julian described it as small and very white, with two bedrooms. There was a balcony they had sat out on and he told me that one night John had seen a UFO from there.

May was like a big sister to Julian: she knew how to talk to him and they were at ease with each other, although

John was still often distant and awkward. However, Julian had started to learn the guitar at school, so John showed him some new chords and they enjoyed playing together.

Julian's Christmas present from John was a drum machine that played all kinds of rhythms. He loved it and they tried it out with their guitars. John, apparently, was gadget mad, and Julian was bowled over by the TV, with its separate box for changing channels – the forerunner of today's remote control. It was the cause of the worst moment in Julian's holiday. One morning while May was out shopping and John was sleeping late in the living room, Julian, bored and waiting for his dad to wake up, crept in and started to play with the channel box. Although he tried to do it quietly the clicks woke John, who leapt up and shouted at him. Julian, shocked and frightened, ran back to his room. From then on he had to watch John's moods constantly to avoid accidentally provoking him.

I was furious with John for putting yet another barrier between himself and his son – it had been a wild over-reaction to a minor incident. He could be astonishingly insensitive and cruel and gave no thought to the consequence for their relationship, which made Julian ever more wary around him. But none of this stopped Julian worshipping him and he was delighted when his father mentioned having him over again in the summer.

Things seemed set for a much more positive future until, soon after Julian's return, we heard that John had gone back to Yoko, who was pregnant. The baby was expected in October. I said little to Julian, other than how exciting it would be to have a brother or sister, and that I hoped he'd still be able to go over as planned. But privately I was worried. Would this be the end of our new-found rapport with John? I was also sad for May, who I imagined must be hurt. Yoko had dismissed her fifteen months with John as his 'lost weekend', and John made a statement saying that his separation from Yoko hadn't been a success. I prayed that this was the right move for him and that he'd stay in touch with Julian.

Initially he did: he phoned Julian as before, every few weeks. But the calls became less frequent, and all too often when Julian tried to phone John he couldn't get through. Yoko, or one of their employees, would tell him that John was sleeping or busy. Discouraged, Julian would wait weeks before trying again.

On John's thirty-fifth birthday, 9 October 1975, Yoko gave birth to their son, Sean. I was impressed by the timing and, despite my misgivings on Julian's behalf, glad for Yoko, who had had several miscarriages since she'd been with John.

Julian was eager to see his baby brother, but the planned summer visit never materialised and it was another two years before he met Sean.

In the meantime I had a new man. My old college friend Helen Anderson, now a dress designer, had introduced me to a TV engineer called John Twist. He was six years younger than me, and before long we had begun to see each other. One day he arrived at my bungalow in his car, with his dog on the back seat, and told me he'd lost his job and his home. I was unsure of what to do. Although I liked him, I wasn't ready to let him move in. Still, never good at saying no, that was exactly what I did. It was not until much later that I learnt he had been married when I met him. I was horrified: I would never have wanted him to do such a thing, but by the time I discovered the truth we had been living together for some time. Still, a seed of mistrust had been planted and although John was charming and always keen to please me, I never felt quite sure of him again.

Before I'd met John, I had bought a beautiful rundown old cottage in Llandurnog, north Wales, for fifteen thousand pounds, planning to do it up. The work took about a year to complete, but eventually John, Julian and I moved in. Although John and I had fun together, part of me knew that he was not the man of my dreams. Not only did I feel I couldn't entirely trust him, but it became clear gradually that he had little genuine rapport with Julian. However, never one to give up on a relationship if I thought there was a chance of making it work, I put my doubts to one side and we married on 1 May 1976 in Glendywr register office. It was only when Helen Anderson said to me at the wedding party afterwards, 'Why on earth have you married him, Cyn?' that I admitted to myself I'd made a mistake. If only she'd said it before the ceremony I might not have gone through with it. John and Yoko had sent us a telegram, saying 'Congratulations, good luck, God bless the three of you, love John and Yoko', and I took this as an encouraging sign.

A few weeks later, John rang and invited Julian over for a holiday and to meet Sean. Julian was thirteen now, at Ruthin School, just into north Wales. His friend Justin was there too, and Julian seemed happy, doing well in art.

He had joined the CCF (Combined Cadet Force) and became cadet of the year. He was delighted by his father's invitation, so I put him on a plane with presents for Sean and reassurances, more for my sake than his, that everything would be fine.

Julian adored Sean, who was almost a year old. On the phone he told me that his baby brother was crawling, and sat on his lap to watch television.

Two young teenage relatives of Yoko's – Julian was never sure who they were – were also visiting. I knew that Yoko still had no contact with her daughter and imagined that perhaps she and John had decided to have the girls at the same time as Julian to make it easier for her. They took all the children to Long Island for a holiday, which seemed to pass relatively smoothly.

However, when he got home Julian told me he'd been mugged in New York. He'd been for a walk, seen a harmonica in a music shop and told John he wanted to buy it. His father had told him to go ahead, but Julian was shy and afraid of going into the shop alone. John, perhaps wanting to toughen him up and encourage him to be independent, wouldn't go with him or send anyone else. He told Julian, 'If you want it, go on your own.' Eventually Julian set off, but as he stood outside the shop, plucking up the courage to go in, he was leapt on by two boys who hit him and stole his money. Luckily I didn't know about it until after he'd come home, or I would have let John know exactly what I thought about him encouraging such a young boy to go out on the streets of New York alone.

Although Julian was glad to see his father again, he felt like an outsider in John and Yoko's home, and they did little to help him feel part of the family. There were many occasions on which he had felt excluded, when, with a little effort, John and Yoko could have put him at ease. I was upset that John was not more sensitive to this, especially as he knew how hard it was to be on the edge of a family unit. When he was Julian's age and had visited his mother's home, he'd felt like an outsider too, because Julia, Bobby and the girls lived together while he was with Mimi and could only visit them. I had hoped that, with this in mind, John would make an extra effort to include Julian, but sadly he didn't.

Neither was he conscious of the difference between Sean and Julian's lifestyles. While Julian lived in a modest Welsh cottage with limited possessions and money, Sean's bedroom was full of the most expensive toys money could buy. John had boasted publicly of splashing out on anything and everything Sean might want, yet he gave Julian only modest presents at birthdays and Christmas. Julian was not a materialistic child, but he was sensitive and he could not help noticing the gifts, time and attention that his father lavished on his brother.

I learnt later, from John's sister Julia who was in touch with him by phone at this time, that John felt guilty about the way he had treated Julian and, determined not to repeat his mistakes, made an extra effort with Sean. But how was Julian supposed to know this unless John told him – or at least attempted to make up for lost time? As it was, he was left only too keenly aware of the difference between his relationship with his father and Sean's. If only John had tried harder when Julian was there and hadn't felt so awkward around him, things might have been different. Julian loved John and longed to be close to him, but as a shy teenager far from home and everything familiar, he needed encouragement. I'm sure John loved Julian, but he was never able to show his elder son the affection he so badly needed.

# 19

EVER SINCE THE DIVORCE SETTLEMENT I had relied on interest from Julian's trust fund to pay his school fees. When Sean was born the fund was cut in half, reducing the amount of interest I could draw, as well as the amount Julian would eventually inherit. This was a blow and I found it hard, given that John was now worth many millions of pounds, that he hadn't left Julian's fund intact and set up a new one for Sean. Also, John and Yoko were still trustees of the fund, with me, and I needed their signatures on every withdrawal. There was always a delay, which often left me overdrawn at the bank as I waited for the money to be released. And the trust was not making the amount of interest I thought it should.

Encouraged by Julian's visit to the States, I decided to appeal directly to John. My letter was cautious, polite and to the point. I explained that, as he and Yoko were out of the country, it was impossibly difficult for me to get at Julian's money:

*Nothing can proceed without your signature – it means I'm forever overdrawn at the bank and have to wait on the convenience of your lawyers . . . I want the best for Julian, and his standard of life shouldn't suffer because of lack of good management on your part, which has been happening since the fund was set up . . . The money, instead of having doubled through good investment, is dwindling through lack of interest on your part . . . It's just so important that this whole arrangement is sorted out without animosity or aggravation . . . The way things are going Julian's financial prospects when he is 25 will be virtually nil and he is going to want to know why . . . It is one thing fighting for your rights but totally ridiculous fighting against your own son's interests, which is what seems to be happening.*

I suggested to John that we appoint trustees in Britain and perhaps separate the two halves of the trust fund to make things simpler. I ended, 'Take care of yourself, Yoko and little Sean – he looks beautiful. And thank Yoko for being so kind to Julian when he was with you last summer, especially for arranging the Concorde flight. It will be something for him to remember for the rest of his life. Love to you all. Cyn.'

To my relief John responded positively, and it was made easier for me to draw on the money. I was glad that he still had enough common sense to grasp that there was no point in making my life or Julian's any harder. I felt I'd taken a small step forward in our dealings, but my sense of satisfaction was short-lived.

Around this time John Twist and I went to live in Ireland for a year. He had persuaded me to go so that I could write a book about my life and we found a cottage in a small village, Kilmacanock, in County Wicklow. A publisher was interested and I went ahead. I was reluctant to delve deep at that stage and it was a superficial, lightweight book. I think John Twist believed it would make our fortune, but all it did was provoke fury in John and Yoko. The *News of the World* published a lurid story announcing its publication and alongside it they ran pictures of John and Yoko. Fearing some huge exposé, John and Yoko tried to get the book stopped. They failed because I wasn't exposing anything. Far from it. I was so concerned for Julian's welfare that I had deliberately avoided putting in anything that might offend them. But the threat of litigation was once again in the air.

While we lived in Ireland Julian boarded at Ruthin School. He'd been reluctant to do so, and had left me countless notes around the house beforehand saying, 'Don't make me do this.' I promised that if he didn't like it I'd come straight back. But he loved it and had a very happy year.

By 1978 we were back anyway because I didn't enjoy being away from Julian or my mother. We bought an old townhouse in Ruthin and did it up, and soon afterwards I spotted a building in the centre of the town that I thought I would turn into a bistro and bed-and-breakfast business. I didn't have enough money to buy it, but we secured a mortgage and started converting it.

Eventually we opened Oliver's Bistro, and above it, the bed-and-breakfast business that John's parents were to run. Above that there was a small flat, where my mother lived. When John and Yoko tried to get my book stopped, she had been upset and suffered a stroke shortly afterwards. Then she began to show symptoms of Alzheimer's.

We invited Angie McCartney to come and work for us at the bistro. She had been married to Paul's brother Mike and had gone through a host of problems, including bulimia and anorexia, all of which meant that when they divorced he had been awarded custody of their three daughters. Angie was a fabulous cook and we wanted to help her get back on her feet. She had a room in the B-and-B and her children stayed with us when they visited her at weekends.

It was a hectic period. With the bistro, the bed-and-breakfast business, Mum becoming increasingly forgetful and Angie's erratic behaviour – she regularly failed to appear for her shifts – life was full and I was often exhausted. I enjoyed being busy and having people around, it suited me better than being alone, but at times I wondered if I'd taken on too much, especially as my marriage, rocky from the start, was foundering.

In the spring of 1979 John invited Julian to the States for his sixteenth birthday in the Easter holidays. The two still spoke over the phone occasionally and Julian wanted to be in touch with John as he hoped to become a musician and already had a fledgling band. I knew that his father's support and advice would mean a great deal to him.

It was with high hopes that Julian flew to the States at the beginning of April. After a few days in New York John announced that they were all going to Florida on holiday. He sent Julian ahead with a young man named Fred Seaman, his personal assistant, who drove them in a Mercedes station-wagon. John told Julian that he'd enjoy the trip and the rest of the family would follow shortly. Fred was friendly and kind, but Julian was keenly aware that his father had fobbed him off.

When John and the others arrived, a day or two later, Fred was giving Julian a driving lesson. John was furious that he had done this without his permission and shouted at both of them. Julian took cover in his bedroom.

For some reason – Julian had no idea why – John decided to celebrate Julian's birthday a few days early by taking the whole family out in a boat with the birthday cake. Things went well until he realised that someone in a nearby boat had recognised him. Instantly the party ended even though the cake hadn't been cut, and they headed for the shore. Julian had his cake at the house, but no alternative celebration. He was delighted with his present, though: a motorbike that John had had delivered to our home in Wales a few weeks earlier.

John's erratic behaviour around Julian continued – fun one moment and violent anger the next. And he could be like this with Sean too, reducing the little boy to tears of terror. Fred Seaman, or sometimes Yoko, would act as a buffer when John lost his temper. Julian was constantly on tenterhooks, sensing that an eruption was coming and retreating to his room in the hope of avoiding it.

One incident in particular did him lasting damage. The whole family had been having fun, making Mickey Mouse pancakes and fooling around, when Julian giggled. John turned on him and screamed, 'I can't stand the way you fucking laugh! Never let me hear your fucking horrible laugh again.' He continued with a tirade of abuse until Julian fled once again to his room in tears. It was monstrously cruel and has affected him ever since. To this day he seldom laughs.

Julian returned home with bleak reports of the eccentric life John and Yoko led in the Dakota building. Yoko ran their business empire from another apartment downstairs, known as Studio One (they owned five in the building). She hardly ever appeared in the family apartment, and often slept in her office, where she had a bed. Meanwhile, Sean was looked after by a nanny, Helen Seaman, Fred's aunt, while John spent much of his time in bed. Julian would lurk outside his room, occasionally plucking up the courage to go in. John would be sitting in bed with his guitar, the phone, a coffee and the TV on. He liked to watch the news and would shout at the TV, 'That's a load of crap,' when he didn't agree with something. After a while Julian would slip away to play with Sean, He saw no evidence that John was a devoted househusband or Sean's carer. Helen looked after Sean, and there were periods when he saw little of either parent. Far from baking bread or playing with Sean, John seemed to live in his own small world in the bedroom. He had relinquished all power to Yoko, who, he told Julian, 'knows best', and he appeared to have little interest in making music or anything else.

Back in Wales Julian continued to ring John, but all too often he was unable to get past Yoko. Julia, John's sister, experienced the same response. When, on rare occasions, Julian managed to reach his father, John seemed glad to hear from him and would chat happily about what he was up to.

By late 1980 Julian began to feel that there was a genuine breakthrough in his relationship with his father. John was at last making a new album, *Double Fantasy*. He began to call Julian more often. It was as though, with his creative juices flowing, he had woken up and realised his son needed him. He even played Julian tracks from the album and asked his opinion, something he had never done before and which gave Julian's confidence an enormous boost.

Then, just as it looked as though they might be forging a closer relationship and that John's life was turning round, he was shot outside the Dakota building returning from a recording session. It was just before eleven, on that terrible 9 December night, when a supposed fan, Mark Chapman, who had asked John for his autograph earlier that day and exchanged friendly words with him, shot him four times. John was hit twice in the back and twice in the shoulder. A fifth bullet missed him. Yoko had been a couple of steps behind him and it must have been appalling for her to witness. John staggered up the six steps outside the Dakota building before he fell. As Yoko screamed for an ambulance, the porter, who knew John well, covered him with his jacket, then pressed a button that connected him straight to the police. They arrived within a couple of minutes, decided there was no time to wait for an ambulance, carried John to their car and sped him to hospital. John died from massive blood loss shortly after they arrived.

That morning Julian woke in our home in Wales and knew that something was wrong. I was away from home, and as he looked at the reporters gathering outside our door, he knew that something must have happened to John. He went to find John Twist, who told him, as I had asked, to wait for me to get home from London. But Julian was devastated.

When I arrived home we cried together, each trying to comfort the other. Julian was determined to go to New York but I was reluctant to let him be on his own. In the end he insisted and I took him to the airport. At seventeen Julian had begun to look uncannily like John, the same aquiline profile, the same slim build. Julian wore his hair long, and in his leather jacket and jeans he could almost have been the young John. Watching him walk towards the plane was heartbreaking. I knew he would be bracing himself for what lay ahead and that he'd find a way to cope. Still, I longed to protect him from the searing pain I understood only too well, not just because I too mourned John but because I knew what it was like to lose your father at seventeen.

Julian described that journey as being in 'the twilight zone'. Nothing was real: all he saw around him were headlines about his father and all he wanted was to be where his father had last been.

Fred Seaman met him at the airport and took him back to the apartment, offering sympathy on the way. Outside the Dakota building hundreds of fans were holding a vigil, chanting John's songs, which were being played on almost every radio station across the country.

As Julian and Fred emerged from the car, Julian was mobbed by crowds of fans and press, wild with excitement at the sight of John's son. There was chaos as bulbs flashed and people screamed at him. Julian, unprepared for this, put his hands in front of his face and was hustled through the crowd to the car-park entrance of the building, fortunately avoiding the spot where John had fallen.

Inside the Dakota there was no sign of Yoko or Sean, who had been taken off by his nanny. Yoko had said that Sean wasn't to be told of John's death until she was ready to talk to him herself. Fred, Julian said, was devastated. He had spent a lot of time with John over the last two years and had loved him very much. At one point Fred drew Julian aside and told him quietly to be prepared for the fact that Yoko was not going to include him in any arrangements. 'She will do anything to keep you in your place,' he said. 'Sean is the only person who matters to her. There's simply no place for you in her world.' Fred's message was pretty brutal but it was proven absolutely true over the next weeks and months.

Lost, confused and shocked, Julian had no idea what to do or where to go. He just sat in the kitchen-cum-living room, waiting, as various employees came and went. Eventually he was invited to see Yoko, who was in bed in the large room that she and John had shared. She was calm and composed. She offered to take Julian down and show him where John had fallen and where he had crawled after he was shot, the spot still stained with his blood. 'Do you want to see John before he is cremated?' she asked. Julian declined both offers and told her he preferred to remember his father as he had known him.

'I don't know how to tell Sean,' Yoko admitted, in a rare moment of vulnerability. Julian advised her to tell him straight, and she asked Julian if he would do it with her. She knew Sean would have to be told soon – he had already seen Julian and was excited that his big brother had arrived but kept asking, 'Why's Julian here? Where's Dad?' Julian, who is hopeless at lying, had no idea what to say to him.

The next day Yoko and Julian sat down together with Sean. 'In the end we both talked to him,' Julian told me. 'We told him Daddy was dead and when he finally understood he burst into tears and I cuddled him. I told him the man who killed our dad would go to court. Sean said, "Do you mean a tennis court or a basketball court?" I said, "No, it's a different kind of court."'

Yoko later issued a statement saying:

*I told Sean what happened. I showed him the picture of his father on the cover of the paper and explained the situation. I took Sean to the spot where John lay after he was shot. Sean wanted to know why the person shot John if he liked John. I explained that he was probably a confused person. Sean said we should find out if he was confused or if he really had meant to kill John. I said that was up to the court. He asked what court – a tennis court or a basketball court? That's how Sean used to talk with his father. They were buddies. John would have been proud of Sean if he had heard this. Sean cried later. He also said, 'Now Daddy is a part of God. I guess when you die you become much more bigger because you're part of everything.'*

*I don't have much more to add to Sean's statement. The silent vigil will take place December 14th at 2 p.m. for ten minutes.*

*Our thoughts will be with you.*

*Love, Yoko and Sean*

This statement, published in newspapers around the world, summed up so much of what Julian had had to go through. There was no mention of his being there with Yoko when she told Sean. Yoko even quoted Julian's words as her own. There was no mention that John's older son had also lost his father. And Julian's name was not added to the signature. The insensitivity of this took my breath away. I knew John would have been just as proud of Julian as of Sean. Julian showed immense courage and composure throughout the terrible days after his father's death, and on top of this he had to endure being excluded from all Yoko's public responses to John's death.

The morning after Sean was told, Julian was asked to go to Yoko's room. 'Would you like to touch it?' she said, indicating an urn that stood over the fireplace. Julian stopped in his tracks and stared in horror. Almost on auto-pilot he did as he was told. It was still warm.

No doubt Yoko meant well – perhaps she thought it would help Julian to say goodbye or feel connected to his father – but for Julian it came as the rudest shock to realise that the urn contained his father's ashes. It was just over forty-eight hours since John had died and although Julian had known there would be no funeral – Yoko had made a public statement to this effect – he wasn't ready to be presented with a jar and told it contained the remains of his father. The shock stayed with him for a long time afterwards.

However, Yoko also suggested he might like a friend to keep him company, an offer he accepted gratefully and asked for Justin. The whole thing was arranged speedily and Justin was soon in New York. Then she came up with the idea that Fred should take Julian and Justin to the house at Cold Spring Harbor on Long Island that she and John had bought the previous year. She thought it might be nicer for them to be away from the crowds surrounding the Dakota building. She and Sean would join them later so that Sean didn't have to see the crowds gathered outside their home for the silent vigil in memory of John on 14 December. A hundred thousand people gathered in Central Park, next to the Dakota; all normal radio and television programmes were suspended so that people around the world could take part. In the Long Island house Yoko forbade anyone to put on the TV or radio to protect Sean from what was going on, Julian slipped quietly to his room and lay on his bed while the vigil was in progress to think about his father.

It took place instead of a funeral: Yoko had felt a funeral would be impossible with so many fans around. She had arranged for the early cremation because she didn't want fans to find out where John's body was and morbid crowd scenes to erupt. It must have been hard for her to have to decide all this after the shock of his death, and I sympathised, although it was upsetting for many of those who had known and loved John to be denied a funeral or memorial service.

In the Dakota building the apartment next to the main one was used solely for John's instruments. It held the white piano on which he had composed 'Imagine', at Tittenhurst, as well as dozens of guitars. Julian and Justin spent a lot of time in there, listening to John's music.

When I asked Justin, who is still a close friend of Julian, for his memories of that time, he wrote:

*It was very surreal being there at the Dakota, there were crowds of people for days across the road on the edge of Central Park, with police barriers etc. I remember looking down from the windows and seeing the vigil, candles, at night and hearing the people singing songs.*

*We would enter the Dakota via an underground parking lot just to the west of the main entrance where it had all happened. I remember how shocking it was to pass by the main entrance.*

*We stayed in the apartment where John, Yoko and Sean had all lived. Julian and I slept on white Japanese futons in a big white room.*

*It seemed like John had just stepped out for a little while, though I had never been there before. His stuff was still around the apartment. I remember looking at the stereo system in the open-plan kitchen/living area where we spent most of our time, thinking that only a few days ago he would have been playing music on it, or watching TV sitting on the couch. You just felt his presence everywhere. Yoko would spend a lot of time in her bedroom, coming out occasionally and we would sit together and have tea, and eat together sometimes. They had a chef.*

*Elliot Mintz [a former disc-jockey who had become a close friend and confidant of Yoko and John's] was around a lot and there were others in and out, mainly staff from John and Yoko's offices downstairs. Basically it seemed people were pretty dazed, but still functioning, trying to make sense of it all and being supportive to Yoko, Julian and Sean. Sean seemed too young to quite understand* what was going on. I remember him as a sweet and gentle boy. His room was near the main kitchen/living area and he would be in there quite a bit, playing, just doing what a normal five-year-old would do, I imagine.

Yoko seemed dazed, understandably, but cognisant. I found her to be a very intelligent woman, she commanded a certain respect. She spoke to us as adults, though we were teenagers.

It was a terribly sad thing to be a part of and one that forever changed me. I loved John's music myself; the first album that was bought for me as a kid was Abbey Road. I remember Julian giving me a copy of Mind Games. His dad had given him some copies on one of Julian's previous visits. I loved that album, especially the title track.

While at the Dakota I remember Julian and I going into the apartment next to theirs that they also owned. John used this one for storage for his musical instruments and to write and make rough demos. All his guitars were in a room or large closet, and I recognised some of them from Beatles footage. There was an upright piano that Julian would tinker on. There was a record player that we placed on the floor in the middle of the room and I remember we put on the Plastic Ono Band album. We sat on the floor and listened over and over to some of the tracks and we cried. I shall not forget that.

At other times Elliot would take us out and show us the sights in New York – I had never been there before. We also went to John and Yoko's other house in Cold

*Spring Harbor, Long Island, at the edge of the sea, for a week or so. It was a very cold winter, the sea was frozen over. We would swim in the outside pool which had a plastic bubble over it to keep the heat in, it was like a sauna. We had fun running from the house through the snow into that. There was a pool table at the house. We'd play pool and listen to the jukebox a lot. I think Fred was staying with us there.*

*So it was all a curious mix of terrible sadness and loss, and in a way adventure, for me and possibly for Julian. He seemed like he was keeping it together and I think Yoko felt that we should try to make the most of the trip, get out and do stuff. I don't know how Julian was feeling inside, though I can only imagine. My feelings then and in retrospect were that it was so sad for him because he was just starting to develop a better relationship with his dad, they were becoming friends. From the outside it seemed that John was wanting to re-connect with Julian and be more available to him. That is conjecture on my part, but I believe it to be true.*

In many ways, Justin's description sums up that time for Julian. He was miserable about his father's death but Justin's presence allowed him to have some fun – they could almost forget, for a few hours, what had happened, and just be kids in America.

After a week or two Julian wanted to come home, but Yoko asked him to stay on to join her at the ground-breaking ceremony for the Strawberry Fields site. This was a section of Central Park, opposite the Dakota building, which was to be dedicated to John's memory. The ceremony would take place in front of a gathering of the world's press and Yoko urged Julian to be there, for the sake of family unity, and even asked him to wear John's cap and scarf. Julian felt under pressure and agreed, hoping it would improve relations between the different sides of John's family. He hated every moment of it and couldn't wait to get inside and tear off the cap and scarf. Wearing his father's things in public was too painful for him and he felt that, in asking him to wear them, Yoko was not only wanting him to look like John but was also trying to make it appear that she and Julian were close. This was far from true – in fact, he felt she had little interest in him.

Before Julian left, Yoko offered both him and Sean one of John's guitars. Julian asked for one he had always loved, a black Yamaha inlaid with a pearl dragon. He remembered John playing Sean songs on it. Yoko told him he couldn't have that one and gave him two others instead, which, sadly, he didn't recognise and which therefore had no meaning for him. These were the only possessions of John's Julian was ever given, yet when he returned to the Dakota building on another occasion he saw that Sean had the full use of all John's musical equipment, including the guitar Julian had wanted.

Julian was home in time for Christmas. I had been anxious about him the whole time he was in New York and was relieved to have him home. But he was despondent and unhappy, not only because his father had died, but because he'd found Yoko's attitude hurtful. Her behaviour over

John's possessions was indicative to him of how little interest Yoko had in him and he felt he had only been asked to go to New York because it would have looked bad otherwise.

He wasn't the only one who appeared to be very much on the outside. After John's death none of his relatives were privately informed or contacted, though the speed of the media reaction did make that difficult. 'Relatives only want you for your money,' Yoko reportedly told Julian. A sad and, in my experience of John's family, untrue perception. I don't think they were ever interested in his money: they just loved him because he was part of them.

We got through Christmas and into the New Year as quietly as we could. The press were still turning up daily and I was still working in the bistro with John Twist and Angie. Julian was at a loose end, spending his days sleeping or going over to Justin's house a couple of miles away. I watched him closely, worried that he would sink into depression. There was no escaping John for any of us: his records filled the charts. *Double Fantasy*, the album released a few weeks before he had died, hadn't been doing particularly well, but after his death it sold six million copies. His songs were on the radio all the time, old interviews were replayed, the papers were full of articles and tributes to him, and souvenirs like ties and mugs appeared in the shops. If he had been popular in life, he was ten times more so in death. That he'd had to die to receive the accolades seemed terrible.

Shortly after John's death Yoko issued a single, 'Walking

On Ice', which contained snippets of John's voice interspersed with her singing. I found it macabre, but it was her most successful single ever. She followed it with an album, *Season of Glass*, which had a picture on the front of John's broken, blood-stained glasses, taken from the scene where he had died. In an interview she said that she had wanted to bring home to people the reality of what had happened, but for Julian it was all too apparent without an unhappy reminder of his father's death in every record store.

In the New Year we received a surprising visitor: Fred Seaman. He had travelled from New York to tell us that John had kept detailed diaries for the last six years before he died. Apparently a few months earlier, on a trip to Bermuda, he had said that if anything ever happened to him he wanted Fred to pass the diaries to Julian.

Fred had taken them from the Dakota building and given them into what he had thought was the safe keeping of a 'friend' in New York. This person had promised to copy them so that Fred could return the originals to Yoko, but had then realised their value, and refused to hand them back to him. He was still trying to retrieve them, and hoped to be able to pass them on eventually to Julian. We appreciated his kindness in coming to let us know and spent a nostalgic evening with him, talking about John.

Julian was terribly excited and touched that John had wanted him to have something so personal. He hoped the diaries would help him understand his father, and that by reading his most private thoughts of the past few years, he would be able to get beneath the surface of the father he'd never really had the chance to know.

Sadly, though, he never read them because Fred never got them back. They were returned, for a fee, to Yoko, who had Fred arrested on a charge of grand larceny. In 1983 he was sentenced to five years' probation, later reduced to three. The diaries remain in Yoko's possession but Julian has the comfort of knowing that John intended them for him.

We knew that John had left a will, but had no idea of its terms. However, I was certain that John would have left Julian a financial legacy. Whatever his faults as a father, he had loved Julian and would have wanted to ensure that he was recognised as his son and provided for. It was inconceivable to me that he would neglect this duty. My only worry was that, because John had had no interest in legal matters and always left them to others, he might not have made sure that things were as straightforward as possible. I remembered the difficulty I'd had with the trust fund.

Eventually we were told that John had left his fortune, then considered to be well in excess of a hundred million pounds (and subsequently increased to several hundred million) to Yoko and the issue of John and Yoko: Julian, Sean and Kyoko. The trustees were Yoko and the Lennons' lawyers.

We expected that it would take some time for the terms of the will to be fulfilled. But when, after a year or more, nothing had happened, we asked our lawyers to investigate. They advised us that, because Yoko and her lawyers were the trustees of the fund, whether or not Julian got any money was in their hands, despite the fact that John's wealth would have been enough to run a small country.

# 20

JOHN'S DEATH HIT ME AND Julian hard. I was grieving for the man I had loved as a husband and partner, Julian for the father he never really knew and needed so badly. For some months he appeared to be spiralling out of control and I worried about him. He was angry with me, blaming me for the break-up with John, for not being stronger, for not making him behave like a proper father. He wanted to push me away and I understood, but it still hurt.

There was little comfort for us in seeing John's killer

jailed. John's death had been so unnecessary and for both of us the hardest part of losing him was the loss of hope – for me that we would one day be friends again, able to laugh together, enjoy past memories and talk about the son we both loved, and for Julian that he and John would become close, understand each other and spend time together. I knew that Julian's loss was by far the greater one. I had had something wonderful with John that I could always look back on. But Julian lost the possibility that John would make up to him for past neglect and become the father he had always wanted.

Not long before he died John gave an interview to *Newsweek* in which he said: 'I hadn't seen my first son grow up and now there's a 17-year-old man on the phone talking about motorbikes. I was not there for his childhood at all. I was on tour. I don't know how the game works, but there's a price to pay for inattention to children. And if I don't give him attention from zero to five then I'm damn well gonna have to give it to him from 16 to 20, because it's owed, it's like the law of the universe.'

I felt sure that, if John had lived, so much might have been possible. He had been making more effort towards Julian before he died and, judging by the *Newsweek* interview, he was aware that he had neglected Julian and intended to make up for it. Also, persistent rumours that John's second marriage was far from happy and that Yoko was considering divorcing him heightened my sense that in time many things might have changed. But time had run out and we had to cope with what was, not what might have been. I had to move on, and most importantly

I had to find a way to help Julian to move on.

By this time my marriage had collapsed and I had to sell the bistro, bring Mum to live with me and make tough decisions about my future. But I wasn't in any state to make decisions. Both Julian and I were in a low state: we had both been knocked off course by John's death and floundered, each trying to find a new sense of direction and purpose.

I knew that I'd get myself back on track, but I wanted to help Julian find some direction in his life. There was nothing for him in north Wales. He had helped out in the bistro for a while but other than that he was hanging around or being chased around town by the local bullies, who singled him out for being John's son and were unable to believe that he wasn't secretly stinking rich.

As he wanted to go into the music business, I rang a few old contacts. I had thought George Martin might be able to give him a job, even if it was just making tea in the studios, but he couldn't. After a few more false starts I remembered Elton John's percussionist, Ray Cooper, and the promise he'd made us on board the SS *France*. 'Send him to me,' Ray said. 'I'll put him up for a few months and introduce him to a few people.'

So Julian, now eighteen, was packed off to London to find his musical direction. Ray was now living with a Greek girl, who had once been married to Magic Alex. Julian had been playing in a band with Justin, and had also begun composing on the piano I'd bought him for his sixteenth birthday – a beautiful, hand-carved upright Steinway. I remember walking down the stairs one day when he was playing and singing, and being startled because he sounded like John all over again. Now Julian's greatest wish was to prove himself good enough to earn a recording contract. I hoped he would do it, but I feared for him because of the inevitable comparisons with his father.

For a while things seemed uncertain. After six months with Ray, who gave him all sorts of valuable introductions, Julian got his own flat. Eventually the press got wind of who he was and stories began to appear in the papers. He was photographed in L'Escargot, a famous London restaurant, dressed in drag with Karen O'Connor – daughter of comedian and TV star Des – pointing a gun at his head. Of course the press had a field day with that one, even though it was just two kids fooling around after a few drinks. He hadn't even seen that Karen was pointing the gun at him because he was looking at the camera.

Nightclubs began to offer him free entry, knowing that he was a magnet for the press, and plenty of other bad-boy stories appeared. The impression was that Julian, if not going off the rails, was certainly enjoying his share of youthful wild living. There were regular items in the papers about him and various girls, mostly blonde models who confessed their love for 'so shy Julian'. I laughed at most of it, knowing it was wildly exaggerated. Julian phoned me regularly to reassure me that he wasn't actually dating every dizzy blonde he was pictured with.

In April 1982 his nineteenth birthday party was held at Stringfellows and I was there to join in the fun. The highlight came when Julian was presented with a white horse by Stephanie LaMotta, daughter of former boxer Jake, whose story was told in the film *Raging Bull*. Everyone was stunned when she made a dramatic entrance with the horse – but, as with so much else, it was a publicity stunt, set up to cash in on Julian's name.

From the stories in the papers the public might have concluded that Julian was John's spoilt millionaire playboy son. In fact, nothing could have been further from the truth. Soon after Julian had given an interview saying he had no money, Yoko agreed to give him an allowance of a hundred dollars a week, which she called 'beer money' and said was quite enough for a young man of his age. She insisted that he would only blow it if he had more. But Julian found it humiliating to take handouts from her when he felt he should already have received a share of his father's estate. I was angry too. It wasn't that I wanted Julian to be rolling in money; if he had received a share of John's I'd have wanted him to invest it for the future or contribute to good causes, rather than just spend it. But the point was not what Julian might or might not do with the money: it was that John had wanted him to have it.

While our lawyers continued in their attempts to resolve this and Julian carried on writing songs in London, the legend of John Lennon grew daily greater. Within two months of his death he had sold two million records in Britain alone, and tribute songs were released by everyone from George Harrison, who had his biggest hit in eight years with 'All Those Years Ago', to Elton John, Paul

Simon, Pink Floyd, Mike Oldfield, Queen and Roxy Music.

I was touched when Paul McCartney wrote his own tribute to John, a lovely song called 'Here Today', which showed his affection for his old friend. The two had never been close after the Beatles' split, but I know they'd met and talked a number of times. John's style was to walk away and stay away – as he did with me: once his mind was made up he didn't go back. But he and Paul had had a deep and enduring affection for each other since they were teenagers and it had never disappeared.

In addition to all the tributes, John's death sparked a dramatic increase in the rock memorabilia market. Anything that had belonged to John went for enormous sums at auction. I was amused when our Rolls-Royce – the one John had painted in psychedelic colours – was sold for $2.2 million at Sotheby's the year after John's death. If only I'd asked him to include it in my divorce settlement! Meanwhile, Yoko was selling limited editions of John's artwork, engraved in marble. A few friends received these as presents, but not Julian. Instead he received Christmas and birthday presents of John's sketches printed on a metallic blue tie and a coffee mug. He was desperately upset that John's work was being commercialised in this tacky way, and insulted that Yoko should give them to him rather than the marble engravings.

Some time after John's death I received a series of letters claiming to be messages from him to me through a psychic. I have an open mind about these things, and I know that the 'psychic' might have been a crackpot, but the warm, loving messages they contained, and 'John's' expressions of regret over the way he ended our relationship were heartwarming. It was nice to believe, just for a moment, that he really had sent them.

For years I had led a quiet life, mostly outside the spotlight of John's fame, so it was unnerving when, after his death, I found myself inundated with requests. There was a huge demand for anything connected with him and I was asked to do all kinds of things, from becoming a TV presenter to being involved with a restaurant named after John, to making a record of my own. No doubt I should have turned down more of these offers than I did because some were a nightmare. But I had a living to make and no one experienced or professional to advise me. And I was happy to honour John's name and his talent, which was what I felt I was doing with the projects I accepted.

One of the most exciting invitations came nearly two years after John's death when I was invited to the States for an exhibition of my artwork in a Long Island gallery. The owner, Gary Lajeski, was a friend of Peter Brown, the former Beatles manager, who had come up with the idea. The exhibition included a series of drawings I had done years earlier of the Beatles and scenes from our life in Liverpool. I was delighted: I felt that, in some way, it was my own tribute to John. While I was in Long Island I found myself alone and at a loose end for much of the time, so I phoned May Pang, John's former girlfriend, and suggested we get together. I had liked her when we met in 1974 and thought it might be nice to catch up. She invited me over to her apartment and we became firm friends.

A few days later it was the second anniversary of John's death. On the spur of the moment, May and I went to the Dakota building, where a vigil for John was going on outside. Thousands of people held candles, sang and chanted for him. May and I stood unrecognised among the crowds, saying our own silent farewells to the man we had both loved.

During that visit May told me much more about her relationship with John. She believed John had loved her – he had told her so frequently, and they'd had some very happy times. When he had gone back to Yoko, May was shocked because he had told her only days previously that he never would. Even when he did, it was not the end of their relationship: he continued to see her as a lover until 1977 and after that they talked from time to time on the phone. She had last spoken to him during the summer before he died.

It was from May I learnt that John's reluctance to see or have contact with me was fuelled by Yoko, who told him constantly that I still loved him and would do anything to get him back. I had already suspected that this was so and, of course, I did still love John, but I had never considered the possibility of us getting back together: I had moved on to other relationships, just as he had. All I ever wanted was for us to be friends, as so many ex-partners are, for the sake of our shared history and our child. I was sad to think that he and I had been denied friendship because Yoko felt threatened by me.

Back in England, after a couple of years alone, in 1982 I started a new relationship, with a man called Jim Christie.

I'd met him when he gave Julian motorbike lessons a few years earlier and we'd become friends. When his marriage and mine both ended we got together and soon afterwards decided to move, with my mother, to Wiltshire, to be nearer to my brother Charles and his wife Penny.

As Julian's twenty-first birthday drew near, he was approached by Tony Stratton Smith, the head of Charisma Records. Tony had heard Julian's demo tapes without knowing who he was, and offered him a contract for an album. Julian was thrilled and it was arranged that Tony, Julian and Julian's band, including Justin, would go to stay in a château in France for several weeks to work on his songs and write some additional material.

The album was recorded in New York and Julian called it *Valotte*, after the French château. Within a few months of its release it had sold 1.5 million copies and gone platinum in the States. His first single from it, 'Too Late For Goodbyes', was a song many thought was about his father, although in fact it was about an ex-girlfriend. It reached number six in the British charts in September 1984. He had written another song about John, 'Well I Don't Know', which posed questions about life after death. Julian's career, it seemed, was assured at last, and I couldn't have been more proud or delighted. He did his first tour of North America at twenty-one, visiting fourteen states and Canada, and played in venues that were filled to capacity. He earned himself some wonderful reviews: critics said they couldn't help but be reminded of John, but his success was down to his own musical talent.

I was happy for him, but watching him perform I felt

a sense of *déjà vu*. Not only was it was like seeing John's story all over again, but Julian had the added burden of being John's son. I was afraid it was too much too soon. Rather than being treated as a young début artist he was seen as the pretender to John's throne, which he had never wanted and set him up to fail. I hoped that he would have the strength and wisdom to cope with all that fame entailed and wouldn't become burned out or turn to drugs.

While Julian was forging ahead musically, I did a TV interview and as a result was offered a job as an interviewer for a Granada Television programme called *Weekend*. I'd never been confident in front of the cameras, but I decided to give it a try. My first interview was with Julian, then riding high in the charts. It was fun for both of us and I went on to interview all sorts of other people. My TV career ended when the programme finished its run. I had no great desire to move on to other TV work, but I was glad I'd done it and was more confident for it.

Another venture I became involved in and enjoyed was designing bed linen for a major company, on a three-year contract. I was also invited to design paper products, such as napkins, for another company. Both ventures gave me a chance to do something artistic again.

Some time before, Jim and I had moved to Penrith in Cumbria. I'd been up there to visit my old friend Phyl and fallen in love with it, so we bought an old house and set about restoring it. A couple of years later we moved again, this time to the Isle of Man where we opened a restaurant, Bunter's. Once again I was cooking, cleaning and serving customers for long hours each day, but I've never been afraid of hard work, and on that quiet little island I found a kind of peace – at least for a while.

Meanwhile, in 1985 Julian decided to move to Los Angeles. It made sense in terms of his musical career, but for me the prospect of him being so far away was hard to come to terms with. Still, I gave him my blessing and he bought a lovely villa where he settled for the next few years. In 1986 he was nominated for the best new artist award at the Grammys and later that year he played at London's Royal Albert Hall, in a charity event hosted by the Duchess of York. Following that, his second album went gold in the States and he embarked on a world tour.

Sadly, at the end of that year, my mother died, at the age of eighty-three. In the end it was a release: Alzheimer's had made her last few years desperately hard for all of us, but I missed her: she had been a staunch support, friend and companion throughout my life.

In January 1989 I was asked by promoter Sidney Bernstein, who had organised the Beatles' first New York concert back in 1964, to lend support to a major charity concert in John's memory, to be called Come Together. The idea was to mark what would have been John's fiftieth birthday, 9 October 1990, with a rock symphony, to be performed in the States and televised around the world. The funds raised would be used to support charitable ventures in John's name. Sid, who had known John well in his New York days, was excited about it and his enthusiasm was infectious, although I knew the venture would involve months of intensive unpaid work. I thought long and hard about it, but in the end it seemed so fitting and so worthwhile that I had to say yes.

By July, after six months of planning, the concert was shaping up brilliantly. I had meetings with both Rudolf Nureyev and Michael Jackson, who had agreed to dance together as one of the highlights of the evening. *Star Wars* producer George Lucas had agreed to do the special effects and other performers who were to appear included Ravi Shankar, the Moody Blues and, to our delight, Paul McCartney, who was to play with the Berlin Philharmonic Orchestra, under the direction of George Martin.

It was an enormous enterprise and I travelled all over Europe and the States, met German chancellor Willi Brandt and many others who wanted to help. I was even invited to the US Senate, and stood beside the Mayor of New York as the peace bell rang at the United Nations. As the concert took shape I felt one thing was needed to seal its success and make it the perfect tribute to John: Yoko's endorsement. I called her and told her I had something to discuss with her and she agreed to a meeting.

So it was that one sweltering August day Jim and I went to the Dakota building and walked in, past the spot where John had been shot, which I found almost unbearable. That evening Julian was due to give a concert at the Beacon Theater in New York. He had invited Sean, who was then almost fourteen, to join him on stage and Yoko and I would be in the audience. It seemed to me the perfect moment for us, in supporting our sons, to make a declaration of friendship and share in the opportunity to honour John.

Jim and I were shown into the kitchen – Jim hadn't

removed his shoes at the door or allowed me to. It was large, light and airy, with light wood units and a central island. To one side there was a low table with cushioned benches along both sides and this was where we sat. I was a little startled to see a plastic replica of the gun that had shot John sitting on a shelf nearby.

With Yoko, when she came in, was the man who had been her partner since a few months after John's death, Hungarian designer Sam Havadtoy. Although she had declared that she would never marry again, rumours persisted that she and Sam had married in secret in Hungary. I had no problem with their relationship, whether they were married or not. Why shouldn't Yoko move on and have another man in her life? Sam seemed pleasant and I hoped he might encourage her to join me.

As I explained about the planned concert the work and ideas that had gone into it, Yoko said nothing. She didn't smile. My heart sank but I ploughed on, showing her the plans and suggesting that it would mean a great deal to our sons and to John's fans if we could support the concert and honour John's memory together.

When I had finished Yoko said she would consider it and get back to me. I raced across New York in time for the start of Julian's concert, but there was no sign of Yoko until the interval, when she made a last-minute entrance and sat with Sam and Sean a few rows behind me. The concert went beautifully – and when Julian called Sean up on stage to play guitar with him the audience went wild, screaming and applauding.

After the show Yoko, Sean, Julian and I held a drink

together at the Hard Rock Café, where photos were taken – the only ones ever taken of me with Yoko. I beamed and Yoko managed a small smile. Neither of us could pretend that we were at ease.

A few days after my meeting with Yoko, Sam Havadtoy phoned me. 'Yoko doesn't want to take part in your concert,' he informed me. 'In fact she has been planning a concert for John for some time.' I was speechless.

Soon after Sam's call Yoko gave an interview to the press, saying that I planned to make millions from the concert, although I had explained to her that it was for charity. When I read it, I knew our concert was doomed. Once the ugly rumour that we were not being honest had started to circulate it would sow the seeds of doubt in potential performers.

Sid was desperately disappointed when we were forced to abandon it, and I was hurt at what I saw as a betrayal. After such high hopes and so much hard work, it seemed a sad conclusion.

Yoko went ahead with a concert, also called Come Together. It was held in May 1990 in Liverpool. Only fifteen thousand people attended, instead of the expected forty-five thousand, and none of John's family was sent a ticket. I imagine Yoko was disappointed by the low turnout but I was just sorry that we hadn't been able to do something together: I believed it might have been a real success.

Yoko and I had one more meeting, a couple of years later, when Aunt Mimi died in 1992. A few months before, John's sister Julia had phoned me to say that Mimi was unwell with heart problems. I rang her and we talked briefly. She wasn't well enough for a long conversation, but I made my peace with her and that felt important.

When she died I went to the funeral, with John's sisters and cousins. Mimi had been the eldest of the five Stanley sisters but was the last to die, having reached the ripe old age of eighty-nine. Her nurse told Julia that Mimi's last words had been, 'I'm afraid of dying. I've been a wicked woman.' When I heard this it sent a chill through me. I felt sorry for Mimi, who had obviously suffered, but her self-assessment had been true.

Yoko was also at the funeral, with Sean, who met most of his English relatives for the first time. After the service, we all had lunch in a local hotel. Sean, who was seventeen, was having a great time, chatting to everyone and knocking back the red wine, but while Yoko made an effort to join in, talking politely to those around her, she was struggling with what was clearly unfamiliar English buffet food. She pushed it gingerly around her plate, smiling, and delicately avoided putting a single forkful into her mouth.

Eventually she abandoned the attempt and went back to smoking her long black cigarettes, crushing each one out after only one puff. John's sister Jacqui looked at the ashtray, overflowing with unsmoked cigarettes, giggled and whispered, 'Who's learning to smoke, then?'

Once more, in an echo of the housewarming party over twenty years earlier, Yoko approached me and began to chat as though we were old friends. First she complimented me on the way I'd brought up Julian, which almost caused me to choke on my drink. I remembered her message to me, which John had delivered, that Julian was a badly behaved child. Then she asked my advice about how she should handle Sean. I kept my smile in place and stuck to pleasantries.

That afternoon Mimi's house was put on the market. Although John and I had bought it for Mimi, intending it to belong to her and to be passed on to her family, it was never put in her name, with unfortunate consequences for his English family. Mimi herself had only found out after John's death that she didn't own her house: 'I can't believe he's left me beholden to Yoko,' she told her family. After that she was so frightened the house might be taken away from her that she changed all her photos of Julian for photos of Sean, in case Yoko or anyone who knew her should visit.

Mimi kept the house, but the second she died it was sold. The house John and I had bought for Harrie and Norman, intending it to pass to Julia and Jacqui as their nest egg, was similarly disposed of after Norman's death. In a public gesture of generosity Yoko gave it to the Salvation Army, who left it empty for four years, then installed one of their officers in it.

John was distant from his family for most of his time with Yoko, and it was possible that he never told her why the house had been bought. I wrote to her, telling her that John had intended the house for his sisters, but whether or not she saw my letter, it made no difference.

After the house had been given away, I wrote to Julia:

*I want to confirm to you and whomsoever it may concern that John and I discussed your wellbeing endlessly. We both thought that the home we purchased for Aunty Harriet and Uncle Norman – who were your guardians at the time – would eventually be yours and ensure your future security. If John had still been with us this situation would never have arisen. The house was bought by us for you and Jacqui . . . The house bought for Mimi was also intended to be for John's family . . . To my mind it is truly sad that John's wishes have not been carried out.*

As for the money for Julia in John's will, when nothing had been forthcoming after six years the matter was put in the hands of lawyers. It took another ten years for Yoko to reach a settlement, by which time she had received a fair amount of bad press over it. I was glad that in the meantime Julian had been a success in his own right and had earned his own money. He was never dependent on his father's legacy, but it was a point of principle that he should receive it.

We hoped then that goodwill could be achieved on all sides and we could get on with our lives. But on 18 May 1998 Julian released *Photograph Smile*, his first album for several years. He had put a great deal of thought, energy and creativity into it and was justifiably proud of it, calling it his first real album. So he was devastated when he discovered that Sean, now a musician too, had released an album on the same day. He never blamed Sean for this, but he knew it couldn't be a coincidence that he and his brother had been pitted against each other so blatantly.

To add to Julian's hurt and anger, Sean chose to announce then that he thought John's death was part of a conspiracy. After this controversial declaration all the press wanted to talk to Julian about was Sean's statement and what he thought of it. The last thing Julian wanted to discuss was his father's death, but questions about it overshadowed any interest in his album. Having tried to get on with his life for so long, Julian had reached his limit. He gave a series of angry interviews condemning Yoko, who he felt, rightly or wrongly, had been behind both the release date of Sean's album and his pronouncement about John's death. In one he said, 'Yoko is very insecure . . . She's got everything she could ever want . . . Any success that I have is a bane in her life and a thorn in her side because I'm John Lennon's blood . . . She always has to be the winner.'

Despite its rocky start, Julian's album was a worldwide success, reaching the top twenty in many countries and getting some excellent reviews in the world press.

I understood Julian's frustration with Yoko. I have always been sorry, and frustrated, that Julian and I could not have had a better relationship with her. I often wonder if her trouble in accepting Julian, or indeed John's life before he met her, was due in part to the loss of her daughter. I was glad to hear that, in her late thirties and by then a mother, Kyoko got in touch with her and re-established their relationship. I hope it brought Yoko some genuine happiness.

There was another startling reappearance a few years back, from John's older half-sister Victoria, the baby Julia had given up for adoption. Now called Ingrid, she announced who she was in a newspaper article. It appeared that she had known all along but had kept quiet until her adoptive mother had died. We were all so glad to learn that her upbringing had been happy and hoped to get to know her. But sadly we never got the chance, as she hadn't wanted to be closely involved with John's family.

Julian left the States in the mid-1990s and travelled before establishing a home in Europe. By this time Jim and I had moved to Normandy but in 1999 we split, after seventeen years together. I was devastated and called Julian, who drove through the night to be with me. The following Christmas he treated me and my dear friend Phyl – now also on her own – to a holiday in Barbados, where we sunned ourselves, swam and had a wonderful time.

It was in Barbados that I met Noel Charles, a former nightclub owner and an old friend of Julian's. His easy wit and gregarious nature charmed me. Not in the least over-awed by my past or my name, he had run nightclubs, mixed with celebrities and lived a vibrant, exciting life. We became good friends and eventually fell in love. And on 7 June 2002 I did something I'd promised myself I would never do again: I got married – for the fourth and last time.

These days, Julian lives in Europe with Lucy, his partner. He cooks like a dream and is a partner in a couple of restaurants, and when the mood takes him he still makes music. He's also put together a wonderful collection of his father's possessions, bought over the years from auctions and carefully preserved. We often see each other and I'm enormously proud of him, for the person he is even more than for the things he has done.

In April 2003 he turned forty. We were both painfully aware that he had now reached the age at which John had died. For a while Julian battled the fear, like so many of those whose parents have died prematurely, that his own time had come. Always remarkably generous, he invited Noel and me, along with friends from his past as well as the present, for a two-week-long party in Barbados to celebrate. And, surrounded by people who loved him, he realised that his time most certainly hadn't come. He has a lot more living to do and his life at forty was very different from John's.

Neither of us will ever escape the Lennon name, and there have been times when it has felt like a millstone round our necks. But we have come through the hard times and learnt to bless all that it has brought us – good and bad. And we both miss John, even after all this time, in so many ways.

I've kept many of the friends John and I had in our Liverpool days and mourned others. Maureen, Ringo's ex-wife and a dear friend to the end, died of leukaemia when she was only forty-seven. Only a few years earlier we had been at her wedding to Isaac Tigrett, owner of the Hard Rock Café. Both Isaac and Ringo were at her bedside when she died, with her three children by Ringo, and Alexandra, the daughter she had with Isaac.

I also felt terribly sad when I read of George Harrison's death from cancer at the age of fifty-eight. I hadn't seen him for many years, but I had fond memories of him and was always grateful to him for the spiritual awakening he encouraged in all of us in the days of peace and love.

While I feel sad when I think of how many people we knew in the Beatles' heyday have died, it's a source of enormous pleasure to me that other old friends are going from strength to strength.

I still meet up with Cilla and we enjoy reminiscing about the old days. Not long ago I went back to Liverpool when the curators of The Beatles Story, the permanent exhibition in the city, bought all the cartoons I had drawn of the boys in the early days. To celebrate we had a party in a restaurant on the Liverpool docks – it was like going back in time. There were friends like Helen Anderson from art college days, Pete Best, looking great, with his younger brother Roag, John's sister Julia and my dear Phyl. We talked, told stories and laughed well into the night. As the wine and memories flowed I couldn't help a moment of sadness, thinking that John had once been part of this crowd and loved them all as much as I did. I'm certain he would have enjoyed the party.

I wondered what he would have thought of his old friends, Paul, Mick and Elton, being knighted. Of course he would probably have been offered a knighthood too, had he lived, even though he'd sent back his MBE. Would he have scoffed and turned it down, or would he have been sentimental and patriotic enough to accept? My money's on the latter.

John could be incredibly sentimental. He once told Julia he wanted to return to Liverpool, sailing up the

Mersey, to see all the people he'd left behind. And not long before he died he asked Mimi to send him all the old photos, ornaments and mementoes of his childhood that she could find so that he could surround himself with them.

It was also in a moment of sentiment, but perhaps of tenderness and love too, that John once said to Julian, 'If anything ever happens to me, look for a white feather and you'll know I'm there, looking out for you.'

I think of John every time I see a white feather.

He was an extraordinary man: talented, flawed, a creative genius who sang movingly about love while often wounding those closest to him.

I never stopped loving John, but the cost of that love has been enormous. Someone asked me recently whether, if I'd known at the beginning what lay ahead, I would have gone through with it. I had to say no. Of course I could never regret having my wonderful son. But the truth is that if I'd known as a teenager what falling for John Lennon would lead to, I would have turned round right then and walked away.

# Photographic Acknowledgments

A = above, b = below, c = centre, l = left, r = right

All photos courtesy of the author with the exception of the following:

Associated Newspapers: 8a; Axel Springer Verlag: 29b; Bob Whitaker/Camera Press; 11a & b; Corbis: 26; Featureflash: 28b; *Manchester Daily Mail*: 28a; Mirrorpix: 19b; Popperfoto: 7b; Redferns: 4a & b (K & K Ulf Kruger OHG), 6a (Astrid Kirchher) 6b (Jurgen Vollmer), 9b, 18b (S & G Press Agency); Star File Photos: 30a; Ronald Traeger: 15a; Geoffrey Wilkinson: 31

# Index

DEAR CYN.

I LOVE YOU I LOVE YOU
I love you I love you I love you
I love I love U I I I I I love U
I LOVE YOU LIKE MAD I DO
I DO LOVE YOU YES YES YES
I DO LOVE YOU CYN YOU I LOVE
I love you Cynthia Powell
John Winston love C. Powell
cynthia cynthia cynthia
I love you I love you I love
you forever and ever is NA
it great? I LOVE YOU LIKE
GUITARS I LOVE YOU LIKE
ANYTHING LOVELY LOVELY
LOVELY LOVELY CYN LOVELY CYN
I LOVE LOVELY CYNTHIA CYNTHIA
I LOVE YOU. YOU ARE WONDER
FUL I ADORE YOU I WANT YOU
I LOVE YOU I NEED YOU DONT
GO I LOVE YOU HAPPY XMAS
MERRY CHRIMBO I LOVE YOU
I LOVE YOU I LOVE YOU CYNTHIA
CYN CYN CYN CYN CYN CYN CYN
IS LOVED BY JOHN JOHN JOHN
JOHN JOHN I LOVE YOU
XXXXX                 LOVE John xx